Applications of X-ray Computed Tomography in the Geosciences

Special Publication reviewing procedures

The Society makes every effort to ensure that the scientific and production quality of its books matches that of its journals. Since 1997, all book proposals have been refereed by specialist reviewers as well as by the Society's Books Editorial Committee. If the referees identify weaknesses in the proposal, these must be addressed before the proposal is accepted.

Once the book is accepted, the Society has a team of Book Editors (listed above) who ensure that the volume editors follow strict guidelines on refereeing and quality control. We insist that individual papers can only be accepted after satisfactory review by two independent referees. The questions on the review forms are similar to those for *Journal of the Geological Society*. The referees' forms and comments must be available to the Society's Book Editors on request.

Although many of the books result from meetings, the editors are expected to commission papers that were not presented at the meeting to ensure that the book provides a balanced coverage of the subject. Being accepted for presentation at the meeting does not guarantee inclusion in the book.

Geological Society Special Publications are included in the ISI Index of Science Books Content, but they do not have an impact factor, the latter being applicable only to journals.

More information about submitting a proposal and producing a Special Publication can be found on the Society's web site: www.geolsoc.org.uk.

It is recommended that reference to all or part of this book should be made in one of the following ways:

Mees, F., Swennen, R., Van Geet, M. & Jacobs, P. (eds) 2003. *Applications of X-ray Computed Tomography in the Geosciences*. Geological Society, London, Special Publications, **215**.

Van Geet, M., Lagrou, D. & Swennen, R. 2003. Porosity measurements of sedimentary rocks by means of microfocus X-ray computed tomography. *In*: Mees, F., Swennen, R., Van Geet, M. & Jacobs, P. (eds) *Applications of X-ray Computed Tomography in the Geosciences*. Geological Society, London, Special Publications, **215**, 51–61.

GEOLOGICAL SOCIETY SPECIAL PUBLICATION NO. 215

Applications of X-ray Computed Tomography in the Geosciences

EDITED BY

F. MEES
Ghent University, Belgium

R. SWENNEN
Physico-chemical Geology, K. U. Leuven, Belgium

M. VAN GEET
SCK-CEN, Belgium

and

P. JACOBS
Ghent University, Belgium

2003
Published by
The Geological Society
London

THE GEOLOGICAL SOCIETY

The Geological Society of London (GSL) was founded in 1807. It is the oldest national geological society in the world and the largest in Europe. It was incorporated under Royal Charter in 1825 and is Registered Charity 210161.

The Society is the UK national learned and professional society for geology with a worldwide Fellowship (FGS) of 9000. The Society has the power to confer Chartered status on suitably qualified Fellows, and about 2000 of the Fellowship carry the title (CGeol). Chartered Geologists may also obtain the equivalent European title, European Geologist (EurGeol). One fifth of the Society's fellowship resides outside the UK. To find out more about the Society, log on to www.geolsoc.org.uk.

The Geological Society Publishing House (Bath, UK) produces the Society's international journals and books, and acts as European distributor for selected publications of the American Association of Petroleum Geologists (AAPG), the American Geological Institute (AGI), the Indonesian Petroleum Association (IPA), the Geological Society of America (GSA), the Society for Sedimentary Geology (SEPM) and the Geologists' Association (GA). Joint marketing agreements ensure that GSL Fellows may purchase these societies' publications at a discount. The Society's online bookshop (accessible from www.geolsoc.org.uk) offers secure book purchasing with your credit or debit card.

To find out about joining the Society and benefiting from substantial discounts on publications of GSL and other societies worldwide, consult www.geolsoc.org.uk, or contact the Fellowship Department at: The Geological Society, Burlington House, Piccadilly, London W1J 0BG: Tel. +44 (0)20 7434 9944; Fax +44 (0)20 7439 8975; Email: enquiries@geolsoc.org.uk.

For information about the Society's meetings, consult *Events* on www.geolsoc.org. uk. To find out more about the Society's Corporate Affiliates Scheme, write to enquiries@geolsoc.org.uk.

Published by The Geological Society from:
The Geological Society Publishing House
Unit 7, Brassmill Enterprise Centre
Brassmill Lane
Bath BA1 3JN, UK

(*Orders*: Tel. +44 (0)1225 445046
Fax +44 (0)1225 442836)
Online bookshop: http://bookshop.geolsoc.org.uk

British Library Cataloguing in Publication Data
A catalogue record for this book is available from the British Library.

ISBN 1-86239-139-4

Typeset by Aarontype Ltd, Bristol, UK
Printed by The Alden Press, Oxford, UK

Distributors

USA
AAPG Bookstore
PO Box 979
Tulsa
OK 74101-0979
USA
Orders: Tel. +1 918 584-2555
Fax +1 918 560-2652
E-mail: bookstore@aapg.org

India
Affiliated East-West Press PVT Ltd
G-1/16 Ansari Road, Daryaganj,
New Delhi 110 002
India
Orders: Tel. +91 11 327-9113
Fax +91 11 326-0538
E-mail: affiliat@nda.vsnl.net.in

Japan
Kanda Book Trading Co.
Cityhouse Tama 204
Tsurumaki 1-3-10
Tama-shi
Tokyo 206-0034
Japan
Orders: Tel. +81 (0)423 57-7650
Fax +81 (0)423 57-7651
E-mail: geokanda@ma.kcom.ne.jp

Contents

Applications of X-ray computed tomography in the geosciences

F. MEES[1], R. SWENNEN[2], M. VAN GEET[2] & P. JACOBS[1]

[1] *Department of Geology and Soil Science, Ghent University, Krijgslaan 281 S8, B-9000 Ghent, Belgium (e-mail: florias.mees@rug.ac.be)*
[2] *Physico-chemical Geology, K.U. Leuven, Celestijnenlaan 200C, B-3001 Heverlee, Belgium*
[3] *SCK-CEN, Waste & Disposal Department, Boeretang 200, B-2400 Mol, Belgium*

Abstract: X-ray computed tomography (CT) is a non-destructive technique with wide applications in various geological disciplines. It reveals the internal structure of objects, determined by variations in density and atomic composition. Large numbers of parallel 2D sections can be obtained, which allows 3D imaging of selected features. Important applications are the study of porosity and fluid flow, applied to investigations in the fields of petroleum geology, rock mechanics and soil science. Expected future developments include the combined use of CT systems with different resolutions, the wider use of related X-ray techniques and the integration of CT data with results of compatible non-destructive techniques.

X-ray computed tomography (CT) is a non-destructive technique that allows visualization of the internal structure of objects, determined mainly by variations in density and atomic composition. It requires the acquisition of one- or two-dimensional radiographs for different positions during step-wise rotation around a central axis, whereby either the source and detector or the sample are moved. This is followed by the reconstruction of two-dimensional cross-sections perpendicular to the axis of rotation.

CT images record differences in the degree of attenuation of the X-rays, which is material- and energy-dependent. The interactions that are responsible for this attenuation are mainly Compton scattering and photoelectric absorption. The contribution of the photoelectric effect depends on the effective atomic number and is especially important at low energies. At high energies, the Compton effect predominates and attenuation is mainly determined by density.

X-ray CT was developed as a medical imaging technique in the early 1970s (Hounsfield 1972, 1973). The possibility of its use in geology and engineering was soon recognised, resulting in large numbers of publications from the early 1980s onwards. Early applications include studies in the fields of soil science (Petrovic *et al.* 1982; Hainsworth & Aylmore 1983), meteoritics (Arnold *et al.* 1982), petroleum geology (Vinegar 1986; Vinegar & Wellington 1986), palaeontology (Haubitz *et al.* 1988), geotechnics (Raynaud *et al.* 1989) and sedimentology (Kenter 1989).

In this introductory paper, we present some general information about the technique and a brief overview of its applications in geology, with references to some recent studies. More detailed information can be found in recent review articles by Duliu (1999) and Ketcham & Carlson (2001).

Acquisition of transmission data

The basic components of X-ray CT scanners are an X-ray source, a detector and a rotation system. Various possible configurations exist, whereby the selected configuration is determined by sample size and the desired resolution, besides availability and access restrictions. Ideally, the X-ray beam should be parallel rather than fan- or cone-shaped (with a finite size of the origin), in which case the resolution is only determined by detector quality. High resolution (10 μm) can also be attained with microfocus X-ray tubes. Another aspect of acquisition is the energy spectrum of the X-ray source. Unless the X-rays are produced by radioactive decay, they are always polychromatic with a wide range in energy. This complicates quantitative analysis and creates artefacts in the CT images, due to the stronger attenuation of X-rays with lower energies. Monochromatisation by diffraction eliminates these problems, but involves a great decrease in intensity and is therefore only feasible for systems with high initial intensities, such as linear electron accelerators and synchrotron installations.

Most X-ray CT studies of geological materials are carried out using medical scanners of different generations, which differ mostly in configuration of the source and detectors. The highest

From: MEES, F., SWENNEN, R., VAN GEET, M. & JACOBS, P. (eds) 2003. *Applications of X-ray Computed Tomography in the Geosciences*. Geological Society, London, Special Publications, **215**, 1–6. 0305-8719/03/$15.

resolution that can be obtained with medical scanners is rather low, in the order of 600 µm × 600 µm × 1 mm.

Scanners that have been developed specifically for material research are now also available, both with fan-beam and cone-beam configurations. The latter allow the acquisition of data for many cross-sections at the same time, but their use involves more complicated reconstruction procedures. The advantages of CT systems for material research include the possibility of using higher X-ray intensities and the possibility of attaining higher resolutions by using microfocus X-ray tubes and by rotating the specimen rather than the source and detector. CT images of the highest quality are obtained with synchrotron radiation, which has the advantage of near-parallelism of the X-ray beam and the possibility of monochromatisation. A drawback of high-resolution systems is that high resolutions can only be obtained for small samples, e.g. <2 mm diameter for a 5 µm resolution in microfocus CT scans.

Reconstruction of CT images

The mathematical principles of backprojection procedures required for reconstruction have been known for a long time. In X-ray CT, a pre-reconstruction transformation of the radiographs is required to enhance contrast and sharpness of the CT images. This is done by applying a digital filter to the obtained signals. Various types of filters can be used for this operation. To be able to perform a reconstruction for a specific cross-section, data for the entire object must be available for this level, in radiographs for all different angles. The object should be completely in the field of view for each of these angle positions, i.e. surrounded by air in each radiograph, which limits the maximum sample size.

Image quality optimization requires a reduction of artefacts, which are inherent to the technique of X-ray CT. Several common types of artefacts can already be reduced to some extent by optimising acquisition conditions, but a further reduction during reconstruction is often needed. This is the case for: (i) misalignment artefacts, related to an imperfect alignment of source, detector and axis of rotation; (ii) ring artefacts, related to detector defects; (iii) star artefacts, related to the presence of dense objects; and (iv) line artefacts, related to the presence of anomalously bright pixels in the radiographs. The most important artefacts in CT images are related to 'beam hardening', which refers to the preferential attenuation of low-energy X-rays when a polychromatic X-ray beam passes through an object. This effect can be reduced to some extent by using a hardware filter, which involves a loss in intensity of the X-ray beam and a lowering of contrast in the radiographs. For beam hardening artefact reduction during reconstruction, linearisation procedures are highly recommended for monomineralic systems (Hammersberg & Mangard 1998). For polymineralic systems, a dual energy approach allows a dramatic improvement in image quality (De Man *et al.* 2001).

The development of simulators is a useful approach for artefact reduction, as they allow optimisation of the choice of acquisition parameters in function of the object to be studied. A major problem with this approach is the fact that the X-ray spectrum is often unknown and very difficult to measure (Hammersberg *et al.* 1998).

Applications in the geosciences

Suitable materials

X-ray CT can be used for qualitative and quantitative analysis of internal features of geological materials, if those features are marked by sufficiently great differences in atomic composition and/or density.

A strong contrast mainly exists between solid phases and the atmosphere. An important application of X-ray CT is, therefore, the study of porosity, including porosity represented by pores with dimensions below the resolution of the CT images. Examples of applications in this field include studies of soil macro-porosity (e.g. Perret *et al.* 1999) and reservoir rock characterization (e.g. Van Geet *et al.* 2000). The contrast with the atmosphere is also used as a means of visualizing the external morphology of objects, e.g. for surface roughness analysis (e.g. Fohrer *et al.* 1999) or for documenting the morphology of fossils (e.g. Rogers 1999).

The contrast between liquid phases and the atmosphere finds wide applications in the study of hydraulic properties, both in soil science (e.g. Mooney 2002) and petroleum geology (e.g. Peters *et al.* 1996). Recent technological improvements allow the discrimination between empty and (water-)filled pores (Roels *et al.* in press). In earlier studies, heavy elements had to be added as tracers (e.g. Mori *et al.* 1999), which has also been done to monitor gas transport (e.g. Karacan & Okandan 2001).

Visualization of the distribution of different solid phases generally requires pronounced variations in atomic composition. The few published studies mainly concern materials with iron-bearing minerals (e.g. Tivey 1998; Spiess *et al.* 2001) or minerals with exceptionally low attenuation values (e.g. Taylor *et al.* 2000). Synchrotron CT with monochromatic X-rays is required to differentiate between less contrasting materials.

For many applications, a correlation with results obtained by more traditional techniques, such as optical microscopy and scanning electron microscopy, is required (e.g. Van Geet *et al.* 2001).

Non-destructive analysis

One of the strengths of X-ray CT is its entirely non-destructive nature, at least at the scale of the sample. This aspect renders the technique highly suitable for monitoring of active processes, such as water and solute movement (e.g. Kasteel *et al.* 2000; Perret *et al.* 2000) and deformation (e.g. Besuelle *et al.* 2000; Wong 2001). These experiments can be carried out at elevated pressures and temperatures, which is a major advantage over more conventional techniques, e.g. in studies of fracture characteristics at reservoir conditions. The non-destructiveness of the technique also allows analysis of valuable or unique specimens, such as fossils (e.g. Rowe *et al.* 2001) and meteorites (e.g. Gnos *et al.* 2002).

3D rendering

CT analysis can produce large numbers of contiguous parallel cross-sections, which allows 3D visualisation of selected features and quantification of 3D volumes. Examples of applications that exploit this aspect of X-ray CT include studies of soil macroporosity (e.g. Capowiez *et al.* 1998; Pierret *et al.* 2002) and rock textures (e.g. Denison & Carlson 1997).

This volume

The opening paper in the present collection, by **Carlson *et al.***, gives an extensive illustration of the possible applications of X-ray CT in igneous and metamorphic petrology, meteoritics and palaeontology. The paper, which is based on studies carried out at the High-resolution X-ray CT Facility of the University of Texas at Austin, demonstrates how CT analyses can contribute to wider investigations by providing information that cannot be obtained by other methods. In a second review, **Akin & Kovscek** discuss applications of CT in petroleum geology, covering measurements of porosity and multiphase fluid flow characteristics. Experimental design and image processing methods are also covered.

CT analyses of pore characteristics are the subject of the next four contributions. **Jones *et al.*** present the results of synchrotron CT studies of various porous materials, carried out at three different research facilities (Brookhaven, Argonne and Grenoble). They address the potential of high resolution (2–10 μm) synchrotron CT analysis for porosity characterization in unconsolidated sediments and sandstone samples, using Wood's metal (a low-melting Bi-Pb-Sn-Cd alloy) for contrast enhancement in one example. The study also illustrates the use of X-ray fluorescence tomography in environmental geochemistry studies. **Van Geet *et al.*** present the results of porosity measurements with microfocus CT, which are compared with results obtained by more traditional methods. The authors describe procedures for artefact reduction, which is necessary to allow quantitative analysis of grey values. They discuss the use of a linearization technique for monomineralic rocks, such as limestone. For heterogeneous materials, a dual energy approach is proposed. **Vandersteen *et al.*** examine different methods for the quantification of fracture apertures by microfocus CT, using phantom objects and optical microscopy for calibration and verification. Their results are of importance in investigations requiring quantitative characterization of planar voids, e.g. in soil research and reservoir studies. **Sellars *et al.*** describe studies of fractures that developed by applying triaxial stress to cubic quartzite blocks with tabular gaps. The fracture patterns revealed by 3D rendering of CT data help to verify the predictions made by numerical models that are currently used to assess fracture development around deep mining excavations. The authors also comment on limitations imposed by the presence of artefacts, which hinder the production of 3D images of fractures.

Vogel & Brown present a geostatistical assessment of variations in gamma ray attenuation in scans with different resolutions. Their study assesses heterogeneities in bulk density that occur in a dolomite formation at a nuclear waste disposal site. The authors provide an evaluation of the effect of resolution and scale on semivariogram parameters.

The next four papers focus on fluid flow in rocks. **Géraud *et al.*** examine the hydraulic properties of a rock with very low porosity by using capillary saturation experiments. Using a fluid with high attenuation values, they are able to document differences in microporosity between different constituents of a granite, which affect flow patterns within the test samples. **Hirono *et al.*** use X-ray CT for direct imaging of fluid flow along faults in heterogeneous deformed rocks. These laboratory experiments, using KI as a tracer, help to clarify the relationship between the texture and hydraulic properties of fault zones.

Rousset-Tournier *et al.* present a study of water migration during drying experiments in sandstones. Their findings show that drying behaviour is influenced by the method of saturation. **Ruiz de Argandoña *et al.*** describe the patterns of water uptake during the total immersion of a porous limestone, showing that CT helps to relate water movement to petrographical characteristics of the rock. Both of the last two studies address issues that are important in the field of cultural heritage conservation.

In a first of three studies of soil behaviour, **Anderson *et al.*** describe the determination of porosity and hydraulic conductivity in soils by X-ray CT measurements during solute breakthrough experiments. Differences between measured values and simulated breakthrough curves derived by a finite element method are explained by the existence of small-scale heterogeneities, which can be assessed by X-ray CT. **Rogasik *et al.*** use X-ray CT to study variations in structure, macroporosity and calculated bulk density for soils with different textures and under different management systems. A quantitative study of compaction around earthworm burrows is also included. **Delerue *et al.*** present a novel quantitative method for 3D image analysis of soil macroporosity using CT images to obtain calculated hydraulic conductivity and permeability values. The method is based on the segmentation of the pore space into elementary objects and quantification of their characteristics, followed by the creation of a network model that allows conductivity calculations.

Two papers describe the results of compression tests using specially designed triaxial cells, both providing detailed information on the experimental set-up. **Karacan *et al.*** describe a triaxial cell that permits permeability measurements during deformation. Using X-ray CT to monitor deformation experiments with this cell, the authors evaluate changes in porosity, permeability and fluid flow under conditions of brittle and ductile failure. **Thomson & Wong** describe a system for performing triaxial tests under undrained conditions, which they use as a means for studying the deformation behaviour of unconsolidated sands. They discuss the observed void ratio redistributions and indicate directions for future research. **O'Neill *et al.*** present a qualitative assessment of a test that simulates the evolution of backfill material of an opencast mine. They use CT to unravel the nature of particulate interactions undergoing creep settlement.

The last group of papers illustrates the applications of X-ray CT in a number of geological disciplines. **Flisch & Becker** investigate sedimentary and deformation structures in lake deposits by CT scanning of sediment cores. They illustrate the potential of CT for core analysis and the study of unconsolidated sediments. The contribution by **Schreurs *et al.*** illustrates an application of X-ray CT in structural geology that involves the CT-monitoring of deformation experiments with sand-box models. These authors document changes in structure in three dimensions through time, which allows analysis of complex geological structures in contractional and extensional settings. A palaeontological study is presented by **Stock & Veis**, who perform micro-focus CT analyses of Jurassic echinoid fragments. Some of the features that they describe are interpreted as diagenetic modifications.

Future developments

One expected trend is the combined use of various CT instruments, whereby a material is scanned at different resolutions. For example, a rock core can be scanned at full size with a medical scanner at moderate resolution, followed by microfocus and synchrotron CT analysis of subsamples of different sizes. Optical microscopy and scanning electron microscopy will continue to be important sources of additional information.

A more widespread use can be expected for certain special procedures that have been developed for medical applications. One technique that has already been used in a number of geological studies is dual energy scanning (e.g. Kalendar *et al.* 1987), which allows quantification of density and effective atomic number. It involves the processing of scans that were recorded for energies at which the relative contributions of Compton scattering and photoelectric absorption are different. A promising technique is 'region of interest' scanning, providing high resolution CT images of a region in the interior of a sample. It requires high resolution acquisition for the region of interest, combined with a second scan of the same plane for the full width of the object, at low resolution; the second scan is used to provide information about the part of the object that is outside the field of view during acquisition for the high resolution scan (e.g. Gentle & Spyrou 1990).

The potential of a number of powerful techniques that are related to X-ray CT is still largely unexplored for geological materials. They mainly require a monochromatic or very intense X-ray beam, which means that they are mainly available at synchrotron installations. Examples are fluorescence X-ray tomography, recording element distributions down to trace level concentrations (e.g. Simionovici *et al.* 2001); phase contrast X-ray tomography, which allows visualisation of discontinuities by phase shifts of a coherent

X-ray beam, in materials with low attenuation coefficients (e.g. Cloetens *et al.* 1997, 1999), and diffraction X-ray tomography, which allows mapping of variations in mineralogical composition (e.g. Hall *et al.* 1998).

An integration of X-ray CT results with data obtained by other non-destructive techniques can be expected to become more common. Neutron tomography is a compatible technique, because neutrons interact with the nuclei of atoms, whereas X-ray CT is based on an interaction with electrons (e.g. Winkler *et al.* 2002). This results in large differences in element sensitivity whereby neutron tomography, in contrast to X-ray CT, is highly sensitive to elements such as hydrogen and carbon. A more widely used compatible technique is Nuclear Magnetic Resonance (NMR) analysis, which allows imaging of fluid flow in porous media and also provides information about pore size distributions (e.g. Watson & Chang 1997; Amin *et al.* 1998; Baraka-Lokmane *et al.* 2001). Its use for geological materials is limited by the rather low resolution and by the problems that can be caused by the presence of iron compounds.

Since medical CT scanners became widely available, CT results for geological materials have been successfully integrated with other analytical data in various geological disciplines. It is expected that in the near future, CT will be used more and more as a routine research tool. In future studies, CT data will be obtained for much greater numbers of samples, allowing a statistical evaluation of a large body of CT results.

References

AMIN, M.H.G., HALL, L.D., CHORLEY, R.J. & RICHARDS, K.S. 1998. Infiltration into soils, with particular reference to its visualization and measurement by magnetic resonance imaging (MRI). *Progress in Physical Geography*, **22**, 135–165.

ARNOLD, J.R., TESTA, J.P.J., FRIEDMAN, P.J. & KAMBIC, G.X. 1982. Computed tomographic analysis of meteorite inclusions. *Science*, **219**, 383–384.

BARAKA-LOKMANE, S., TEUTSCH, G. & MAIN, G. 2001. Influence of open and sealed fractures on fluid flow and water saturation in sandstone cores using Magnetic Resonance Imaging. *Geophysical Journal International*, **147**, 263–271.

BESUELLE, P., DESRUES, J. & RAYNAUD, S. 2000. Experimental characterisation of the localisation phenomenon inside a Vosges sandstone in a triaxial cell. *International Journal of Rock Mechanics and Mining Sciences*, **37**, 1223–1237.

CAPOWIEZ, Y., PIERRET, A., DANIEL, O., MONESTIEZ, P. & KRETZSCHMAR, A. 1998. 3D skeleton reconstructions of natural earthworm burrow systems using CAT scan images of soil cores. *Biology and Fertility of Soils*, **27**, 51–59.

CLOETENS, P., PATEYRON-SALOMÉ, M., BUFFIÈRE, J.Y., PEIX, G., BARUCHEL, J., PEYRIN, F. & SCHLENKER, M. 1997. Observation of microstructure and damage in materials by phase sensitive radiography and tomography. *Journal of Applied Physics*, **81**, 5878–5886.

CLOETENS, P., LUDWIG, W., BARUCHEL, J., VAN DYCK, D., VAN LANDUYT, J., GUIGAY, J.P. & SCHLENKER, M. 1999. Holotomography: quantitative phase tomography with micrometer resolution using hard synchrotron radiation X-rays. *Applied Physics Letters*, **75**, 2912–2914.

DE MAN, B., NUYTS, J., DUPONT, P., MARCHAL, G. & SUETENS, P. 2001. An iterative maximum-likelihood polychromatic algorithm for CT. *Institute of Electrical and Electronics Engineers Transactions on Medical Imaging*, **20**, 999–1008.

DENISON, C. & CARLSON, W.D. 1997. Three-dimensional quantitative textural analysis of metamorphic rocks using high-resolution computed X-ray tomography. 2. Application to natural samples. *Journal of Metamorphic Geology*, **15**, 45–57.

DULIU, O.G. 1999. Computer axial tomography in geosciences: An overview. *Earth-Science Reviews*, **48**, 265–281.

FOHRER, N., BERKENHAGEN, J., HECKER, J.M. & RUDOLPH, A. 1999. Changing soil and surface conditions during rainfall. Single rainstorm/subsequent rainstorms. *Catena*, **37**, 355–375.

GENTLE, D.J. & SPYROU, N.M. 1990. Region of interest tomography in industrial applications. *Nuclear Instruments and Methods in Physics Research Section A*, **299**, 534–537.

GNOS, E., HOFMANN, B., FRANCHI, I.A., AL-KATHIRI, A., HAUSER, M. & MOSER, L. 2002. Sayh al Uhaymir 094: A new martian meteorite from the Oman desert. *Meteoritics and Planetary Science*, **37**, 835–854.

HAINSWORTH, J.M. & AYLMORE, L.A.G. 1983. The use of computer-assisted tomography to determine spatial distribution of soil water content. *Australian Journal of Soil Research*, **21**, 435–443.

HALL, C., BARNES, P., COCKCROFT, J.K., COLSTON, S.L., HAUSERMANN, D., JACQUES, S.D.M., JUPE, A.C. & KUNZ, M. 1998. Synchrotron energy-dispersive X-ray diffraction tomography. *Nuclear Instruments and Methods in Physics Research Section B*, **140**, 253–257.

HAMMERSBERG, P. & MANGARD, M. 1998. Correction of beam hardening artefacts in computerized tomography. *Journal of X-ray Science and Technology*, **8**, 75–93.

HAMMERSBERG, P., STENSTRÖM, M., HEDTJÄRN, H. & MANGARD, M. 1998. Measurements of absolute energy spectra for an industrial micro focal X-ray source under working conditions using a Compton scattering spectrometer. *Journal of X-ray Science and Technology*, **8**, 5–18.

HAUBITZ, B., PROKOP, M., DÖHRING, W., OSTROM, J.H. & WELLNHOFER, P. 1988. Computed tomography of *Archeopterix*. *Paleobiology*, **14**, 206–213.

HOUNSFIELD, G.N. 1972. A method of and apparatus for examination of a body by radiation such as X- or gamma-radiation. British Patent No 1.283.915, London.

HOUNSFIELD, G.N. 1973. Computerized transverse axial scanning (tomography). Part 1: Description of system. *British Journal of Radiology*, **46**, 1016–22.

KALENDAR, W., BAUTZ, W., FELSENBERG, D., SÜSS, C. & KLOTZ, E. 1987. Materialselektive Bildge-bung und Dichtemessung mit der Zwei-Spektren-Methode. I. Grundlagen und Methodik. *Digitale Bilddiagnose*, **7**, 66–72.

KARACAN, C.O. & OKANDAN, E. 2001. Adsorption and gas transport in coal microstructure: investigation and evaluation by quantitative X-ray CT imaging. *Fuel*, **80**, 509–520.

KASTEEL, R., VOGEL, H.J. & ROTH, K. 2000. From local hydraulic properties to effective transport in soil. *European Journal of Soil Science*, **51**, 81–91.

KENTER, J.A.M. 1989. Applications of computerized tomography in sedimentology. *Marine Geotechnology*, **8**, 201–211.

KETCHAM, R.A. & CARLSON, W.D. 2001. Acquisition, optimization and interpretation of X-ray computed tomographic imagery: applications to the geosciences. *Computers and Geosciences*, **27**, 381–400.

MOONEY, S.J. 2002. Three-dimensional visualization and quantification of soil macroporosity and water flow patterns using computed tomography. *Soil Use and Management*, **18**, 142–151.

MORI, Y., IWAMA, K., MARUYAMA, T. & MITSUNO, T. 1999. Discriminating the influence of soil texture and management-induced changes in macropore flow using soft X-rays. *Soil Science*, **164**, 467–482.

PERRET, J., PRASHER, S.O., KANTZAS, A. & LANGFORD, C. 1999. Three-dimensional quantification of macropore networks in undisturbed soil cores. *Soil Science Society of America Journal*, **63**, 1530–1543.

PERRET, J., PRASHER, S.O., KANTZAS, A. & LANGFORD, C. 2000. A two-domain approach using CAT scanning to model solute transport in soil. *Journal of Environmental Quality*, **29**, 995–1010.

PETERS, E.J., GHARBI, R. & AFZAL, N. 1996. A look at dispersion in porous media through computed tomography imaging. *Journal of Petroleum Science and Engineering*, **15**, 23–31.

PETROVIC, A.M., SIEBERT, J.E. & RIEKE, P.E. 1982. Soil bulk density analysis in three dimensions by computed tomographic scanning. *Soil Science Society of America Journal*, **46**, 445–450.

PIERRET, A., CAPOWIEZ, Y., BELZUNCES, L. & MORAN, C.J. 2002. 3D reconstruction and quantification of macropores using X-ray computed tomography and image analysis. *Geoderma*, **106**, 247–271.

RAYNAUD, S., FABRE, D., MAZEROLLE, F., GÉRAUD, Y. & LATIÈRE, H.J. 1989. Analysis of the internal structure of rocks and characterisation of mechanical deformation by a non-destructive method: X-ray tomodensitometry. *Tectonophysics*, **159**, 149–159.

ROELS, S., VANDERSTEEN, K. & CARMELIET, J. in press. Measuring and simulating moisture uptake in a fractured porous medium. *Advances in Water Resources*.

ROGERS, S.W. 1999. Allosaurus, crocodiles, and birds: evolutionary clues from spiral computed tomography of an endocast. *Anatomical Record*, **257**, 162–173.

ROWE, T., KETCHAM, R.A., DENISON, C., COLBERT, M., XU, X. & CURRIE, P.J. 2001. Forensic palaeontology: The Archaeoraptor forgery. *Nature*, **410**, 539–540.

SIMIONOVICI, A., CHUKALINA, M., GUNZLER, F., SCHROER, C., SNIGIREV, A., SNIGIREVA, I., TUMMLER, J. & WEITKAMP, T. 2001. X-ray microtome by fluorescence tomography. *Nuclear Instruments and Methods in Physics Research Section A*, **467**, 889–892.

SPIESS, R., PERUZZO, L., PRIOR, D.J. & WHEELER, J. 2001. Development of garnet porphyroblasts by multiple nucleation, coalescence and boundary misorientation-driven rotations. *Journal of Metamorphic Geology*, **19**, 269–290.

TAYLOR, L.A., KELLER, R.A., SNYDER, G.A., WANG, W.Y., CARLSON, W.D., HAURI, E.H., MCCANDLESS, T., KIM, K.R., SOBOLEV, N.V. & BEZBORODOV, S.M. 2000. Diamonds and their mineral inclusions, and what they tell us: A detailed 'pull-apart' of a diamondiferous eclogite. *International Geology Review*, **42**, 959–983.

TIVEY, M.K. 1998. Documenting textures and mineral abundances in minicores from the TAG active hydrothermal mound using X-ray computed tomography. *In*: HERZIG, P.M., HUMPHRIS, S.E., MILLER, D.J. & ZIERENBERG, R.A. (eds) *Proceedings of the Ocean Drilling Pogram, Scientific Results*, **158**, 201–210.

VAN GEET, M., SWENNEN, R. & WEVERS, M. 2000. Quantitative analysis of reservoir rocks by microfocus X-ray computerised tomography. *Sedimentary Geology*, **132**, 25–36.

VAN GEET, M., SWENNEN, R. & DAVID, P. 2001. Quantitative coal characterisation by means of microfocus X-ray computer tomography, colour image analysis and back scatter scanning electron microscopy. *International Journal of Coal Geology*, **46**, 11–25.

VINEGAR, H.J. 1986. X-ray CT and NMR imaging of rocks. *Journal of Petroleum Technology*, **38**, 257–259.

VINEGAR, H.J. & WELLINGTON, S.L. 1986. Tomographic imaging of three-phase flow experiments. *Review of Scientific Instruments*, **58**, 96–107.

WATSON, A.T. & CHANG, C.T.P. 1997. Characterizing porous media with NMR methods. *Progress in Nuclear Magnetic Resonance Spectroscopy*, **31**, 343–386.

WINKLER, B., KNORR, K., KAHLE, A., VONTOBEL, P., LEHMANN, E., HENNION, B. & BAYON, G. 2002. Neutron imaging and neutron tomography as non-destructive tools to study bulk-rock samples. *European Journal of Mineralogy*, **14**, 349–354.

WONG, R.C.K. 2001. Strength of two structured soils in triaxial compression. *International Journal for Numerical and Analytical Methods in Geomechanics*, **25**, 131–153.

Applications of high-resolution X-ray computed tomography in petrology, meteoritics and palaeontology

W. D. CARLSON, T. ROWE, R. A. KETCHAM & M. W. COLBERT

Department of Geological Sciences, University of Texas at Austin, Austin, Texas 78712, USA (e-mail: wcarlson@mail.utexas.edu)

Abstract: High-resolution and ultra-high-resolution X-ray computed tomography are rapid, non-destructive and extremely powerful techniques for three-dimensional examination and measurement of a great variety of geological materials and specimens with sizes from several millimetres to several decimetres. A review of recent applications in petrology, meteoritics and palaeontology, which utilized an instrument optimized for geological studies (High-Resolution X-ray Computed Tomography Facility of the University of Texas at Austin), documents an abundance of novel scientific results and illuminates the potential for still broader application of these techniques in the earth sciences.

Many earth science investigations require examination or measurement of the internal features of specimens in three dimensions, tasks to which X-ray computed tomography (CT) is well-suited. A variety of different X-ray CT instruments and techniques are available. The scale of the object to be studied and the spatial resolution required in the images commonly dictate which is employed (Table 1). High-resolution X-ray CT (HRXCT) and ultra-high-resolution X-ray CT (UHRXCT) are techniques suitable for studying objects with dimensions from a few millimetres to a few decimetres. Into this range of scales falls a wide variety of intriguing geological problems. This paper provides a selective review of recent HRXCT and UHRXCT work done at the University of Texas at Austin, in the form of very brief synopses of individual projects with references to the published literature for more complete exposition. Our purpose is to stimulate further exploitation of these techniques in the earth sciences by providing an overview of the broad range of geological questions that can be productively addressed using this rapid, non-destructive, visually powerful and fully digital tool.

Table 1. *General classification of X-ray computed tomography*

Type	Scale of observation	Scale of resolution
Conventional (medical)	m	mm
High-resolution	dm	100 μm
Ultra-high-resolution	cm	10 μm
Microtomography	mm	μm

Advantages of X-ray computed tomography for geological investigations

The best-known advantage of X-ray CT is its ability to quickly and non-destructively image the interior of opaque solid objects in three dimensions. For rare or irreplaceable specimens that cannot or should not be destructively sectioned – including most meteorites and many fossils – X-ray CT may be the only practical means of gaining information on internal materials and geometries or other features hidden from external view. Even if a specimen is regarded as expendable, X-ray CT eliminates the extreme laboriousness of serial sectioning.

The digital character of a CT dataset facilitates computer visualization and animation, allowing a user to interact with the data and to better understand the features and interrelationships among elements of the dataset. Similarly, digital data can be interrogated to quickly obtain quantitative measurements of dimensions, angles, volumes or nearly any metrical feature of interest. Finally, these digital data provide an unrivalled means of archiving and exchanging information. For example, few of the many palaeontologists interested in dinosaurs will ever have a chance to handle the skull of the world's oldest dinosaur, *Herrerasaurus* – only a single complete specimen is known and it resides in a museum in Argentina – but a high quality CT dataset provides unlimited virtual access, while offering information on its anatomy that is impossible to see even if one were holding the specimen in one's hands.

From: MEES, F., SWENNEN, R., VAN GEET, M. & JACOBS, P. (eds) 2003. *Applications of X-ray Computed Tomography in the Geosciences*. Geological Society, London, Special Publications, **215**, 7–22. 0305-8719/03/$15.

X-ray CT at the University of Texas at Austin

This article does not attempt to provide a comprehensive review of geological applications of HRXCT and UHRXCT in the published literature. Instead it focuses on selected projects conducted at the High-Resolution X-ray Computed Tomography Facility of the University of Texas at Austin, projects chosen to illustrate the wide range of geological objects and problems to which these techniques are applicable. Many projects in anthropology, biology, and engineering are likewise amenable to CT study, but these fields were considered to lie beyond the scope of this article.

The X-ray CT laboratory at the University of Texas is dedicated to applications in the geosciences and related fields. Because partial operational funding is provided by the US National Science Foundation, the laboratory functions as a national shared multi-user facility, welcoming external researchers. Most of the projects described below were undertaken in collaboration with colleagues from other institutions, as the cited literature clearly reflects.

The industrial scanner at the University of Texas was custom designed and built by Bio-Imaging Research, Inc., of Lincolnshire, Illinois. The principal design requirement was sufficient flexibility to provide imagery for a very broad range of geologic specimens and materials, that is, across a wide span of combinations of spatial resolution, density discrimination and penetrating ability. The objective was to produce an instrument that would complement existing facilities employing modified medical scanners (capable of penetrating specimens of moderate density, decimetres to metres in size, with spatial resolution on the order of a millimetre) and specialized facilities employing synchrotron radiation (capable of penetrating low- to moderate-density specimens, up to several millimetres in size, with spatial resolution on the order of a micrometre). This objective was met with a modular design that incorporates, within a single radiation-safety enclosure, two X-ray sources and three detectors. These sources and detectors can be used in various combinations to optimize trade-offs among penetrating ability, spatial resolution, density discrimination, imaging modes and scan times.

The scanner is comprised of two subsystems. One yields ultra-high-resolution data on small objects that can be penetrated by relatively low-energy X-rays. The other yields high-resolution data on larger or denser objects that can be penetrated only by higher-energy X-rays.

Ultra-high-resolution tomography of specimens up to a few cm in diameter employs a 200 kV micro-focal X-ray source, in combination with a specimen-positioning stage capable of ~1 μm reproducibility in vertical positioning and a conventional medical image-intensifier detector. The system's primary magnification, which increases with the specimen's proximity to the X-ray source, combines with the fixed pixel size of the video image to determine the limits of spatial resolution. The great flexibility of this system allows imaging of specimens from several cm to a few mm in diameter with spatial resolution from *c.* 250 μm to as low as *c.* 5 μm in favourable cases.

High-resolution tomography of large specimens employs a 420 kV tungsten X-ray source, a rotating turntable that can accommodate samples up to 50 kg in weight and either of two available high-energy detectors. One detector is a 512-channel cadmium-tungstate solid-state linear array, which provides superior sensitivity because of its high absorption efficiency. Its vertical aperture (slice thickness) ranges from 5 mm down to 0.25 mm, with a horizontal channel pitch of 0.31 mm. The other high-energy detector is a 2048-channel gadolinium oxysulfide radiographic line scanner. This detector, although less sensitive, provides higher in-plane spatial resolution, with a channel pitch of 0.05 mm. It can be used either in a high-resolution mode, with a vertical aperture of 0.25 mm, or (by sacrificing some sensitivity) in a thick-slice, rapid-scan mode using a vertical aperture that can vary from 0.5 to 5 mm. Operating in translate-rotate mode, the high-energy subsystem can image specimens up to 500 mm in diameter. The source and detectors are capable of 750 mm of vertical motion, but the unobstructed sample volume is 1500 mm tall, so specimens up to a metre and a half in maximum dimension (e.g. segments of drill cores) can be scanned by first imaging one half, then inverting the specimen to scan the other half. At the opposite end of the size spectrum, this subsystem can also be used to scan any smaller specimen for which *c.* 250 μm spatial resolution is sufficient.

Adding to the flexibility of this scanner is its ability to acquire data in several different modes, to further optimize performance trade-offs. Both subsystems can collect data in third-generation geometry (rotate-only, centred and variably offset) and the high-energy subsystem can also be operated in second-generation geometry (translate-rotate), which allows for increased resolution within subvolumes of the specimen by selective reconstruction of the raw absorption data. On both subsystems, complete control over the positioning of the specimen ensures that the maximum magnification (hence, maximum resolution) can always be achieved. A 'multi-slice' mode on the ultra-high-resolution subsystem permits simultaneous acquisition of data for a central slice and several neighbouring slices above and below. This markedly reduces scan times at a small cost in data quality owing to slight off-axis distortion. This distortion is caused in part by utilization of standard filtered back-projection reconstruction, which assumes all data are coplanar. A 'cone-beam' imaging mode, using the Feldkamp reconstruction algorithm, is also implemented on the ultra-high-resolution subsystem. This allows simultaneous collection of data throughout the full 3D volume of small specimens, further reducing scan times for applications in which maximal resolution and dimensional accuracy are not required.

Contrast resolution in CT images depends on the differential X-ray attenuation characteristics of the phases present, which are complex functions of X-ray energy, mass density and composition. Using the lower-energy X-rays on the UHRXCT system, which provides greater sentivity to differences in atomic number, it is possible to differentiate phases with virtually identical mass densities, such as quartz and orthoclase (cf. Fig. 4 in Ketcham & Carlson 2001).

Additional background information about X-ray CT methods applied to geological materials and further details of image acquisition, optimization and interpretation, are presented elsewhere (Ketcham & Carlson 2001).

Applications in petrology

Most of the applications in this section exploit the digital character of X-ray CT data, which makes it practical to obtain precise quantitative measurements of the proportions, shapes, sizes and locations of particles, crystals, segregations or voids. Because such measurements are commonly numbered in the thousands or even tens of thousands, it is possible with such data to succeed with statistical analyses that would be otherwise impractical, if not impossible, to undertake.

Quantitative analysis of metamorphic textures

Three-dimensional CT data on the sizes and locations of crystals (Fig. 1) permit discrimination between competing hypotheses for the atomic-scale mechanisms controlling the growth of porphyroblasts. Early work demonstrated that reaction kinetics limited by rates of intergranular diffusion were common for garnet in regionally metamorphosed rocks (Carlson & Denison 1992; Denison & Carlson 1997; Denison *et al.* 1997).

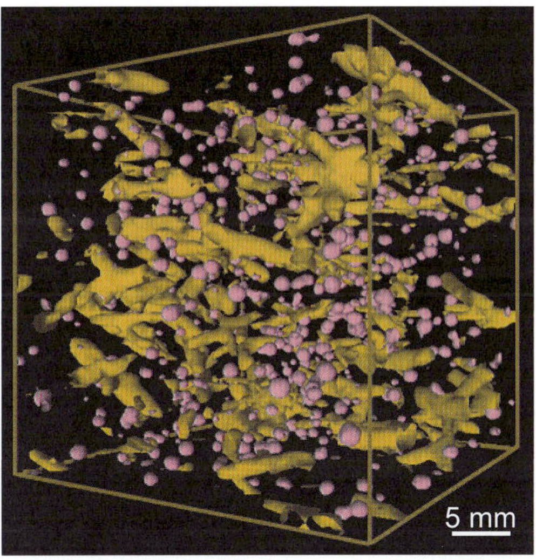

Fig. 1. Isosurface image of garnet (rendered violet) and staurolite (rendered yellow-brown) in pelitic schist from Picuris Mountains, New Mexico. Bounding box for image is approximately 45 × 45 × 42 mm.

Numerical simulations of such processes, constrained by the X-ray CT data, allowed first-ever estimates of key kinetic parameters such as the absolute rates of garnet nucleation and intergranular diffusion of aluminium (the species whose transport is rate-limiting) (Carlson *et al.* 1995). Compositional studies based on serial-sectioning guided by X-ray CT data (Chernoff & Carlson 1997, 1999) demonstrated the importance of this reaction control: sluggish intergranular diffusion appears to account for the observation that in some metamorphic rocks, Ca and many trace elements fail to equilibrate at the thin-section scale during garnet growth, severely restricting the confidence that can be placed in thermobarometric methods that assume otherwise. X-ray CT data made possible a statistical analysis of textures in a prograde sequence of garnetiferous pelites, which revealed that increases in grain size and decreases in crystal number densities are principally the consequence of higher nucleation temperatures, rather than the result of post-crystallization annealing (Carlson 1999; Hirsch & Carlson 2001).

Metasomatism in Earth's mantle and the origin of diamonds

The character of mantle metasomatism and the origin of diamonds – fundamental enigmas in mantle petrology – have been addressed by complete three-dimensional characterization of

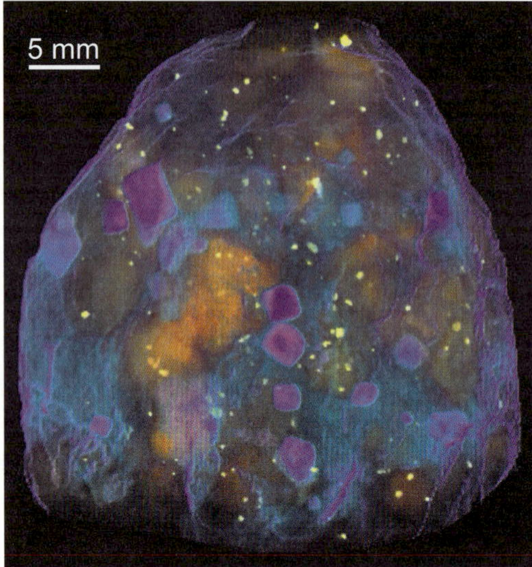

Fig. 2. Three-dimensional reconstruction from X-ray CT data of a diamondiferous eclogite from Udachnaya, Siberia, approximately 35 mm in diameter. Matrix of partially altered clinopyroxene has been rendered semi-transparent in shades of green; large crystals of garnet are semi-transparent in shades of red-orange; diamonds are opaque in shades of blue to violet; sulfides are yellow and the outer surface of the specimen is semi-transparent in shades of blue. Specimen courtesy of Lawrence Taylor, University of Tennessee, Knoxville.

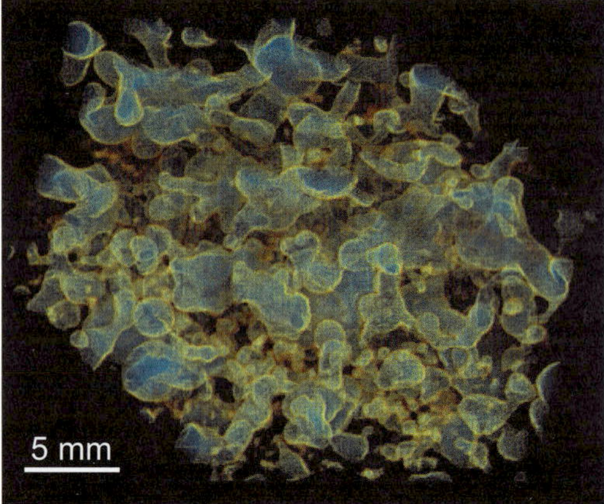

Fig. 3. Three-dimensional reconstruction from X-ray CT data of a 15 mm cube of vesicular basalt. Vesicle boundaries have been rendered semi-transparent in shades of green to yellow. Volumes of such irregular vesicles cannot be determined accurately by any two-dimensional sectioning technique, but are rapidly computed from X-ray CT imagery. Specimen courtesy of Dork Sahagian, University of New Hampshire. Reprinted from Carlson *et al.* (1999), with permission.

textural relationships in extraordinary diamondiferous eclogites from Udachnaya, Siberia (Taylor *et al.* 2000). CT imagery revealed that none of the diamonds were enclosed within either garnet or clinopyroxene, but instead all were spatially associated with 3D networks of subplanar zones of altered clinopyroxene that penetrate the specimens (Fig. 2). The implication is that the diamond post-dates formation of the eclogitic assemblage and that it was introduced into the rock in conjunction with metasomatic input(s) of carbon-rich fluids.

Palaeoaltimetry and continental uplift from size distributions of basalt vesicles

Differences in the modal vesicle sizes of samples from the bottom and top of basaltic lava flows preserve a quantitative record of palaeoelevations at the time of eruption (Sahagian & Maus 1994). When vesicles are nearly spherical, modal vesicle sizes in three dimensions can be recovered by laborious stereological techniques in thin section, but deformation or coalescence of bubbles in the melt renders stereology erroneous and requires collection of true three-dimensional data (Fig. 3). Such data are acquired easily, rapidly and accurately by means of X-ray CT (Proussevitch *et al.* 1998). When combined with geochronology on the flows, determinations of palaeoelevation yield quantitative rates of continental uplift. Application of this technique to basaltic lavas of the SW USA has defined the timing and extent of uplift of the Colorado Plateau (Sahagian *et al.* 2002).

Topology of melt flow paths in migmatites

Processes and rates of melt flow in the anatectic zone depend on several factors related to the geometry and topology of the flow paths. CT imagery of leucosome-melanosome-mesosome relationships in migmatites (Fig. 4) can be used to make quantitative determinations of features relevant to the permeability of such rocks to melt. This includes the effective porosity, ratios of pore surface to volume, mean pore radius, surface roughness and the connectivity and tortuosity of flow paths. The leucosome network geometry of two contrasting migmatites has been determined by X-ray CT and serial sectioning (Brown *et al.* 1999). This analysis demonstrated the 3D connectivity of leucosome in both samples, highlighted differences between the two in effective porosity and tortuosity, and provided constraints on flow-path topology that can be used in quantitative models simulating melt flow.

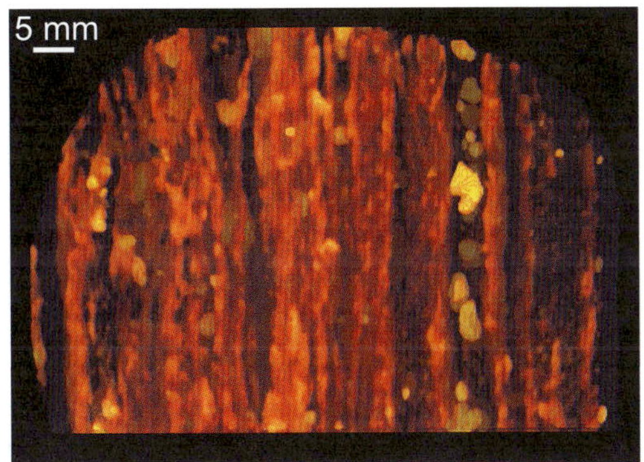

Fig. 4. Three-dimensional reconstruction from X-ray CT data of stromatic migmatite from southern Brittany, France (maximum dimension 63 mm). Leucosome has been rendered mostly transparent; melanosome and mesosome are shown as opaque in reddish shades and garnet crystals are rendered opaque in shades of yellow. Specimen courtesy of Michael Brown and Mary Anne Brown, University of Maryland. Reprinted from Brown *et al.* (1999), with permission.

Magmatic differentiation from crystal-mush compaction

The recent discovery that plagioclase crystals cluster together to form a continuous 3D network early in the crystallization history of slowly cooled basaltic magma (Philpotts *et al.* 1998) offers a means of determining the degree of the compaction of cooling flows and thus permits evaluation of this potential mechanism for their chemical differentiation. Measurements from thin sections and CT data (Fig. 5) on the thick Holyoke flood basalt of Connecticut (USA) revealed a pattern of chain compaction and dilation that matches exactly the pattern indicated by the chemical variation in this flow (Philpotts *et al.* 1999). The origin of these networks in the upper solidification front of the flow, combined with their preservation in lower parts of the flow, is clear evidence for convective transfer of dense crystal mush from the roof to the floor of the magma sheet, a mechanism long postulated but previously never unequivocally demonstrated (Philpotts & Dickson 2000).

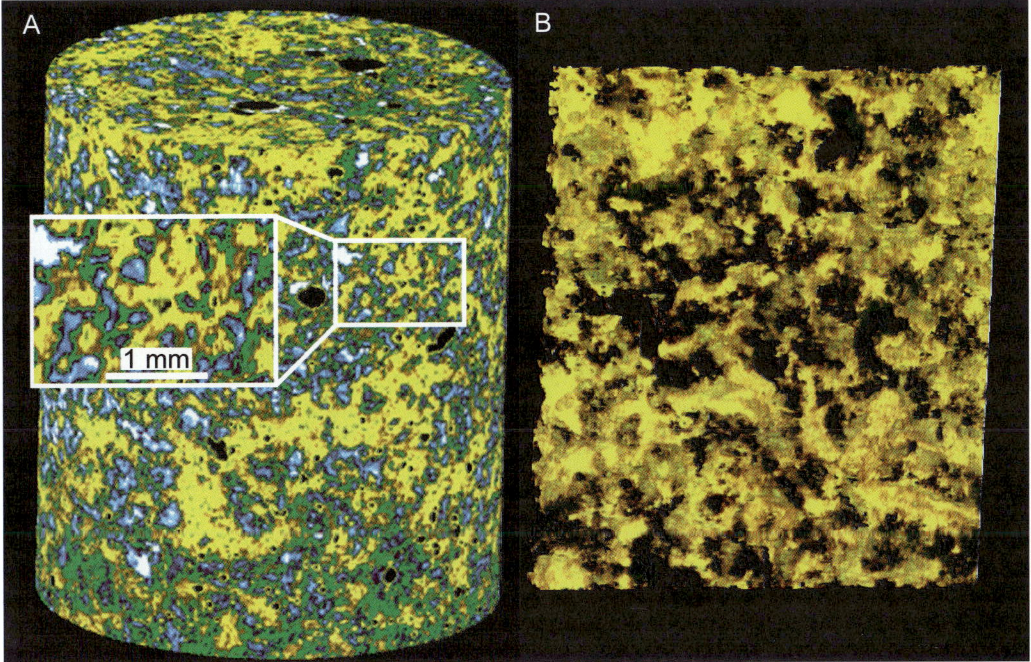

Fig. 5. (**a**) Three-dimensional reconstruction from X-ray CT data of a 1 cm tall core of Holyoke basalt subjected to 50% partial melting. Plagioclase is rendered yellow, pyroxene is blue-to-white, glass is green, and voids are black. (**b**) Three-dimensional reconstruction of subsampled region, in the shape of a thin slab from the centre of the core, with all components except plagioclase rendered transparent. Specimen courtesy of Anthony Philpotts, University of Connecticut. Reprinted from Philpotts *et al.* (1999), with permission.

Modelling of fluid flow through vuggy pore space

Medical CT data has been employed extensively to determine porosity and permeability and to visualize fluid flow in sediments and core specimens (Wellington & Vinegar 1987; Withjack 1988; Withjack & Akervoll 1988). However, in some cases the flow characteristics of reservoir lithologies are dominated by large-scale features, such as fractures or vugs, that make them impossible to analyse either with medical CT or traditional flow methods because the size of sample required to adequately represent these features is too large. Industrial CT scanners that use higher-energy X-rays are able to image larger and denser samples, allowing flow properties to be studied and modelled effectively in many cases. In one such study, high-resolution CT of a 28 × 36 cm caprinid reef block from the Pipe Creek Formation in Texas revealed that the vuggy porosity caused by rudist fossils was highly but not completely connected (Fig. 6). Computer models based on the CT data indicate high effective permeability, poor sweep efficiency and large anisotropy, all caused by the large but tortuous fluid pathways.

Fig. 6. Three-dimensional reconstruction from X-ray CT data of porosity dominated by rudist fossils in a large block (28 cm diameter, 36 cm high) of 'Pipe Creek Reef', a caprinid build-up in the Albian-aged Glen Rose Formation, Texas. Green shades indicate air-filled vugs and blues represent vugs with mud infilling, in addition to smaller-scale macroporosity. Specimen courtesy of James Jennings, University of Texas at Austin Bureau of Economic Geology.

Applications in meteoritics

Because of the extreme rarity of meteorites, destructive analysis of them is avoided whenever possible. For these precious specimens, X-ray CT permits imaging of three-dimensional relationships that are otherwise unobservable. The applications presented in this section exploit this ability of X-ray CT to reveal internal features non-destructively, as well as its ability to quantify proportions of materials and to measure particle volumes.

Melt generation, segregation and migration in lodranites

Initial curatorial slabbing of the lodranite meteorite GRA95209 produced the surprising discovery of abundant metal segregated from the mixed silicate-metal matrix. X-ray CT revealed a sheath of metal (obscured by fusion crust) extending along the exterior surface of much of the specimen and into the interior, where it connected to several veins and stringers of metal within the matrix. The matrix was seen to include metal-poor silicate-rich regions. The coalescing veins and dykelets revealed in 3D by X-ray CT (Fig. 7) are readily interpreted as pathways of melt migration, and measurements from CT data of the relative volumes of metal and the metal-depleted matrix permit mass-balance calculations that rule out a completely local origin for the melt (Carlson & McCoy 1998; McCoy & Carlson 1998). The images, therefore, record a complicated and intriguing picture of melt generation and migration in the lodranite parent body, a primitive planetisimal that is one of the best existing models for processes occurring during core-mantle segregation in the terrestrial planets.

Impact-induced melting and metamorphism of chondrites

Veinlets of metal apparent on the exterior surface of the Portales Valley meteorite prompted a CT study to determine the textural character of metal-silicate relationships and to guide curatorial efforts at cutting the meteorite for distribution to scientists for research purposes. The scans revealed that the specimen was strongly brecciated, with metal filling the interstices between angular

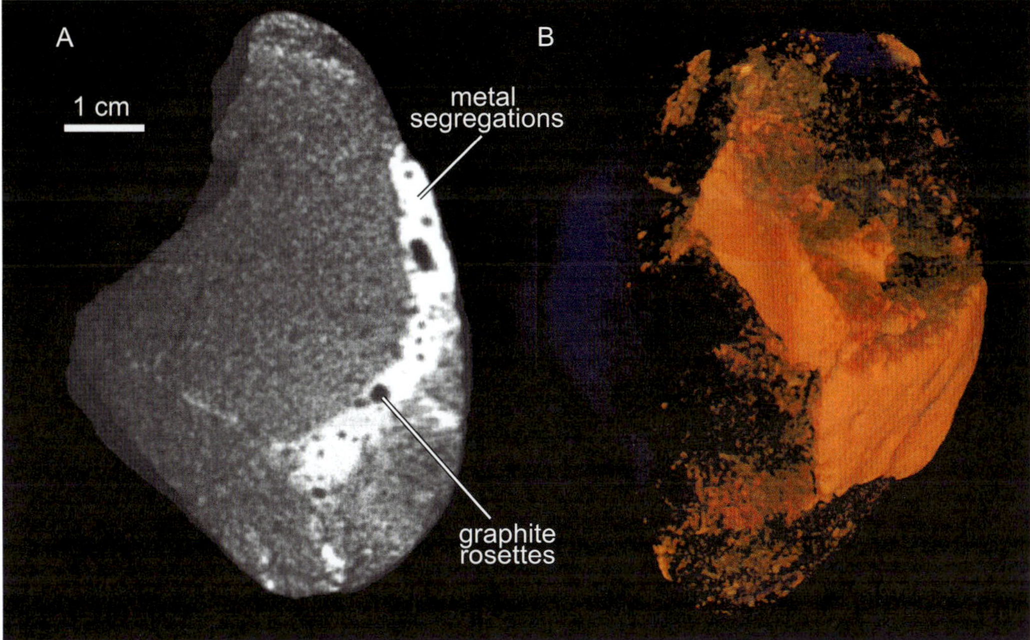

Fig. 7. (**a**) Perspective view of X-ray CT images of lodranite meteorite GRA95209 (maximum dimension *c.* 72 mm), producing a 'virtual sawcut' through the specimen exposing its internal features. Homogeneous bright regions are metal segregations; they contain dark, roughly spherical spots identified optically as rosettes of graphite. Mottled regions contain mixture of fine-grained metal and silicates, in varying proportions as expressed by overall grey-scale level. (**b**) Three-dimensional volume rendering of the same data to show the distribution of metal. Metal particles are rendered opaque in shades of orange; all other materials are nearly transparent in shades of dark blue. Specimen courtesy of Timothy McCoy, Smithsonian Institution, and Marilyn Lindstrom, NASA-Johnson Space Center. Reprinted from Carlson & McCoy (1998), with permission.

silicate clasts (Fig. 8). Subsequent study suggested an origin for this chondrite in which metamorphism, metal fusion and brecciation resulted from impact. The CT data suggest the possibility of size grading of the silicate clasts within larger pockets of metal, which requires the presence of an appreciable gravity field, implying that the impact likely took place while the specimen was part of its parent body (Rubin *et al.* 2001).

Particle sorting in the solar nebula

Particle sizes in chondritic meteorites potentially contain information on turbulence and sorting processes in the solar nebula. Hypotheses that differentiate between mass sorting in a quiescent nebula and aerodynamic sorting in a turbulent nebula can be tested if the sizes of both chondrules and metal-troilite particles can be measured. Although chondrules may be equant enough to allow size determination by stereological methods in thin-section, metal-troilite particles are so irregularly shaped that their true volumes cannot be recovered from any sectional measurement (Fig. 9). A high-resolution X-ray CT study of particle sizes in three different meteorites (Kuebler *et al.* 1999) provided key data regarding the processes of particle sorting in the solar nebula. Because in all three cases the metal-troilite particles, in comparison to the chondrules, have different masses but similar aerodynamic stopping times, these data are supportive of a model involving accretion in eddies in a turbulent solar nebula.

Applications in palaeontology

The rarity of many fossils, combined with their morphological complexity, makes them ideal candidates for examination using X-ray CT. Many of the applications in this section benefit principally from the ability to digitally manipulate X-ray CT data for enhanced visualization, although the ability to non-destructively examine features hidden from external view and to make precise morphological measurements are also prominent advantages to palaeontological applications.

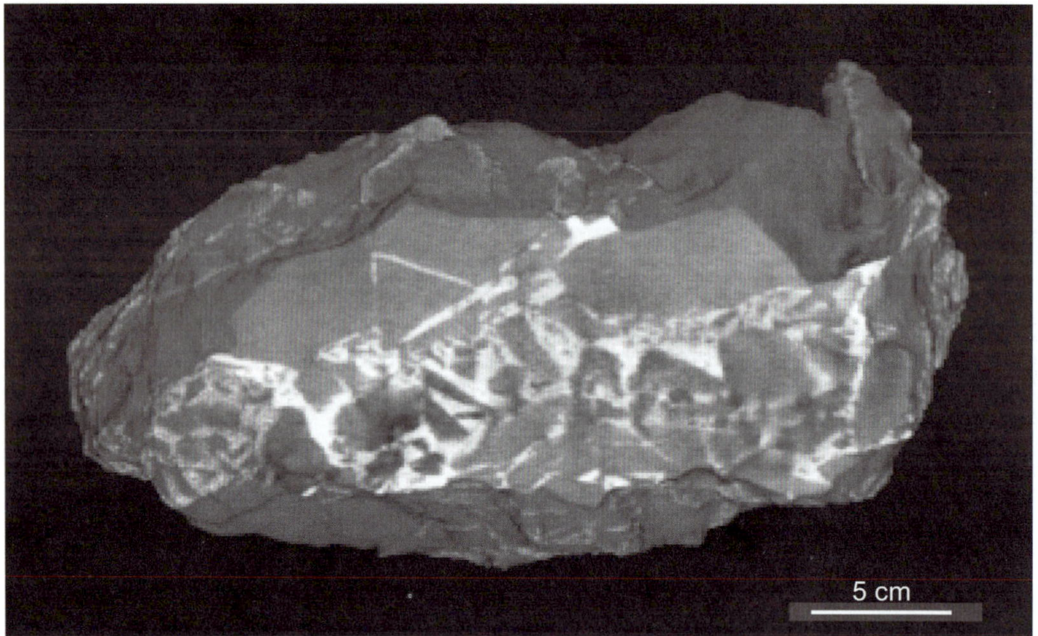

Fig. 8. 'Virtual sawcut' through Portales Valley meteorite (maximum dimension ~30 cm) produced by creating a perspective view of CT images. Brightest whites (metal) lie in the planar surface of the page, and the surrounding material in darker shades is a rendering of the exterior surface below that level. Small silicate particles (grey) in the wider portions of the metallic region tend to concentrate near the 'top' of the metal veins, suggesting segregation in a gravity field (oriented vertically in this image). Specimen courtesy of John Wasson and Alan Rubin, University of California, Los Angeles. Reprinted from Rubin *et al.* (2001), with permission.

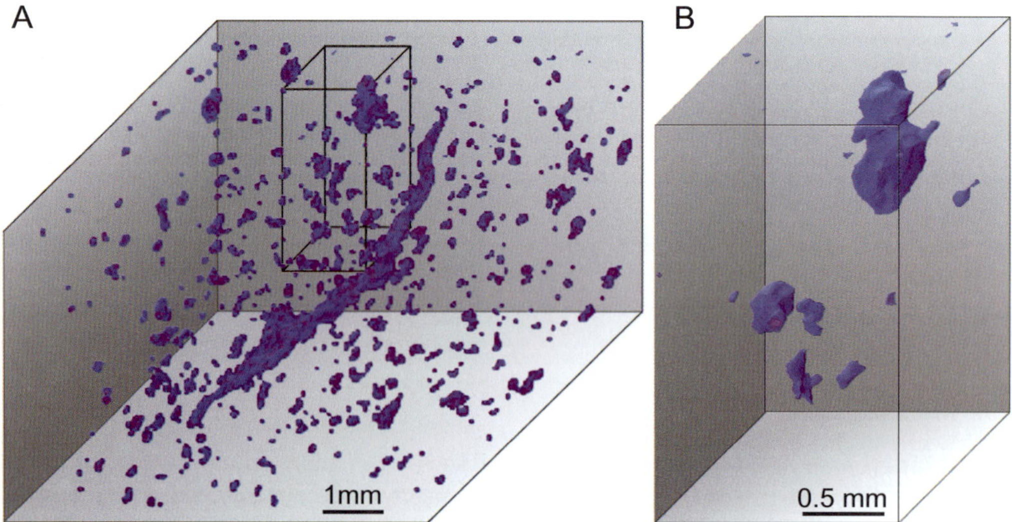

Fig. 9. Three-dimensional representation of shapes, sizes and positions of metal-troilite particles in Hammond Downs chondritic meteorite from X-ray CT data. (**a**) Particles in cubic volume roughly 7 mm on edge include one extraordinarily large vein-like object; dashed prism locates volume shown in part (**b**) of this figure. (**b**) Enlargement of a portion of same dataset, emphasizing that the irregular shapes of these grains make it impossible to determine their volumes accurately by any two-dimensional sectioning method. Specimen courtesy of Harry McSween, Jr, University of Tennessee, Knoxville. Reprinted from Kuebler *et al.* (1999), with permission.

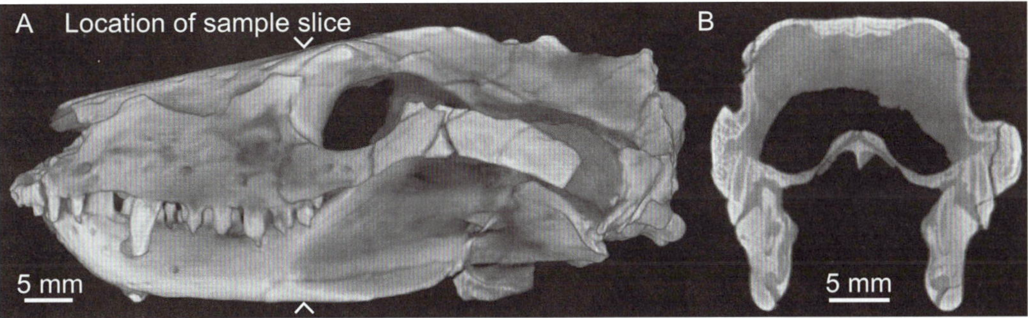

Fig. 10. (**a**) Three-dimensional reconstruction from X-ray CT data of the Early Triassic fossil skull of *Thrinaxodon* (maximum dimension 7.2 cm). (**b**) Single data slice illustrating the complex internal structures that can be imaged by UHRXCT. Bright areas are fossilised bone, darker areas are the silicate matrix. Specimen courtesy of the University of California Museum of Paleontology (UCMP #40466), University of California at Berkeley.

Mammalian origins: the endocranium of Thrinaxodon

The first application of HRXCT to palaeontology was the scanning of a 280 million-year-old fossilized skull of a distant mammalian relative known as *Thrinaxodon* (Fig. 10), an animal that has been at the centre of many studies of early mammalian evolution (e.g. Kemp 1982; Gauthier *et al.* 1988; Rowe 1988, 1993). Early mammalian history is principally represented by tiny animals, and their fossils are very rare and scattered widely among collections (Rowe 1999). *Thrinaxodon* is one of the larger species, being the size of a house cat, and it is known from only a handful of specimens. Non-destructive examination of *Thrinaxodon* using CT imagery was aimed at answering several long-standing problems surrounding the origin of mammals by extracting new information on the internal structure of the skull and on the size and morphology of the brain as recorded in its endocast. This work offered the first look at the endocranial cavity and permitted the first digital endocast of a specimen to be extracted from a digital dataset (Rowe *et al.* 1993).

Mammalian origins: the neocortex

The modern marsupial *Monodelphis domestica*, a popular lab opossum that grows to the size of a rat, is one of the least specialized of living mammals and its brain, compared to its body size, is one of the smallest of any living mammal (Fig. 11). Scanning a growth series of skulls allowed measurement of endocranial volume throughout life, revealing how the brain and skull affect each other as they grow. Comparisons of *Monodelphis* with *Thrinaxodon* and several other fossils showed that the origin of Mammalia (Rowe 1988) coincided with the appearance of the neocortex, a hugely inflated part of the forebrain that is unique to mammals. A consequence of this great cerebral expansion was enhancement of the senses of smell and hearing, heightened motor-sensory integration and probably also an increased basal metabolic rate (Rowe 1996*a*, *b*).

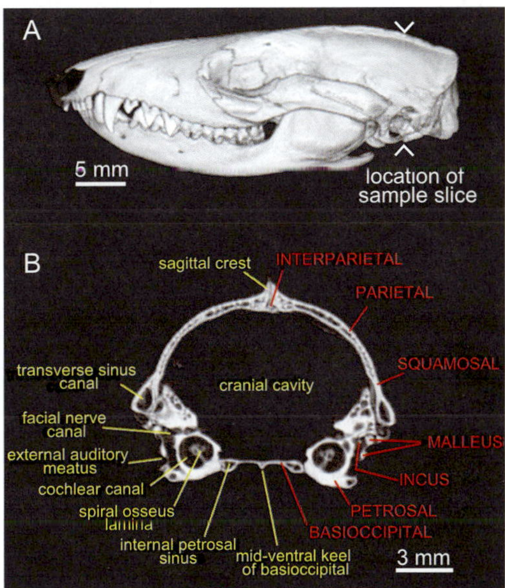

Fig. 11. (**a**) Three-dimensional volumetric reconstruction from X-ray CT data of the skull of the extant opossum *Monodelphis domestica* (maximum dimension 4.0 cm). (**b**) Individual CT slice with structures labelled to show the richness of detail obtainable by UHRXCT, even on very small specimens. Specimen courtesy of the Vertebrate Paleontology Laboratory, University of Texas at Austin.

Mammalian origins: the middle ear

Comparisons of *Monodelphis* and *Thrinaxodon* to other fossils revealed that an inflation in brain volume causes profound remodelling at the back of the skull, which answered one of the oldest problems in mammalian origins. Mammals are distinctive in having three separate bones that conduct sound through the middle ear. Other vertebrates with a sound-transducing ear have only a single middle ear bone. In these species, the homologs of the extra two mammalian middle ear bones remain as parts of the lower jaw. How the bones of the mammalian middle ear became detached from the jaw and repositioned to their definitive mammalian configuration was considered highly problematic. Measurements from the CT data (Fig. 12) demonstrated that different relative growth of the brain and ear ossicles – both in early mammalian evolution and during early development in living species – is responsible for the trajectory and direction of movement of the ear ossicles (Rowe 1996*a*, *b*).

Jaws of fossil marsupials: tooth replacement, reproductive patterns, taxonomy

UHRXCT imagery of an 80-million-year-old fossil jaw of *Alphadon* provided the first evidence of tooth replacement in a Mesozoic marsupial (Cifelli *et al.* 1996) and demonstrated that the modern pattern of marsupial tooth replacement is very ancient. Correlations between tooth development and reproductive pattern permitted further inferences regarding the antiquity of the

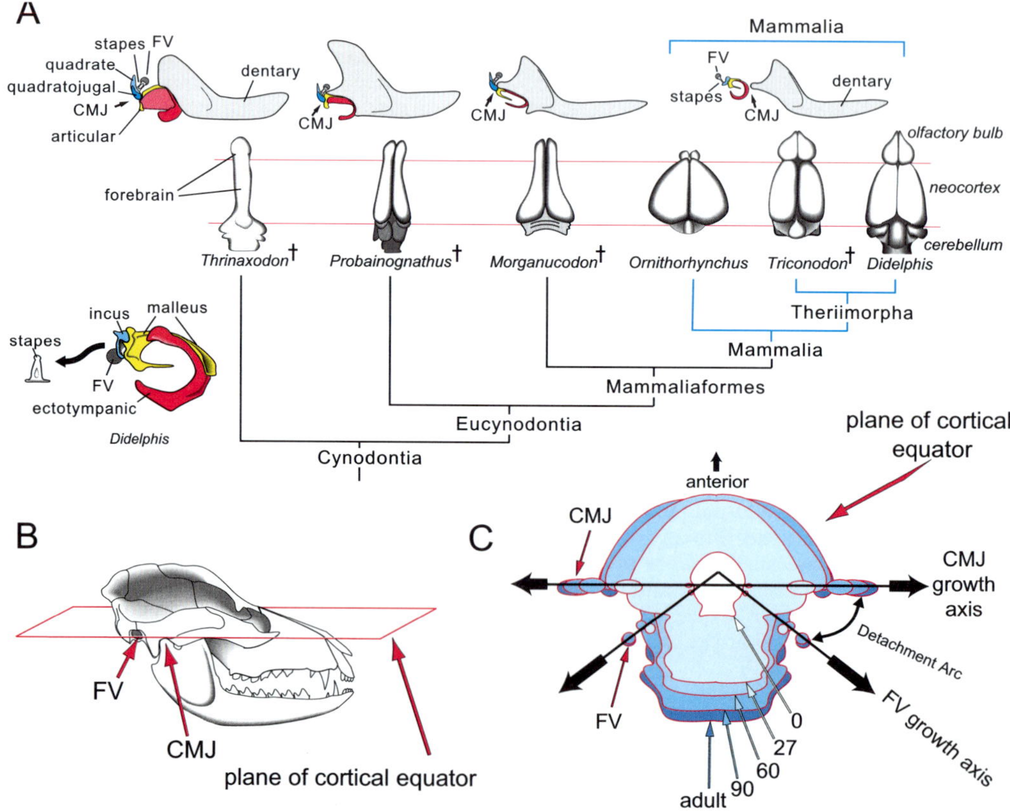

Fig. 12. (**a**) Co-evolution of the mammalian mandible and middle ear (in right lateral view) and the brain (illustrated by dorsal views of endocasts), plotted together on a phylogeny of selected mammals and their closest extinct relatives (Gauthier *et al.* 1988; Rowe 1988, 1993). Abbreviations: CMJ – craniomandibular joint; FV – fenestra vestibuli of inner ear. (**b**) Location of cortical equator on didelphid skull. (**c**) Superimposed projections of cortical equator, from X-ray CT imaging of *Monodelphis domestica*, showing the increase in equatorial circumference with age. Divergent trajectories of the CMJ and FV growth axes define an *arc of detachment* whose growth leads to detachment and caudal displacement of the auditory chain. Reprinted from Rowe (1996*b*), with permission.

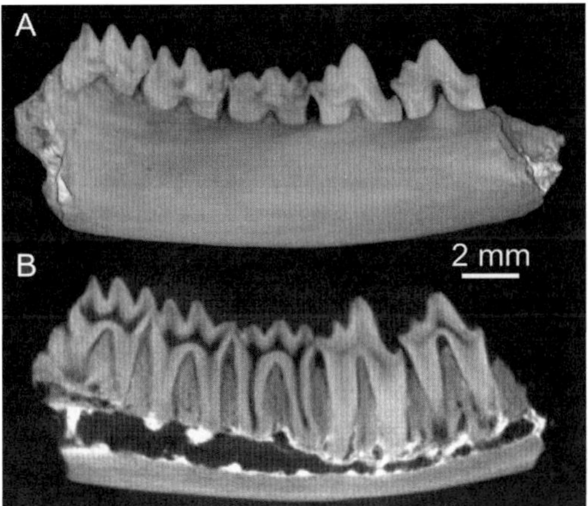

Fig. 13. The lower jaw of *Arundelconodon*, a Late Cretaceous marsupial from the eastern seaboard of the United States (19.3 mm length). (**a**) Three-dimensional volumetric rendering from X-ray CT data. (**b**) A sample CT slice, demonstrating that this specimen retains the primitive condition of non-interlocked tooth roots (Cifelli *et al.* 1999). Specimen courtesy of the Smithsonian Institution, National Museum of Natural History. Reprinted from Cifelli *et al.* (1999), with permission.

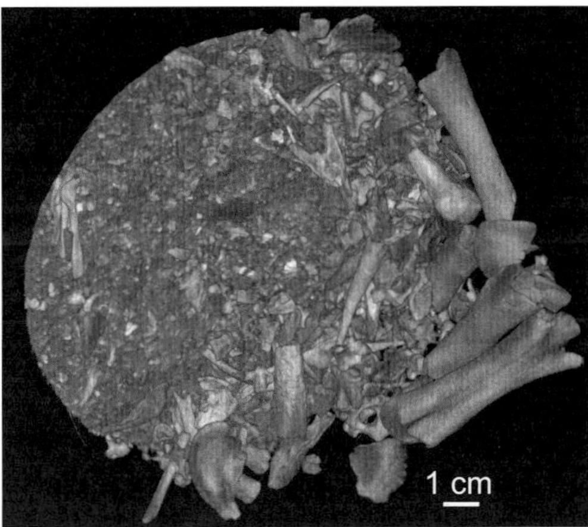

Fig. 14. Three-dimensional volumetric rendering from X-ray CT data of the embryonic remains contained inside an intact egg of *Aepyornis maximus*, the elephant bird (maximum dimension 15.2 cm). Specimen courtesy of the National Geographic Society.

distinctive marsupial reproductive patterns. Successful analysis of this tiny jaw prompted other high-resolution CT studies of tiny Mesozoic fossils (Cifelli & de Muizon 1998; Cifelli *et al.* 1999). In another study, UHRXCT aided in the description of *Arundelconodon*, a new species of Mesozoic mammal known only from a tiny jaw (Cifelli *et al.* 1999), by revealing otherwise unavailable information on taxonomically important aspects of the structure of its tooth roots (Fig. 13).

Elephant bird

HRXCT has successfully imaged the embryo of the extinct elephant bird of Madagascar, *Aepyornis maximus* (Weintraub 2000). The elephant bird was the largest bird known, weighing roughly 440 kg and standing nearly 4 m tall (Amadon 1947), and it laid the largest eggs ever discovered. Although it became extinct in historic times, only a few intact eggs are known. Conventional X-radiography revealed the presence of an embryo in at least one of them (Wetmore 1967), but the imagery provided no useful information on its skeletal structure. HRXCT scanning of a complete egg revealed details of the embryonic skeleton; image processing of the CT dataset permitted digital 'extraction' of the disarticulated embryo from the egg (Fig. 14). Detailed analysis is still underway (Balanoff 2001).

Forensic palaeontology: identifying forgeries

HRXCT data provide unique criteria by which to judge the authenticity of fossils. A high-profile example is '*Archaeoraptor*', which *National Geographic* magazine had heralded as a 'missing link' between birds and extinct dinosaurs (Sloan 1999). Analysis of the CT data revealed that the forgery (Fig. 15) had been assembled from at least two, and perhaps as many as five, different specimens and species (Rowe *et al.* 2001).

Mammalian cranial architecture: visualizing cranial cavities

X-ray CT makes it possible to non-destructively examine and measure the interior cavities of vertebrate skulls (Brochu 2000; Marino *et al.* 2000; Colbert 2001; Franzosa 2001; Maisey 2001). These cavities are typically hidden from external view and are difficult or impossible to study by traditional means. They include not only the endocranial cavity that houses the brain and related structures, but also sinus cavities, inner ear labyrinths or any non-bony space. While it has long been recognized that endocranial casts provide a close representation of mammalian brain morphology, the architectural relationship between the brain, sinus cavities and overall cranial shape has been difficult to evaluate. The

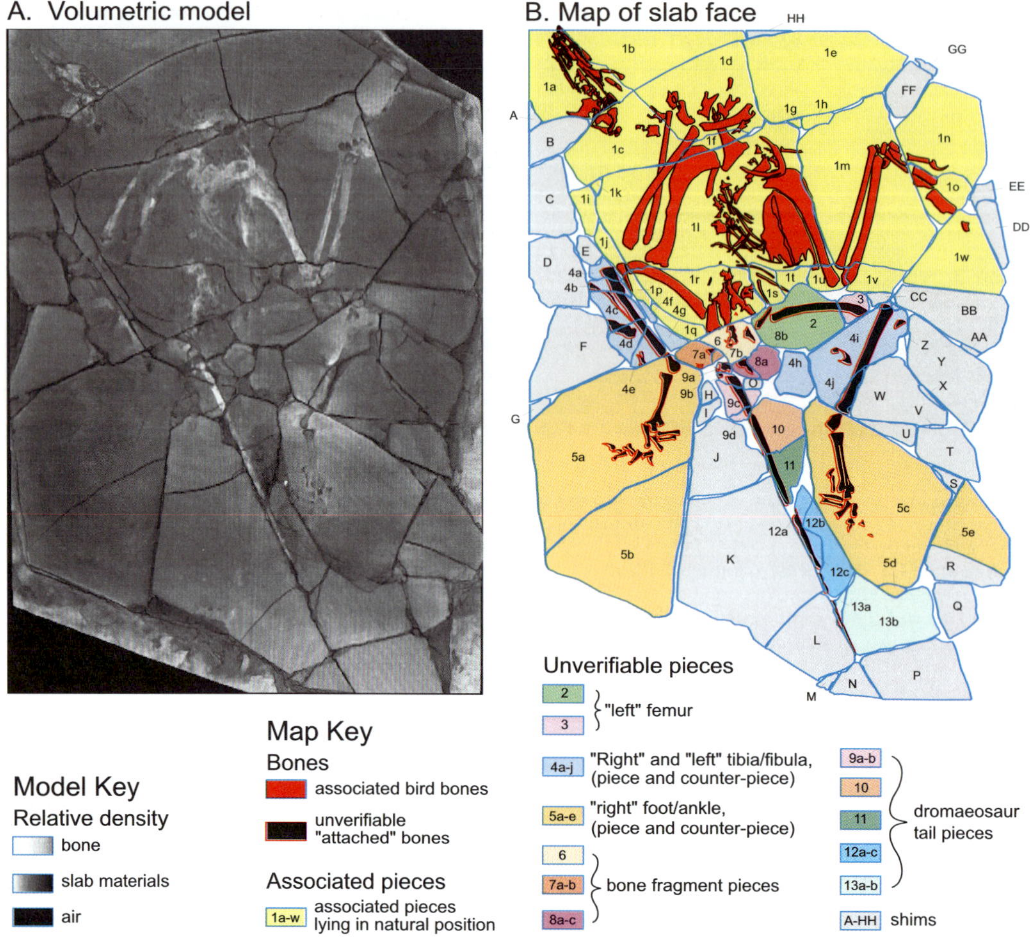

Fig. 15. Two computer-generated models of the face of the *Archaeoraptor* slab as it was presented for X-ray CT scanning (27 cm width). The specimen was scanned from top to bottom in planes perpendicular to this view. (**a**) Volumetric model generated from superimposed CT slices. (**b**) Map of the slab face, colour-coded to indicate the nature of the associations between its 88 constituent pieces. Reprinted from Rowe *et al.* (2001), with permission.

changing architectural contribution of these cavities to skull morphology during evolution was revealed by a series of CT scans of extant tapirs and of a new fossil tapiroid from the Middle Eocene of North America (Colbert 1999). Comparison of CT-based endocasts reveals that the unique posterior 'telescoping' of the skull of Recent tapirs is also accompanied by the development of enlarged frontal sinuses that surround the olfactory bulbs and frontal lobes of the brain and that accounts for the elevation of the frontal bones (Fig. 16). These sinuses are not present in the new Eocene fossil tapiroid, which also lacks elevated frontals, an indication of the importance of these sinuses in the overall architecture of the skull.

Non-destructive examination of fossils in amber

Insects and other small animals trapped in amber provide a rare glimpse of the 'in-life' morphology of ancient fauna. Attempts to extract such specimens from their encasement are inevitably destructive. Although such specimens can often be examined productively using microscopy and photography, ultra-high-resolution CT has proven useful for some more difficult cases (Grimaldi *et al.* 2000). In one case, the surface morphology of a scorpion (*Arthropoda: Buthidae*) was extracted from a specimen in which the amber was too clouded for standard light-based techniques (Fig. 17). In another, the skeleton of a

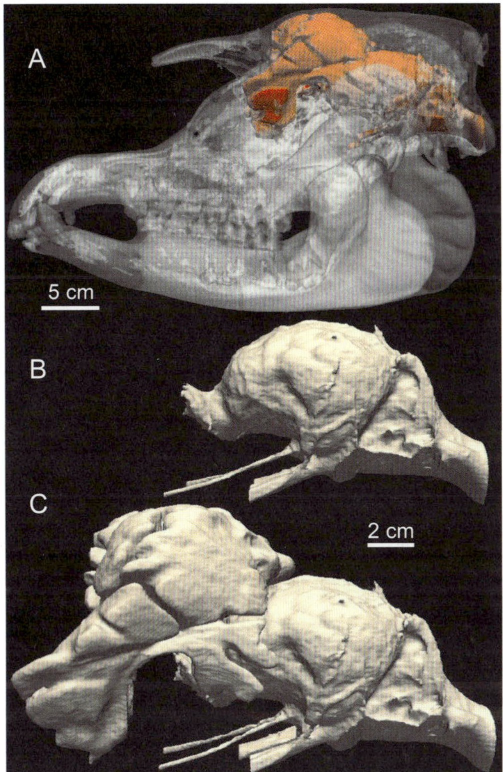

Fig. 16. Skull and endocasts of the Malayan Tapir (*Tapirus indicus*) rendered from X-ray CT data. Maximum dimension of skull is 42 cm. (**a**) Relationship of the sinus cavity and endocranial cavity to the exterior of the cranium. Cavities rendered in red, bone in semitransparent white. (**b**) Cranial cavity endocast. (**c**) Relationship of the sinus endocast to the cranial cavity endocast, showing that the frontal sinus surrounds much of the anterior endocranial cavity. Specimen courtesy of American Museum of Natural History.

Fig. 17. Three-dimensional rendering from X-ray CT data of the remains of a scorpion trapped in Miocene-aged Dominican amber. Optical examination of this specimen was impossible because of the opacity of the amber in which it was embedded. Specimen courtesy of David Grimaldi, American Museum of Natural History.

Sphaerodactylus gecko, which was otherwise obscured by the skin, was sufficiently preserved to permit a number of morphological elements to be identified and extracted, allowing comparison with more commonplace fossilized material.

Structural elements responsible for density banding in scleractinian corals

Density bands that form annually in some corals are a valuable tool for reconstructing past environmental and climatic conditions, because they allow isotopic analysis of material precipitated at known times in the past. Better understanding of these data and considerable additional detail may be available if the skeletal elements responsible for banding can be identified, which may then be related to specific organismal processes. X-radiographs have long been used to reveal density banding within a skeletal slab, but they provide insufficient detail to examine the underlying structures. High-resolution X-ray CT imagery of

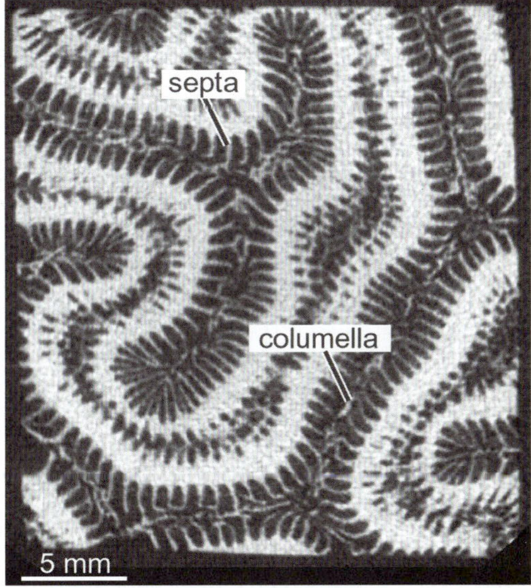

Fig. 18. High-resolution X-ray CT image of the coral *Diploria strigosa* (edge length 25 mm), revealing annual density bands that are associated with enlargement of the thinner structural elements (septa and columella) in the image. Specimen courtesy of Richard Dodge and Kevin Helmle, Nova Southeastern University.

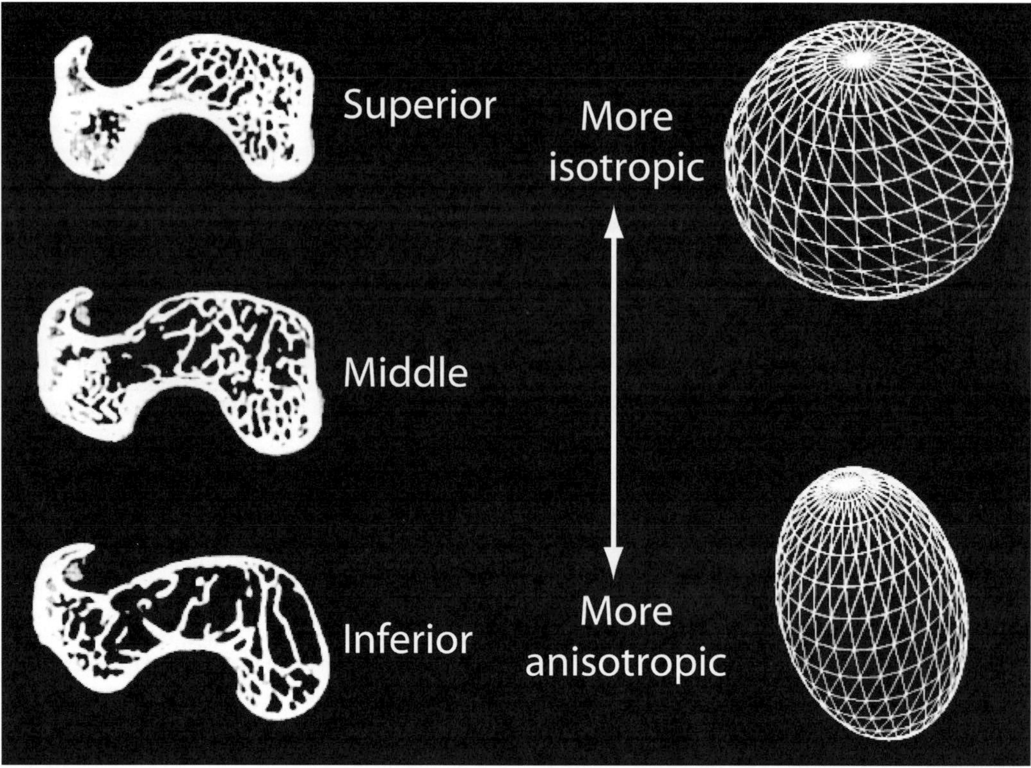

Fig. 19. X-ray CT images of trabecular bone of the femoral head of *Galago senegalensis* (maximum dimension 10 mm) with ellipsoidal representations of the degree of anisotropy of the three-dimensional trabecular structure. Anisotropy increases with increasing distance from the superior end of the bone. Specimen courtesy of the Smithsonian Institution, National Museum of Natural History.

the coral *Diploria strigosa* (Fig. 18) reveals that density bands in this species are associated with thickening of septal and columellar structures (Helmle *et al.* in press).

Trabecular architecture in primates as an indicator of locomotor patterns

It has been hypothesized for over 100 years that trabecular bone plays an important structural role in the musculoskeletal system of animals and that it dynamically responds to applied loads through growth. High-resolution X-ray CT data permit quantitative determination of the architecture of trabecular bone near limb joints in small primates through determination of bone volume fraction and a fabric tensor that indicates the material principal axes and degree of anisotropy (Fig. 19). A systematic study of the trabecular bone in femoral heads of Recent small primates shows that significant differences exist between leaping and non-leaping species, with the former having relatively anisotropic bone (Ryan 2000; Ryan & Ketcham 2002). These results for extant primates permit inferences about the locomotor patterns of extinct species to be drawn from fossil specimens in which trabecular bone patterns are preserved.

Conclusions

The advent of high-resolution and ultra-high-resolution instruments has opened up a wide realm of new applications of X-ray CT to geological problems. Given the advantages of these techniques – their speed, their non-destructive character and their ability to produce data in a digital form that facilitates visualization and quantification – one should expect only further expansion of applications into an even wider range of fields in the coming years. The projects selected for inclusion in this review are indicative of much of the technique's potential, but the full range of possibilities – which remains largely unexplored – is surely as diverse as the geological sciences are broad.

The authors acknowledge, with gratitude, the support of the W. M. Keck Foundation and the Geology Foundation of the University of Texas at Austin, who together with the US National Science Foundation (NSF EAR-9406258) made possible the initial purchase of our CT scanner. We are also grateful for continued support of the facility's operational costs by the US National Science Foundation (NSF EAR-9816020 and NSF EAR-0004082). C. Denison played a significant role in data acquisition and image processing for several of the projects included in this review.

References

AMADON, D. 1947. An estimated weight of the largest known bird. *The Condor*, **49**, 159–165.

BALANOFF, A.M. 2001. Unscrambling the egg: digital extraction of an elephant bird (*Aepyornis*) embryo from an intact egg. *Journal of Vertebrate Paleontology*, **21 (suppl. to No. 3)**, 31A.

BROCHU, C.A. 2000. A digitally rendered endocast for *Tyrannosaurus rex*. *Journal of Vertebrate Paleontology*, **20**, 1–6.

BROWN, M.A., BROWN, M., CARLSON, W.D. & DENISON, C. 1999. Topology of syntectonic melt flow networks in the deep crust: inferences from three-dimensional images of leucosome geometry in migmatites. *American Mineralogist*, **84**, 1793–1818.

CARLSON, W.D. 1999. The case against Ostwald ripening of porphyroblasts. *Canadian Mineralogist*, **37**, 403–413.

CARLSON, W.D. & DENISON, C. 1992. Mechanisms of porphyroblast crystallization: results from high-resolution computed X-ray tomography. *Science*, **257**, 1236–1239.

CARLSON, W.D., DENISON, C. & KETCHAM, R.A. 1995. Controls on the nucleation and growth of porphyroblasts: Kinetics from natural textures and numerical models. *Geological Journal*, **30**, 207–225.

CARLSON, W.D., DENISON, C. & KETCHAM, R.A. 1999. High-resolution X-ray computed tomography as a tool for visualization and quantitative analysis of igneous textures in three dimensions. *Electronic Geosciences*, **4**, 3.

CARLSON, W.D. & MCCOY, T.J. 1998. High-resolution X-ray computed tomography of lodranite GRA 95209. Lunar and Planetary Science XXIX, Abstract #1541, Lunar and Planetary Institute, Houston (CD-ROM).

CHERNOFF, C.B. & CARLSON, W.D. 1997. Disequilibrium for Ca during growth of pelitic garnet. *Journal of Metamorphic Geology*, **15**, 421–438.

CHERNOFF, C.B. & CARLSON, W.D. 1999. Trace-element zoning as a record of chemical disequilib rium during garnet growth. *Geology*, **27**, 555–558.

CIFELLI, R.L. & DE MUIZON, C. 1998. Marsupial mammal from the Upper Cretaceous North Horn Formation, central Utah. *Journal of Paleontology*, **72**, 532–537.

CIFELLI, R.L., LIPKA, T.R., SCHAFF, C.R. & ROWE, T.B. 1999. First Early Cretaceous mammal from the eastern seaboard of the United States. *Journal of Vertebrate Paleontology*, **19** 199–203.

CIFELLI, R.L., ROWE, T., LUCKETT, W.P., BANTA, J., REYES, R. & HOWES, R.I. 1996. Fossil evidence for the origin of the marsupial pattern of tooth replacement. *Nature*, **379**, 715–718.

COLBERT, M.W. 1999. *Patterns of evolution and variation in the Tapiroidea (Mammalia, Perissodactyla)*. PhD thesis, University of Texas at Austin.

COLBERT, M.W. 2001. Digital visualization of cranial cavities: examples from fossil and recent mammals. *Journal of Vertebrate Paleontology*, **21 (suppl. to No. 3)**, 41A.

DENISON, C. & CARLSON, W.D. 1997. Three-dimensional quantitative textural analysis of metamorphic rocks using high-resolution computed X-ray tomography. Part II: application to natural samples. *Journal of Metamorphic Geology*, **15**, 45–57.

DENISON, C., CARLSON, W.D. & KETCHAM, R. 1997. Three-dimensional quantitative textural analysis of metamorphic rocks using high-resolution computed X-ray tomography. Part I: methods and techniques. *Journal of Metamorphic Geology*, **15**, 29–44.

FRANZOSA, J.W. 2001. Constructing digital endocasts of theropods using a high-resolution X-ray computed tomography scanner. *Journal of Vertebrate Paleontology*, **21 (suppl. to No. 3)**, 51A.

GAUTHIER, J., KLUGE, A.G. & ROWE, T. 1988. Amniote phylogeny and the importance of fossils. *Cladistics*, **4**, 105–209.

GRIMALDI, D., NGUYEN, T. & KETCHAM, R. 2000. Ultra-high-resolution X-ray computed tomography (UHR CT) and the study of fossils in amber. *In*: GRIMALDI, D. (ed.) *Studies on Fossils in Amber, with Particular Reference to the Cretaceous of New Jersey*. Backhuys, Leiden, 77–91.

HELMLE, K.P., DODGE, R.E. & KETCHAM, R.A. 2003. Skeletal architecture and density banding in *Diploria strigosa* by X-ray computed tomography. *In*: *Proceedings of the 9th International Coral Reef Symposium*, 365–372.

HIRSCH, D.M. & CARLSON, W.D. 2001. Causes of variation in textures along a regional metamorphic field gradient (abstr.). *In*: *Eleventh Annual V.M. Goldschmidt Conference*, Abstract #3203, (CD-ROM).

KEMP, T.S. 1982. *Mammal-like Reptiles and the Origin of Mammals*. Academic Press, London.

KETCHAM, R.A. & CARLSON, W.D. 2001. Acquisition, optimization and interpretation of X-ray computed tomographic imagery: applications to the geosciences. *Computers & Geosciences*, **27**, 381–400.

KUEBLER, K.E., MCSWEEN, H.Y., JR, CARLSON, W.D. & HIRSCH, D.M. 1999. Sizes and masses of chondrules and metal-troilite grains in ordinary chondrites. *Icarus*, **141**, 96–106.

MAISEY, J.G. 2001. Stone deaf? CT scanning and low frequency phonoreception in extinct sharks. *Journal of Vertebrate Paleontology*, **21 (suppl. to No. 3)**, 51A.

MARINO, L., UHEN, M.D., FROHLICH, B., ALDAG, J.M., BLANE, C., BOHASKA, D. & WHITMORE, F.C., JR. 2000. Endocranial volume of mid-late Eocene archeocetes (Order: Cetacea) revealed by computed tomography: Implications for cetacean brain evolution. *Journal of Mammalian Evolution*, **7**, 81–94.

MCCOY, T.J. & CARLSON, W.D. 1998. Opaque minerals in the GRA 95209 lodranite: a snapshot of metal segregation. Lunar and Planetary Science XXIX, Abstract #1675, Lunar and Planetary Institute, Houston (CD-ROM).

PHILPOTTS, A.R., BRUSTMAN, C.M., SHI, J., CARLSON, W.D. & DENISON, C. 1999. Plagioclase chain networks in slowly cooled basaltic magma. *American Mineralogist*, **84**, 1819–1829.

PHILPOTTS, A.R. & DICKSON, L.D. 2000. The formation of plagioclase chains during convective transfer in basaltic magma. *Nature*, **406**, 59–61.

PHILPOTTS, A.R., SHI, J. & BRUSTMAN, C.M. 1998. Role of plagioclase crystal chains in the differentiation of partly crystallized basaltic magma. *Nature*, **395**, 343–346.

PROUSSEVITCH, A., KETCHAM, R.A., CARLSON, W.D. & SAHAGIAN, D. 1998. Preliminary results of X-ray CT analysis of Hawaiian vesicular basalts. *EOS*, **79**, 360.

ROWE, T. 1988. Definition, diagnosis and origin of Mammalia. *Journal of Vertebrate Paleontology*, **8**, 241–264.

ROWE, T. 1993. Phylogenetic systematics and the early history of mammals. *In*: SZALAY, F.S., NOVACEK, M.J. & MCKENNA, M.C. (eds) *Mammalian Phylogeny*. Springer Verlag, New York, 129–145.

ROWE, T. 1996*a*. Brain heterochrony and evolution of the mammalian middle ear. *In*: GHISELIN, M. & PINNA, G. (eds), *New Perspectives on the History of Life*. California Academy of Sciences Memoir **20**, 71–96.

ROWE, T. 1996*b*. Coevolution of the mammalian middle ear and neocortex. *Science*, **273**, 651–654.

ROWE, T. 1999. At the roots of the mammalian tree. *Nature*, **398**, 283–284.

ROWE, T., CARLSON, W.D. & BOTTORFF, W. 1993. *Thrinaxodon*: Digital Atlas of the Skull [CD-ROM]. University of Texas Press.

ROWE, T., KETCHAM, R., DENISON, C., COLBERT, M., XU, X. & CURRIE, P.J. 2001. The *Archaeoraptor* forgery. *Nature*, **410**, 539–540.

RUBIN, A.E., ULFF-MOLLER, F., WASSON, J.T. & CARLSON, W.D. 2001. The Portales Valley meteorite breccia: evidence for impact-induced melting and metamorphism of an ordinary chondrite. *Geochimica et Cosmochimica Acta*, **65**, 323–342.

RYAN, T. 2000. Quantitative analysis of trabecular bone structure in the femur of lorisoid primates using high-resolution X-ray computed tomography (abstr). *American Journal of Physical Anthropology*, **Suppl. 30**, 266–267.

RYAN, T.M. & KETCHAM, R.A. 2002. The three-dimensional structure of trabecular bone in the femoral head of strepsirrhine primates. *Journal of Human Evolution*, **43**, 1–26.

SAHAGIAN, D., PROUSSEVITCH, A. & CARLSON, W. 2002. Timing of Colorado Plateau uplift: initial constraints from vesicular basalt-derived paleoelevations. *Geology*, **50**, 807–810.

SAHAGIAN, D.L. & MAUS, J.E. 1994. Basalt vesicularity as a measure of atmospheric pressure and paleoelevation. *Nature*, **372**, 449–451.

SLOAN, C.P. 1999. Feathers for *T. Rex*? New birdlike fossils are missing links in dinosaur evolution. *National Geographic*, **196**, 98–107.

TAYLOR, L.A., KELLER, R.A., SNYDER, G.A., WANG, W., CARLSON, W.D., HAURI, E.H., MCCANDLESS, T., KIM, K.R., SOBOLEV, N.V. & BEZBORODOV, S.M. 2000. Diamonds and their mineral inclusions, and what they tell us: a detailed 'pull-apart' of a diamondiferous eclogite. *International Geology Review*, **42**, 959–983.

WEINTRAUB, B. 2000. New 'pictures' of a giant bird. *National Geographic*, **198**, viii.

WELLINGTON, S.L. & VINEGAR, H.J. 1987. X-ray Computerized Tomography. *Journal of Petroleum Technology*, 885–898.

WETMORE, A. 1967. Recreating Madagascar's giant extinct bird. *National Geographic*, **132**, 488–493.

WITHJACK, E.M. 1988. Computed tomography for rock property determination and fluid flow visualization. *Society of Petroleum Engineers Formation Evaluation*, **3**, 696–704.

WITHJACK, E.M. & AKERVOLL, I. 1988. Computed tomography studies of 3-D miscible displacement behavior in a laboratory five-spot model. *In*: *Proceedings of the Society of Petroleum Engineers Annual Technical Conference and Exhibition, Houston, Texas, 2–5 October*, Paper **SPE 18095**.

Computed tomography in petroleum engineering research

S. AKIN[1] & A. R. KOVSCEK[2]

[1] *Petroleum and Natural Gas Engineering Department, Middle East Technical University, 06531 Ankara, Turkey*
[2] *Petroleum Engineering Department, Stanford University, Stanford, CA 94305-2220, USA (e-mail: kovscek@pangea.stanford.edu)*

Abstract: Imaging the distribution of porosity, permeability, and fluid phases is important to understanding single and multiphase flow characteristics of porous media. X-ray computed tomography (CT) has emerged as an important and powerful tool for non-destructive imaging because it is relatively easy to apply, can offer fine spatial resolution and is adaptable to many types of experimental procedures and flow conditions. This paper gives an overview of CT technology for imaging multiphase flow in porous media, the principles behind the technology and effective experimental design. By critically reviewing prior work using this important tool, we hope to provide a better understanding of its use and a pathway to improved analysis of CT-derived data. Because of the wide variety of image processing options, they are discussed in some detail.

The use of X-ray computed tomography (CT) for observing single and multiphase fluid flow in rock and non-destructively viewing porous medium interior is a relatively new technique in petroleum engineering and the associated geological sciences. Radiological imaging using computed tomography was first developed by Hounsfield (1972). In comparison to conventional X-ray radiography, CT scanners generate cross-sectional images of the object by measuring the attenuation of a beam of X-rays as it is rotated around the object at angular increments within a single plane. From a set of these measurements, back projection algorithms employing Fourier transform algorithms are used to reconstruct a cross-sectional image. Sedgwick & Dixon (1988) compared three X-ray-based techniques for measuring fluid saturations: single-beam attenuation, xeroradiography and CT. They concluded that the methods gave similar results, but that the advantage of CT was fine resolution on the millimetre scale. Additionally, three-dimensional images can be generated by interpolating among cross-sectional images.

CT scanners are classified according to the placement of the source-detector combination. First generation scanners used a source and three to six detectors. The images of the specimen were acquired in a translate-rotate fashion. The source and detectors were fixed and the sample was rotated at angular intervals about its central axis in order to acquire data for a cross-sectional image. It usually took several minutes to finish a single scan. Second generation scanners improved the scan time and image quality by keeping the same geometry but incorporating up to 70 detectors. Third and fourth generation CT scanners further improved the scan time and the image quality by rotating the X-ray source in a circular path rather than the sample. Third generation scanners use an arc of detectors, whereas fourth generation machines use a ring of up to 1440 detectors. Scan and processing times are in the range of 30 to 40 seconds. With the introduction of fifth generation scanners, a ring of sources and detectors are used and the scan time is significantly reduced. Thus, there is no need for a rotation or translation and the scan times are so fast that even a heartbeat can be scanned. Second to fourth generation scanners are used in petroleum engineering research. Advanced CT scanners, such as spiral CT systems (Klingenbeck-Regn *et al.* 1999) and fifth generation scanners, are not reported as having been utilized in petroleum engineering research yet.

So-called industrial scanners are intended for analysis of inanimate objects only and have also been used for flow imaging. They differ from a medical scanner in several aspects. Notably, the sample is rotated in this type of scanner as opposed to the X-ray source, as in third and fourth generation medical scanners. Because they do not examine living subjects, the energy of their X-ray sources is generally higher; this allows the penetration of metal and relatively

From: MEES, F., SWENNEN, R., VAN GEET, M. & JACOBS, P. (eds) 2003. *Applications of X-ray Computed Tomography in the Geosciences*. Geological Society, London, Special Publications, **215**, 23–38. 0305-8719/03/$15.

thick objects. Thus, laboratories housing industrial scanners generally require greater shielding. However, the intensity of these sources is commonly less and therefore longer scan times are required, compared to a medical scanner, to achieve similar accuracy. This makes the monitoring of fast displacement processes or frontal advance somewhat more difficult than with a medical scanner. These factors probably contribute to industrial scanners being utilized for flow imaging to a much lesser extent.

The remainder of this review is organized in the following fashion. First X-ray absorption theory as it relates to CT scanning is discussed. Next, imaging artefacts are discussed, as is experimental design. The various applications and image processing options are then given. Our goal is to provide a comprehensive overview of the use of CT for imaging porous media and multiphase fluid flow. However, we do not attempt a complete compilation of every paper that reports using CT for imaging porous media.

Theory

When a CT scanner is operated, X-rays penetrate a thin volumetric slice of an object at different angles as the X-ray source rotates around the object. A series of detectors then records the transmitted X-ray intensity. Thus, many different X-ray attenuations are made available for mathematical reconstruction and enhancement. The basic quantity measured in each volume element (voxel) of a CT image is linear attenuation coefficient as defined by Beer's law:

$$\frac{I}{I_0} = \exp^{-\mu h} \tag{1}$$

where I_0 is the incident X-ray intensity, I is the intensity remaining after the X-ray passes through a thickness h of homogeneous sample, and μ is the linear attenuation coefficient. For a heterogeneous medium, the energy transmitted along a particular ray path is

$$\ln\left(\frac{I}{I_0}\right) = \int_0^L \mu(h(x, y))\,\mathrm{d}h \tag{2}$$

where $h(x, y)$ are the co-ordinates of the attenuation coefficient in two dimensions, L is the path length from source to detector and $\mathrm{d}h$ is a distance along this path length.

Beer's law assumes a narrow X-ray beam and monochromatic radiation. In practice, the actual beam is polychromatic consisting of, for instance, a spectrum of photons ranging from 20 keV to 120 keV. These facts contribute to the introduction of imaging artefacts. Detectors also have an associated efficiency that is energy dependent. The true situation can be represented by an Eq. 3:

$$I = \int_{e_l}^{e_h} \frac{\mathrm{d}I_0}{\mathrm{d}E}\,\varepsilon(E)\exp\left(-\int_0^h \mu(x, y, E)\,\mathrm{d}L\right)\mathrm{d}E \tag{3}$$

where $\mathrm{d}I_0/\mathrm{d}E$ represents the spectral distribution of the incident radiation, $\varepsilon(E)$ is the efficiency of the detector at a particular energy E and e_l and e_h represent the relevant spectrum of energy with subscripts l and h signifying low and high, respectively. The attenuation coefficient is now a function of position and energy.

In practice, Eq. (2) is used to reconstruct images and it is assumed that some particular effective energy characterizes the X-ray beam as a whole. Essentially, Eq. (2) is discretized into n volume elements each with unknown attenuation coefficient. Measurement of the attenuation of multiple ray projections provides sufficient data to solve multiple equations for attenuation coefficient. Images are usually reconstructed with a filtered back propagation method. The method projects a uniform value of attenuation over each ray path so that the calculated uniform value is proportional to the measured attenuation. Each matrix element receives a contribution from each ray passing through it. Images obtained are blurred because of the assumption that attenuation is uniform over the entire length of the ray. A convolution or filtering process is then used to modify the ray sum data and improve images. These filter functions are complex, depend on many parameters given the final result desired and are usually proprietary. As an example, a filter function might enhance edges to sharpen an image.

After image reconstruction, relative values of linear attenuation coefficient are known for each pixel. The computer converts attenuation coefficients into corresponding numerical values, or CT numbers, by normalizing with the linear attenuation coefficient of water, μ_w as shown below

$$CT = 1000\,\frac{(\mu - \mu_w)}{\mu_w} \tag{4}$$

The units of Eq. 4 are Hounsfield (H) units. Each Hounsfield unit represents a 0.1% change in density with respect to calibration density scale. Table 1 lists the CT numbers for some common materials. Note the increase in CT number with mass density.

The linear attenuation coefficient is a function of both the electron density (bulk density), ρ,

Table 1. *CT numbers for common materials*

Material	CT (H) 130 keV	CT (H) 100 keV	Density (kg/m^3)
Air[1]	−1000		1.82
N-hexane[1]	−285		660
N-decane[1]	−283		730
Water[1]	0		1000
1 wt% KBr in water[1]	70	91	1005
2 wt% KBr in water[1]	142	183	1013
8 wt% KBr in water[1]	565	725	1058
PVC[1]	620		1400
Quartz[2]	1589	1836	2190
Berea sandstone[2]	1608	1835	2120
Colton sandstone[2]	1629	1840	2270
Navaho sandstone[2]	1858	2102	2360
Red Navaho sandstone[2]	1912	2156	2390
Indiana limestone[2]	1531	1750	2220
Alumina[2]	2478	2866	2820

[1] Values measured on a Picker 1200 SX instrument, with a tube current of 65 mA and a 16 cm field of view.
[2] CT numbers for minerals and rocks measured on a Philips Tomoscan instrument, with a tube current of 250 mA and a 16 cm field of view.

and effective atomic number, Z in the following form (Vinegar & Wellington 1987):

$$\mu = \rho(a + bZ^{3.8}/E^{3.2}) \quad (5)$$

where 'a' is the Klein–Nishina coefficient and 'b' is a constant. By scanning at dual energy levels, E (high and low), one image proportional to density and one image proportional to effective atomic number can be obtained. For X-ray energies above 100 keV, X-rays interact with matter mainly by Compton scattering which depends on electron density. For energies well below 100 keV, the interaction is dominated by photoelectric absorption, which depends on the effective atomic number. Eq. 5 states that the heavier elements have a greater photoelectric cross-section and the fraction of photoelectric contribution increases rapidly as the X-ray energy is lowered.

Assessment of porosity and *in-situ* phase saturation is possible once CT numbers are measured. Various expressions employed for saturation determination by CT are given later in connection with the review. All derive from the idea of subtraction. The CT numbers measured during an experiment for a partially saturated porous medium (say with oil and water) consist of contributions from the rock, oil and water phases. Subtraction allows us to isolate contributions from a particular phase. For example, by subtracting an image of rock containing both oil and water from an image of water-saturated rock, the contribution of rock is removed from the resultant image. Then, normalizing by the difference in CT numbers between fully water- and oil-saturated rock gives the fraction of the pore space filled with water and scales quantities to lie between 0 and 1.

Types of error

Measurements with X-ray CT are subject to a variety of errors and image artefacts including beam hardening, star-shaped or so-called X-artefacts, positioning errors, and machine errors. These are discussed briefly (see also Van Geet *et al.* 2001).

The majority of CT scanners were developed for medical purposes and were originally intended for qualitative imaging and not for quantitative analysis. Because the X-ray source delivers a spectrum of X-ray energies (polychromatic) rather than monochromatic energy, the lower energy, or soft, portions of the X-ray spectrum are absorbed preferentially at the air/sample interface, as well as in the sample itself. Thus, the X-ray spectrum attenuates toward the lower energy portions of the spectrum. This introduces an error in linear attenuation measurement called beam hardening or 'cupping' because the remaining high-energy photons shift the average energy of the beam toward 'harder'

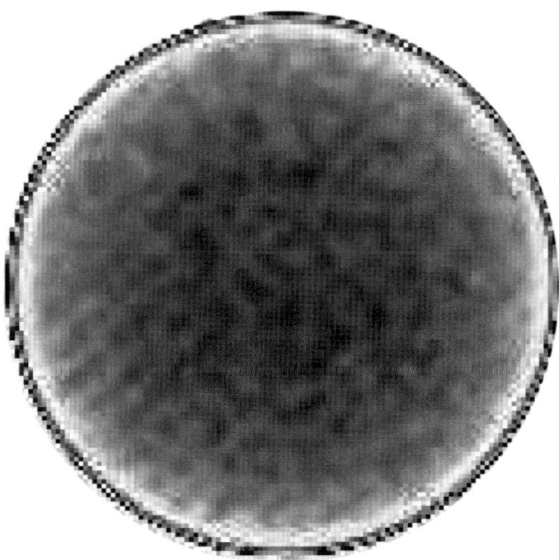

Fig. 1. Example of beam hardening in CT image of a 10.2 cm diameter homogeneous sandpack. Note the lighter shading just inside the core holder.

X-rays. It is realized more often in relatively large objects and is manifested as dark bands around the periphery of objects. For example, the image in Figure 1 displays typical beam hardening effects. It results in falsely high CT numbers (Herman 1980, pp. 76–83 & 161–179).

The artefact amplitudes increase with scanned volume or slice width. Object scatter can be reduced in third generation CT geometry by collimating the detector elements. For fourth generation CT geometry, only poor anti-scatter collimation is possible and a numeric correction is necessary (Ohnesorge *et al.* 1999). A number of methods can be applied directly to reduce beam hardening effects. Reconstruction algorithms provide corrections for beam hardening. Thus, reconstructive processing and beam filtering prior to scanning reduce the occurrence of low energy photons entering the sample. These methods, however, are not optimized for dense materials, such as reservoir rocks. Special core designs, such as surrounding the core holder with a cylindrical water jacket (Schembre *et al.* 1998; Akin *et al.* 2000) or with a crushed rock jacket (Kuru *et al.* 1998) can minimize beam hardening effects. Aluminium (Akin & Demiral 1998) and composite carbon fibre core holders are also used for such purposes (Withjack 1988; Akin & Demiral 1997).

Beam hardening can also be reduced by simply moving to higher energy X-ray sources. With fewer low energy photons, the degree of attenuation of the X-ray beam is reduced. Correspondingly, the average energy of the beam shifts little at material boundaries. In this regard, an industrial scanner might be superior to a medical scanner. Another method to reduce beam hardening is to calibrate the machine to a CT number larger than that of water (see Eq. (4)). For this purpose, a doped or spiked water solution (e.g. with potassium iodide or potassium bromide) or quartz sample can be used. The weight percentage of the solution should be selected to match the CT number of the rock (refer to Table 1). In turn, an image results where CT numbers of a homogeneous sample do not vary with the depth of penetration (i.e. a flat image). Quartz is useful for this purpose if the rock under consideration is a sandstone. Vinegar & Wellington (1987) discuss choice of dopants and calibration in detail.

Object shape can also lead to artefacts. The cross-sectional geometry of the scanner gantry is circular and the machine delivers the best images of objects that are also circular and symmetrical in cross-section. When square or rectangular cross-sections of objects are presented to the scanner, X-shaped artefacts are observed in the images obtained. An example is given in Figure 2. Note that the rock is homogeneous but that the image contains lightly shaded portions in the shape of the letter X. Dashed white lines in Figure 2 mark these portions to guide the eyes.

X-shaped artefacts originate from the image reconstruction and back propagation process. During processing to obtain CT numbers of individual voxels, it is assumed that an average attenuation can be applied along each ray path. In square or rectangular images, the length of the diagonals is greater than the length of the sides. Thus, as the beam is moved in angular intervals around the object, the amount of material encountered varies. The average attenuation then varies depending on the ray path, even though the material is homogeneous. Because the diagonals represent the greatest amount of material, CT numbers are largest there.

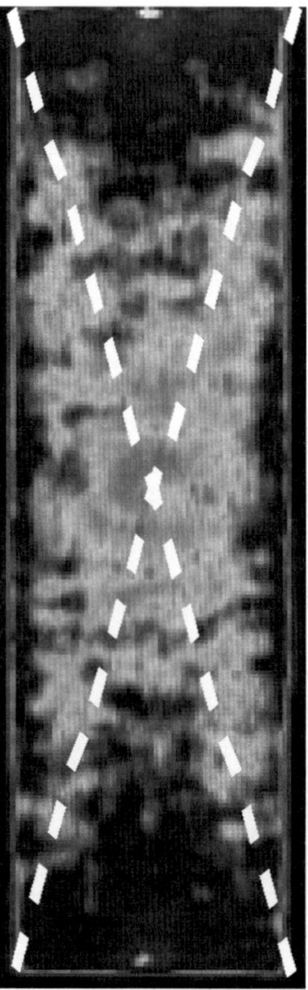

Fig. 2. Example of an X-artefact in scanning a homogeneous diatomite core. Dashed white lines are drawn as a guide. The cylindrical core is 2.5 cm in diameter by 9.5 cm.

Positioning errors can be introduced during the process of image subtraction to obtain fluid-phase saturations or porosity. Essentially, the position of an object within subtracted images must be constant. Objects being scanned must either remain perfectly stationary or be connected to a positioning and alignment system for repeatable placement in a given position. An alternative is to employ image processing routines to centre and align objects of interest prior to subtraction. Because CT numbers depend slightly upon where the object is placed relative to the centre of the gantry, this option works best for objects with cylindrical cross-sectional volume and images where the object is only slightly displaced, relative to other images.

Experimental design and image quality

In designing an experimental set-up to be used in a CT scanner, the errors discussed above must be considered in combination with other factors such as timing, spatial resolution and image quality. If the measurements are of static properties such as density, atomic number, porosity or steady-state saturation, the timing of scans is not important. Dynamic experiments, such as unsteady state relative permeability experiments and corefloods, follow the evolution of a phase as a function of position and time. Special designs are sometimes required to track rapidly moving fronts accurately. To monitor precisely front position and shape, apparatus have been designed that allow the entire length of a core to be visualized in a single scan (Schembre *et al.* 1998; Akin & Kovscek 1999; Akin *et al.* 2000). Figure 3 illustrates the unconventional core holder set-up. Note the horizontally oriented core holder in the centre of the cylindrical cross-section. On either end of the core holder, there are endcaps for fluid distribution and maintaining position. The core holder/endcap assembly is housed within a cylindrical water jacket for temperature control and minimization of image artefacts.

If conventional scanning of cross-sections along the length of a core is employed, one can estimate the location of the flood front and the time it takes to scan the front. A greater number of scans can then be collected at the location of the front. Another method is to stop the experiment at a particular moment to take scans (Siddiqui *et al.* 1996; Closmann & Vinegar 1993). However, capillary and viscous forces cause redistribution of fluids within the core and can effect the results dramatically. Thus, this procedure is not generally recommended.

Spatial resolution is particularly important if the CT data will be used to measure fracture aperture (Hunt *et al.* 1987). The minimum values that medical scanners can resolve are between 1 to 2 mm in spatial resolution. In the study of Hunt *et al.* (1987), the minimum fracture aperture that could be resolved quantitatively was on the order of 0.5 mm. An alternative to direct measurement is the missing mass technique suggested by Johns *et al.* (1993). This technique, unlike conventional measurements, allows fracture apertures as low as 0.01 mm to be inferred by using a medical CT scanner. Jasti *et al.* (1993) describe a specially constructed industrial scanner where the scanned object is viewed on an apparent spatial resolution of 0.01 mm. In contrast to a medical scanner, where a two-dimensional fan-shaped X-ray beam illuminates an object, their high resolution system uses a cone-shaped microfocal X-ray beam. Images obtained are three-dimensional because reconstruction is performed onto a specified three-dimensional grid. This eliminates the need for interpolating between two-dimensional images.

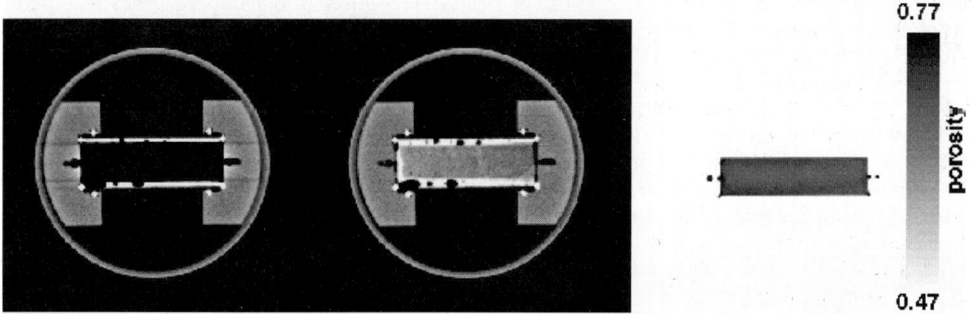

Fig. 3. Scans of an imbibition cell that allows imaging of the entire core length. Dry image (left), water-saturated image (middle) and porosity image (right). The cylindrical core is 2.5 cm in diameter by 9.5 cm.

Such an instrument might be useful for examining some of the relatively large features of the microstructure of a porous medium, a detailed examination of porosity and analysis of static fluid distribution within porous media. Scan times are relatively long (20 min), making frontal tracking difficult.

Perhaps the most important factor in terms of image quality is the geometry of the experimental set-up. Since the geometry of the scanner and reconstruction is circular, it is better to have round objects and to centre the apparatus within the scanner gantry. This minimizes 'X-artefacts'. Note the circular cross-section that the apparatus presents to the scanner in Figure 3. The core holder is immersed in a cylindrical water jacket to minimize beam hardening. Similar designs without a water jacket were reported by Hove *et al.* (1987, 1990). Another factor that affects the image quality is the use of the patient table of medical scanners to carry experimental set-ups. Although the patient table is transparent to X-rays, it causes a wave-like artefact to appear in the images (i.e. superimposed on the object) if the scan diameter is small. The table causes the CT numbers to be lower and noisier than those without the table. The best way to minimize the effect is to remove the table from the scan plane.

Flow characterization

CT scanning has the distinct ability to visualize many core phenomena that are otherwise undetectable by standard practices, such as front tracking and flow profiling. Visualization can be grouped into quantitative and qualitative categories. Several authors presented CT studies that include coreflood front tracking and displacement efficiency during miscible and immiscible displacement experiments (Frenshan & Jelen 1986; Wellington & Vinegar 1987; Hove *et al.* 1987, 1990; Withjack 1988; Withjack & Akervoll 1988; Liu *et al.* 1990; Hicks *et al.* 1990; Peters & Hardham 1990; Peters & Gharbi 1993; Hicks & Deans 1994; Burger *et al.* 1994; Walsh & Withjack 1994; Bertin *et al.* 1999; Schembre *et al.* 1998; Rangel-German *et al.* 1999; Akin *et al.* 2000; Apaydin & Kovscek 2001), mud invasion visualization (Auzerais *et al.* 1991; Krilov *et al.* 1991; Kuru *et al.* 1998), and production of sand along with oil (Tremblay *et al.* 1999).

Quantitative studies include the study by Wang *et al.* (1985) who reported local oil saturations using an X-ray CT scanner. They also presented images of viscous fingering, time derivatives of local composition and residual oil distribution for the case of water displacing oil from a porous medium. Vinegar & Wellington (1987) reported measurements of three-phase saturations utilizing X-ray CT. They also described the image processing system, X-ray transparent high-pressure flow equipment, choice of fluid dopants and X-ray energies for scanning of coreflood experiments. Examples were given of tertiary miscible carbon dioxide displacements in Berea sandstone. Withjack (1988) recorded the use of CT for saturation measurements to be used in the computation of steady state two-phase relative permeabilities. CT saturation results agreed with conventional measurements within 2 saturation %. He also reported that the CT scanner provided an improved understanding of miscible corefloods.

Some studies concentrated on the use of CT-derived multiphase saturation data for steady or unsteady state relative permeability determination. Fresnhan & Jelen (1986) proposed a method to determine saturation changes during the course of a waterflood using a second generation CT. They used production history to calibrate the change in attenuation with a corresponding change in saturation. Relative permeabilities obtained by trial and error were used to simulate the saturation profile along the core. The model predicted overall recovery, but it overestimated recovery in the first one third of the core. The authors attributed this problem to improper modelling of end effects. Mohanty & Miller (1988) obtained porosity and saturation profiles during steady state relative permeability tests. They calculated relative permeability curves for mixed-wet cores using the JBN, unsteady-state method (Honarpour *et al.* 1986) at three different flooding rates (low, intermediate and high). For all cases the flood front was not piston-like, as observed in CT images. MacAllister *et al.* (1990) measured two-phase steady state gas-water relative permeabilities using a CT scanner to calculate three-dimensional fluid saturation distributions during experiments. Similarly, Kamath *et al.* (1995) analysed three-dimensional saturation profiles generated from CT scanning of waterfloods, oil floods and miscible corefloods. They calculated relative permeability end points of the low permeability mixed-wet diatomaceous mudstones using the steady-state method. From the axially non-uniform saturation distribution images at the end of the floods, they concluded that end effects were present.

Cuthiel *et al.* (1993) performed a series of steam corefloods. Trial and error steam relative permeabilities were used to fit experimental saturation and pressure profiles to model data. Clossman & Vinegar (1993) measured steam and water relative permeabilities in natural cores at steam drive residual saturation under *in-situ* conditions. They found that steam relative permeabilities were in agreement with published experimental results. In related work, Ambusso *et al.* (1996) and Satik (1998) determined steady-state steam-water relative permeabilities in sandstone. They used a CT scanner to identify portions of the core with uniform saturation profiles that were, thus, appropriate for computation of relative permeability.

Siddiqui *et al.* (1996), Akin & Demiral (1997) and Sahni *et al.* (1998) used CT-derived three-phase saturation data to estimate three-phase relative permeabilities, whereas Barbu *et al.* (1999) examined three-dimensional phase-saturation patterns in three-phase flow. Similarly, DiCarlo *et al.* (2000*a*, *b*) used CT to elucidate the effect of porous medium wettability on three-phase relative permeability in a gravity drainage mode.

Porosity and core characterization

Porosity can be measured using conventional methods such as Boyle's Law porosimetry (American Petroleum Institute 1960) or thin-section analysis (Van Golf-Recht 1982), but it can be measured more descriptively on a local basis using CT methods. Withjack (1988) performed CT porosity measurements from two scans at the same location obtained with different fluids saturating the porous medium. The following equation, based on Beer's law, is used to determine the porosity for each volume element (voxel):

$$\phi = \frac{CT_{wr} - CT_{ar}}{CT_{w} - CT_{a}}. \quad (6)$$

The subscripts w and a represent water-phase and air-phase CT numbers, whereas wr and ar refer to water- and air-saturated rock, respectively. Close agreement (± 1 porosity %) was reported between the CT-derived porosities and those determined volumetrically.

On the right of Figure 3 a sample porosity image obtained by applying Eq. 6 to a homogeneous diatomite core is shown. Average porosity is in excess of 60%. Note that the endcaps used for positioning the core and delivering and producing fluids are evident in the raw dry and water-saturated images on the left and middle of Figure 3. Also, bubbles are apparent in the epoxy-filled annular region between the core and acrylic sleeve. The image of porosity on the right contains no image remnants of gas bubbles or endcaps because dry and water-saturated images are perfectly aligned and free of beam hardening and X-shaped artefacts.

Lu *et al.* (1992) used a dual scan with single energy level technique that was proposed by Moss *et al.* (1990) to calculate porosity based on equation (6). Using xenon gas as a contrast agent, they determined porosity images for conventional core samples and samples that exhibited dual-porosity features. More recently, Watson & Mudra (1994) investigated gas storage in Devonian shales using a CT scanner and utilized equation (6). Alternatively, Vinegar & Kehl (1988) determined porosity using a method in which the bulk density of the material, ρ_b, must be calculated using several calibration coefficients obtained prior to scanning. The fluid and grain density of the material (ρ_f, ρ_g) must be known as well.

In standard CT porosity measurement, the core sample should be 100% satu [illegible] contrast agent such as xenon gas or [illegible] brine. Special equipment for cleani [illegible] rating samples is sometimes neede [illegible] (1996) proposed a dual energy s [illegible] measurement method at energy le [illegible] based on estimation of a matrix [illegible] CT_M. In this technique, there is [illegible] clean and saturate the samples wit [illegible] agent. The equations to solve on a [illegible] basis are:

$$\phi = \frac{\Delta U}{\Delta N}$$

$$\Delta U = (CT_B^2 - CT_M^2)(CT_{ph1}^1 - CT_{ph2}^1) + (CT_{ph2}^2 - CT_{ph1}^2)(CT_B^1 - CT_M^1)$$

$$\Delta N = CT_{ph1}^2(CT_M^1 - CT_{ph2}^1) + CT_{ph2}^2(CT_{ph1}^1 - CT_M^1) + CT_M^2(CT_{ph1}^1 - CT_{ph2}^1) \quad (7)$$

where the subscript B refers to a given voxel and subscripts ph1 and ph2 refer to the CT numbers of pure phases.

The distribution of porosity and even raw CT images help greatly to characterize the nature (homogeneous versus heterogeneous) of a porous medium (Bergosh & Lord 1987; Peters & Afzal 1992; Karacan & Okandan 1999). Figure 4 presents a set of raw CT images taken of a carbonate core at 1 cm spacing. The presence and location of vugs is evident. Note that dark shading corresponds to high density regions and white to low. As will be discussed later, 3D image reconstructions can be used to examine these and other phenomena further.

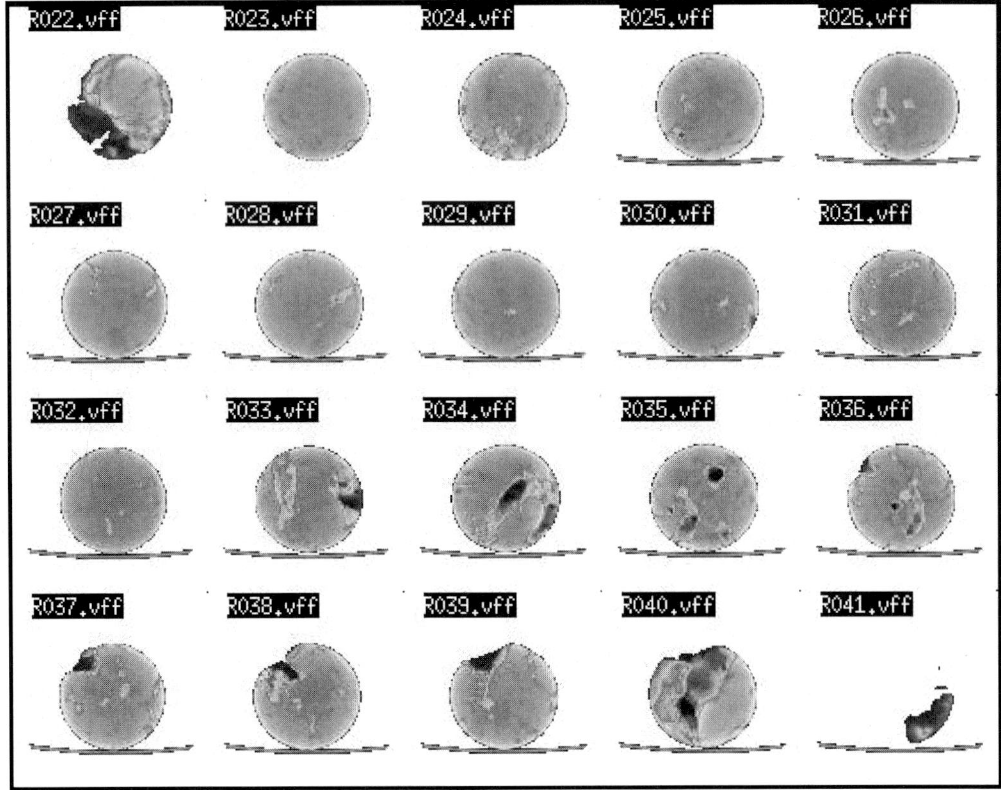

Fig. 4. Scans of a carbonate core at 1 cm spacing. Dark shading corresponds to high density regions and white to low density regions.

Permeability distribution

Permeability cannot be measured directly but its distribution may be estimated with the help of a CT scanner. Withjack *et al.* (1991) presented a permeability distribution determination based on measuring *in-situ* flood-front velocities by CT scanning. In this technique, a core is assumed to be formed by a bundle of streamtubes with negligible variation of cross-sectional area along the length of each streamtube. The streamtube permeability of each voxel can be calculated from the following equation with the knowledge of average core permeability (k_{average}), individual streamtube porosities (ϕ_i), and the time for the solvent to flow along a given length of core (t_i)

$$k_1 = \frac{k_{\text{average}} A_T}{A\left(1 + \frac{\phi_3 t_1}{\phi_1 t_2} + \frac{\phi_3 t_1}{\phi_1 t_3} + \cdots + \frac{\phi_n t_1}{\phi_1 t_n}\right)} \tag{8}$$

where A represents area and the subscript T denotes tube. Other permeabilities can then be computed from

$$\frac{k_n}{k_1} = \frac{\phi_n t_1}{\phi_1 t_n}. \tag{9}$$

This technique relies on accurate estimation of individual voxel porosities and thus should be used with care. Another technique for *in-situ* permeability distribution determination is based on measurement of *in-situ* tracer concentrations with a CT scanner (Mohanty & Johnson 1991; Johns *et al.* 1993). In this technique, a density-matched viscous flood is conducted and monitored with a CT scanner. Complete water- or oil-saturated core scans are taken and scans are then taken at different intervals in time as the displacement is conducted with the same oil doped with a contrast agent. The concentration of the dopant, C_s, is then calculated from the Eq. 10:

$$C_s = \frac{\text{CT}_{\text{ost}} - \text{CT}_{\text{or}}}{\text{CT}_{\text{sr}} - \text{CT}_{\text{or}}}. \tag{10}$$

The subscripts osr, or and sr refer to the CT numbers of rock containing oil and solvent, oil-saturated rock and solvent-saturated rock, respectively. *In-situ* macroscopic dispersivity values can also be computed by scanning the core at 10% and 90% concentrations and measuring the distance between the concentration contours (Mohanty & Johnson 1991). In related work, Peters *et al.* (1996) also examined dispersion using CT. Yet another way of computing the permeability field is to use an empirical permeability-porosity correlation, as suggested by Hicks & Deans (1994).

Determination of two-phase saturations

Many methods have been used for determining fluid saturations during multiphase coreflood experiments. These include transparent models, resistivity, X-ray absorption, nuclear magnetic resonance and X-ray and gamma ray attenuation. Honarpour *et al.* (1986) summarize these methods. All impose restrictions and provide only localized average values for the saturations. In comparison to the above-mentioned techniques, CT measurement of saturation is fast, accurate, easy to calibrate and offers fine spatial resolution. There are several different CT methods for *in situ* saturation determination.

Linear interpolation between pure states

In this technique, it is assumed that the CT number of the core lies on the straight line connecting complete saturation by phase 1 (say water) to complete saturation by phase 2 (say oil). Thus, a single energy scan is sufficient to measure two-phase saturations as shown below:

$$CT_{owr} = (1 - \phi)\mu_r + \phi S_o \mu_o + \phi S_w \mu_w \qquad (11)$$

$$1 = S_w + S_o \qquad (12)$$

where the subscript owr refers to rock containing both oil and water phases. Here, μ_r, μ_o, and μ_w are the attenuation coefficients for the rock matrix, core fully saturated with oil and water, respectively and S_o and S_w are oil and water saturations, respectively. Thus the saturation of oil in each voxel is:

$$S_o = \frac{CT_{wr} - CT_{owr}}{CT_{wr} - CT_{or}}. \qquad (13)$$

One way to obtain the value of CT_{or} after scanning the water-saturated core (to obtain CT_{wr}) is complete cleaning and drying of the core followed by saturation of the core with oil. Because this procedure usually requires removal of the core from its original position, it is subject to positioning errors.

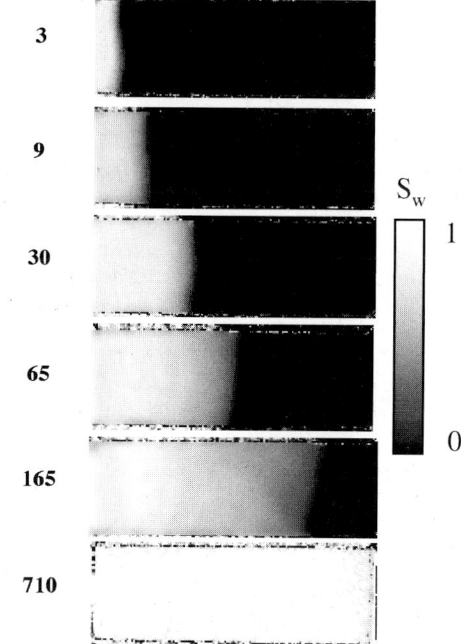

Fig. 5. CT-derived water saturation (S_w) images of spontaneous imbibition in diatomite, with water displacing air. The sample is oriented horizontally during the experiment. Time in minutes is given next to each image.

Figure 5 presents water saturation maps acquired for different diatomite cores during countercurrent water imbibition. The completely wet and dry images presented in Figure 3 are used in the denominator of Eq. 13. The water saturation scale is given on the right. Note that no effect of gravity on the imbibition front is found because capillarity dominates fluid saturation patterns. Wang *et al.* (1985) and MacAllister *et al.* (1990) also demonstrated the use of Eq. 13 for various laboratory corefloods.

Another way to obtain the value of CT_{or} was reported by Alvestad *et al.* (1991). The CT value of the rock completely saturated with oil is interpolated from the dry CT images and water-saturated CT images with knowledge of pure phase CT numbers using the following equation:

$$CT_{or} = CT_{dry} + \frac{CT_o - CT_a}{CT_w - CT_a}(CT_{wr} - CT_{dry}). \qquad (14)$$

Fluid CT numbers

If the porosity distribution of the rock is already available, a different form of Eq. 13 that does not need complete water- and oil-saturated images can be used:

$$S_w = \frac{CT_{ow} - CT_{or}}{\phi(CT_w - CT_o)}. \quad (15)$$

In order for Eq. 15 to yield accurate results, the images must be perfectly flat (i.e. free of beam hardening effects). The CT numbers for water and oil (or air) should be obtained inside the core holder without the core for best results. It should be noted that if the saturation values computed with Eq. 15 decrease towards the centre of the slice, beam hardening effects are present and the results are incorrect.

Figure 6 is a 3D reconstruction of S_w during a hot water flood computed using Eq. 15 (Akin *et al.* 1998). White indicates high water saturation and colours correspond to S_w, as indicated by the colour bar in the figure. Each reconstruction is a different time and flow is from right to left in each image. Temperature increases from 70°F (21°C) for images on the left to 122°F (50°C) in the middle and 150°F (66°F) on the right. The decrease in S_o and corresponding increase in S_w is quite clear.

Linear regression

The procedure described previously assumes that no information is available except for the two pure states. Usually, average saturations at residual states are known from other laboratory techniques such as Dean-Stark extraction. Therefore, complete drying of the core and resaturation with oil can be avoided by scanning the core at connate water saturation, S_{wc}, which can be easily obtained from material balance. At connate water saturation, Eq. 13 can be written as

$$S_o = \frac{(CT_{ow} - CT_{wr})(1 - S_{wc})}{CT_{swcr} - CT_{wr}} \quad (16)$$

where the subscript swcr denotes a CT measurement of a core at connate water saturation. This approach assumes that the connate water saturation is uniform throughout the core. This assumption will lead to errors if a saturation gradient exists. Ganapathy *et al.* (1991) showed that in naturally heterogeneous sandstone cores the oil saturation profile at the start and the end of a coreflood was not uniform. Similarly, Qadeer *et al.* (1994) showed that there were large saturation gradients along the length of a Berea sandstone core for several corefloods monitored with a CT scanner. Similar non-uniform saturations were reported for Middle Eastern carbonates by Sprunt *et al.* (1991) and MacAllister *et al.* (1990). Thus, the use of Eq. 16 is limited and must be undertaken with caution.

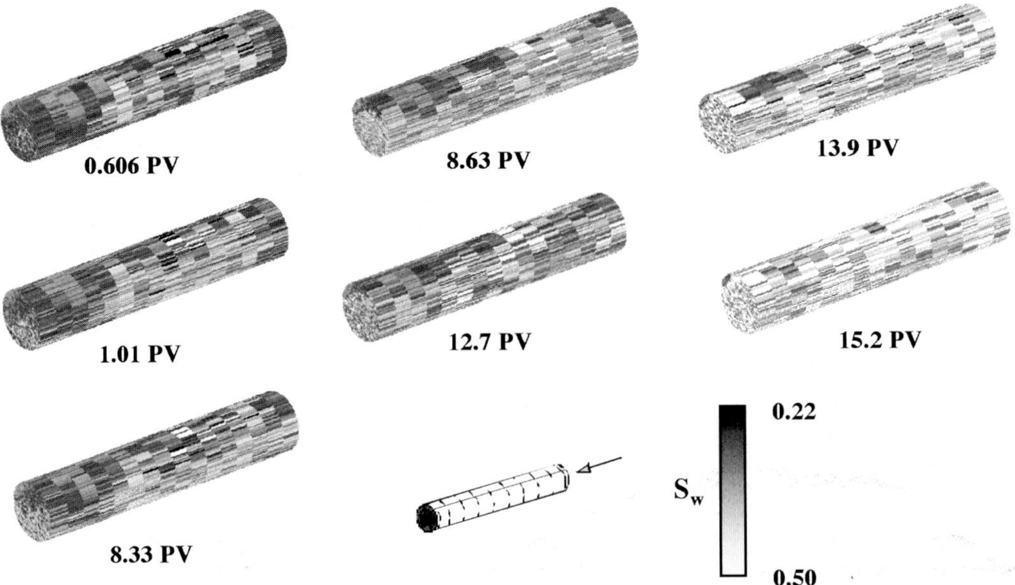

Fig. 6. 3D reconstruction of water saturation in a sandpack undergoing hot-water flood (after Akin *et al.* 1998). Flow is from right to left. Times are given as pore volumes of water injected. Water temperature is increased during the experiment. Left images: 70°F (21°C). Centre: 122°F (50°C). Right: 150°F (66°C).

Determination of three-phase saturations

Several different methods exist for calculating three-phase saturations using a CT scanner. These techniques can be categorized into four major groups: one immobile phase methods, matched CT-fluids methods, linear interpolation methods and dual-energy scan methods.

One immobile phase

In this technique, it is assumed that one phase is immobile and its value is constant throughout the core. With this assumption, the CT number of the immobile phase is included in the rock CT number in Eq. 13. Thus, it is assumed that only changes in saturation of the remaining phases alter the CT values. The saturations of the mobile phases can be obtained by implementing the two-phase CT saturation methods given by Eqs 11 to 13.

Linear regression

In this method, a two-phase flood is conducted until irreducible water saturation, S_{wirr}, is achieved and a scan, CT_{oirr}, is taken at this saturation. Then, three-phase flow is initiated and the water and oil saturations are obtained using Eqs 17 and 18, and assuming that the saturations are linearly related to CT numbers. The remaining saturation can be obtained by material balance. This method was used by Siddiqui *et al.* (1996) to obtain three-phase saturations of water, benzyl alcohol and decane.

$$S_w = S_{wirr} + \left(\frac{CT - CT_{oirr}}{CT_{wr} - CT_{oirr}}\right)(1 - S_{wirr}) \quad (17)$$

$$S_o = \left(\frac{CT_{wr} - CT}{CT_{wr} - CT_{oirr}}\right)(1 - S_{wirr}) \quad (18)$$

Dual energy scan

Three-phase saturations can be most accurately obtained by scanning the core at two energy levels that are linearly independent from each other. Vinegar & Wellington (1987) presented a dual-energy method where fluid CT numbers can be used to obtain three-phase saturations. They eliminated the term for the gas-phase scan by assuming that the attenuation of the gas is zero. Eq. 13 can be adapted for three phases and two energy levels as shown below:

$$CT_1 = (1 - \phi)\mu_{r1} + \phi S_o\mu_{o1} + \phi S_w\mu_{w1} + \phi S_g\mu_{g1} \quad (19)$$

$$CT_2 = (1 - \phi)\mu_{r2} + \phi S_o\mu_{o2} + \phi S_w\mu_{w2} + \phi S_g\mu_{g2} \quad (20)$$

$$1 = S_o + S_w + S_g. \quad (21)$$

Here, subscripts 1 and 2 refer to measurements at high and low energy levels. It should be noted that the difference in linear attenuation coefficients or CT numbers should be large enough to ensure linear independence of the above equations. In other words, the high and low energy levels must measure different physical properties (Compton scattering and photoelectric absorption), as discussed in an earlier section. Therefore, at least one of the phases must be doped with a strong photoelectric absorber such as potassium bromide. The equations to obtain the saturation of each voxel are:

$$S_o = [(CT_{w1} - CT_{g1})(CT_2 - CT_{g2}) - (CT_{w2} - CT_{g2})(CT_1 - CT_{g1})] / [(CT_{o2} - CT_{g2})(CT_{w1} - CT_{g1}) - (CT_{o1} - CT_{g1})(CT_{wr2} - CT_{g2})] \quad (22)$$

$$S_w = [(CT_{o2} - CT_{g2})(CT_1 - CT_{g1}) - (CT_{o1} - CT_{g1})(CT_2 - CT_{g2})] / [(CT_{o2} - CT_{g2})(CT_{w1} - CT_{g1}) - (CT_{o1} - CT_{g1})(CT_{wr2} - CT_{g2})]. \quad (23)$$

where CT_1 and CT_2 refer to data for the three-phase system at high and low energy. It should be noted that the rock is not considered in the denominator of Eqs 22 and 23. Use of this method is demonstrated by Akin and Demiral (1997, 1998) in a Berea sandstone plug.

A different form of the above equation that includes images of gas-, water- and oil-saturated rock can also be used. Thus, the set of equations that use rock influence instead of pure fluid CT numbers are:

$$S_o = [(CT_1 - CT_{gr1})(CT_{wr2} - CT_{gr2}) - (CT_2 - CT_{gr2})(CT_{wr1} - CT_{gr1})] / [(CT_{or1} - CT_{gr1})(CT_{wr2} - CT_{gr2}) - (CT_{or2} - CT_{gr2})(CT_{wr1} - CT_{gr1})] \quad (24)$$

$$S_w = [(CT_2 - CT_{gr2})(CT_{or1} - CT_{gr1}) - (CT_1 - CT_{gr1})(CT_{or2} - CT_{gr2})] / [(CT_{or1} - CT_{gr1})(CT_{wr2} - CT_{gr2}) - (CT_{or2} - CT_{gr2})(CT_{wr1} - CT_{gr1})]. \quad (25)$$

The experimental procedure is somewhat complicated, because three end-point calibration scans are needed at each energy level. This method has been used to measure three-phase saturations during gravity drainage experiments (Sahni *et al.* 1998; DiCarlo *et al.* 2000*a*, *b*).

It should be noted that multiphase liquid saturations need to be corrected using room temperature values for the case of steam injection or high temperature corefloods. This can be achieved by either obtaining the reference CT values at the corresponding temperatures or correcting the saturations after calculation. Closmann & Vinegar (1993) discuss this issue in more detail.

Discussion

The accuracy of CT-derived parameters, such as porosity and multiphase fluid saturations, can be obtained theoretically. Error analyses for different porosity measurement techniques are presented in detail by Akin *et al.* (1996). They propose two techniques. In the first technique, measurements from two scans at identical positions are compared to determine the random error. In the second technique, multiple scans of a single core position that comprise a representative statistical population of the same pixel are analysed. They reported that the conventional CT porosity measurement technique is subject to a 3.8% error if the CT number measurement has an error of 1.6%. Withjack (1988), in establishing the correctness of Eq. 6, measured porosity of Berea sandstone and dolomite samples with CT and volumetric methods. He reported agreement within 1%.

Error analysis for CT-derived multiphase saturation measurements based on Eqs 13, 22 and 23 are presented by Sharma *et al.* (1997). They reported that two-phase saturation errors were between 0.7% and 2.1%. On the other hand, three-phase saturations measured during a steam injection experiment were subject to an error of up to 18.7% in magnitude. Absolute errors were roughly the same for all values of phase saturation. Therefore, the largest percentage errors occur for measurements of low saturation.

Experiments in medical CT scanners have largely been conducted in a horizontal fashion. While most scanners are designed to allow some degree of inclination, with respect to the vertical, it is difficult to find scanners where the gantry can be oriented in a true vertical fashion. Chiefly, this is a mechanical design problem in that the bearings supporting the gantry will not allow it to rotate while in vertical mode. Notable exceptions are the work of Kantzas *et al.* (1988), Sahni *et al.* (1998) and DiCarlo *et al.* (2000*a*, *b*), for which true vertical positioning was employed.

This overview of previous work suggests many best practices for obtaining high-quality data from CT. In brief, apparatus that present circular cross-sections to the scanner eliminate or greatly reduce X-artefacts originating from unequal X-ray path lengths through scanned materials. With some care in the design of coreholders, beam hardening effects can be reduced greatly along with X-artefacts and flat artefact-free images obtained. In this regard, it is strongly suggested to design positioning systems that do not rely on standard medical CT patient couches. These couches are prone to errors in positioning apparatus repeatedly and introduce wave-like artefacts in images. At the very least, patient couches should be removed from the scan plane.

For quantitative representation of porosity or fluid-phase saturation, image processing equations that use fully saturated porous media as end states give the most accurate and least ambiguous results. Specifically, Eq. 6 should be used for porosity, Eq. 13 for determination of two-phase saturation and Eqs 24 and 25 for three-phase saturation. If it is difficult or impossible to obtain one of the necessary fully-saturated images, then pure fluid CT numbers and the local porosity field should be used to back-calculate the fully saturated porous media image. For example, Eq. 15 could be used for two-phase saturation and Eq. 22 and 23 for three-phase saturation.

Frontiers

There are a number of frontiers that remain in CT scanning of porous media. The first is complete volumetric monitoring of dynamic experiments. Current first to fourth generation scanners collect data from thin volumetric sections. Generally, coverage of a porous medium with such scans is incomplete because of the need to sample the entire core at a relatively rapid rate. Interpolation between, essentially, two-dimensional cross-sections is required to infer three-dimensional information. Hence, detail is lost. Spiral CT (Klingenbeck-Regn *et al.* 1999) provides the opportunity for subsecond acquisition times of multiple slices. This development puts complete volumetric scanning of 'fast' displacement processes within reach. To date, there have been no reports of experiments utilizing this tool.

Another frontier is routine scanning at spatial resolutions on the order of 0.1 to 0.01 mm. With such detailed information, fracture networks and other types of fine-scale heterogeneity could be characterized and flow in them could be resolved (Van Geet & Swennen 2001). This objective will be quite difficult to obtain with a medical scanner, as there is little driving force in the medical community to obtain resolutions finer than about 0.25 mm with these devices. In this regard, industrial microfocus CT scanners (μCT) may provide a pathway forward (Kalukin *et al.* 2000; Van Geet *et al.* 2001). Machines have been introduced with a resolution of around 0.1 mm. Problems to overcome for measurement of *in-situ* saturations might include shorter scan times, as discussed earlier, accurate repeatable positioning of the object to be scanned and positioning of the source/receiver assembly, which is variable in some industrial scanners. Synchrotron microtomography for imaging of porous media offers spatial resolution on the order of 5 μm and promising results (Spanne *et al.* 1994; Coles *et al.* 1996, 1998). In these studies, porous medium sample sizes to date have only been about 2.5 cm in both diameter and length.

CT scanning has advanced sufficiently to provide information on convective transport properties and multiphase fluid saturation structures in porous media. This is not the case, however, with diffusive transport processes. Recent work showed how CT imaging could be used to measure diffusion coefficients of ions in water-saturated rocks and clays (Nakashima 2000). So far the technique is limited to relatively heavy ions, such as iodine, whose mass attenuation coefficient is large. Extension to other and perhaps lighter ions could provide advantages over other conventional laboratory techniques for measuring ion diffusion coefficients.

Image processing remains an area where progress could be made. The requirements of medical and petrophysical scanning are not identical. To date, most image processing to convert raw attenuation data to CT numbers has relied upon the proprietary algorithms developed by scanner manufacturers. These image processing routines are not optimized for dense materials such as rock and the aluminium or plastic used in coreholders. Improvements might be made by simply considering image reconstruction for flow in porous media applications separate from the medical community, as is the case with synchrotron microtomography.

Summary

In this paper, a review of CT scanning as a qualitative and quantitative tool in petroleum engineering research is given. The development and application of X-ray computed tomography for the determination of rock properties and the study of multiphase fluid flow dynamics are discussed in detail. Multiphase saturation, porosity and permeability determination using different methods with a CT scanner are presented. The advantages and disadvantages of these methods as well as their accuracy are discussed. Finally, factors affecting experiments are discussed and techniques to handle such problems are suggested.

Appendix

a	Klein–Nishina coefficient
A	area
b	constant, 9.8×10^{-24}
C	concentration
CT	CT number
e	energy, limits of integration
E	energy level
h	thickness
I	X-ray intensity
k	permeability
L	length
S	saturation
t	time
Z	atomic number

Greek

ε	efficiency
ϕ	porosity
μ	linear attenuation coefficient
ρ	density

subscripts and superscripts

0	incident radiation
1, 2	low and high energy levels
avg	average
b	bulk
e	experimental
f	fluid
g	gas or grain
l, n	index
o	oil
ow	oil + water
os	oil + solvent
ph1	phase 1
ph2	phase 2
r	rock
s	solvent
swc	connate
T	tube
w	water
wrsc	connate water + rock
B	base
M	matrix

References

AKIN, S., DEMIRAL, B. & OKANDAN, E. 1996. A novel method of porosity measurement utilizing computerized tomography. *In Situ*, **20**, 347–365.

AKIN, S. & DEMIRAL, B. 1997. Effect of flow rate on imbibition three-phase relative permeabilities and capillary pressures. *In*: *Proceedings of the Society of Petroleum Engineers Annual Technical Conference and Exhibition, San Antonio, Texas, 5–8 October*, Paper **SPE 38897**.

AKIN, S. & DEMIRAL, B. 1998. Application of computerized tomography to the determination of three phase relative permeabilities. *In*: *Proceedings of the 8th International Symposium on Flow Visualization, Sorrento, Italy, 1–4 September*, Paper **122**.

AKIN, S., CASTANIER, L.M. & BRIGHAM, W.E. 1998. Effect of temperature on heavy-oil/water relative permeabilities. *In*: *Proceedings of the 1998 Society of Petroleum Engineers Annual Technical Conference and Exhibition, New Orleans, Louisiana, 27–30 September*, Paper **SPE 49021**.

AKIN, S. & KOVSCEK, A.R. 1999. Imbibition studies of low-permeability porous media. *In*: *Proceedings of the Society of Petroleum Engineers Western Regional Meeting, Anchorage, Alaska, 26–28 May*, Paper **SPE 54590**.

AKIN, S., SCHEMBRE, J.M., BHAT, S.K. & KOVSCEK, A.R. 2000. Spontaneous imbibition characteristics of diatomite. *Journal of Petroleum Science and Engineering*, **25**, 149–165.

ALVESTAD, J., GILJE, E., HOVE, A.O., LANGELAND, O., MALDAL, T. & SCHILLING, B.E.R. 1991. Coreflood experiments with surfactant systems for IOR: Computer tomography studies and numerical modelling. *In*: *Proceedings of the 6th European IOR Symposium, Stavanger, Norway, 21–23 May*, 789–800.

AMBUSSO, W., SATIK, C. & HORNE, H. 1996. Determination of relative permeability for steam-water flow in porous media. *In*: *Proceedings of the Society of Petroleum Engineers Annual Technical Conference and Exhibition*, Denver, Colorado, 6–9 October, Paper **SPE 36682**.

AMERICAN PETROLEUM INSTITUTE 1960. *API Recommended Practice for Core Analysis Procedures.* American Petroleum Institute, Dallas, Texas, Report **40**.

APAYDIN, O.G. & KOVSCEK, A.R. 2001. Surfactant concentration and end effects on foam flow in porous media. *Transport in Porous Media*, **43**, 511–536.

AUZERAIS, F.M., DUSSAN, E.V. & REISCHER, A.J. 1991. Computed tomography for the quantitative characterization of flow through a porous medium. *In*: *Proceedings of the 66th Annual Technical Conference and Exhibition of the Society of Petroleum Engineers, Dallas, Texas, 6–9 October*, Paper **SPE 22595**.

BARBU, A., HICKS, P.J. JR & GRADER, A.S. 1999. Experimental three-phase flow in porous media: Development of saturated structures dominated by viscous flow, gravity, and capillarity. *Society of Petroleum Engineers Journal*, **4**, 368–379.

BERGOSH, J.L. & LORD, G.D. 1987. New developments in the analysis of cores from naturally fractured reservoirs. *Proceedings of the 62nd Annual Technical Conference and Exhibition of the Society of Petroleum Engineers, Dallas, Texas, 27–30 September*, Paper **SPE 16805**.

BERTIN, H.J., APAYDIN, O.G., CASTANIER, L.M. & KOVSCEK, A.R. 1999. Foam flow in heterogeneous porous media: Effect of crossflow. *Society of Petroleum Engineers Journal*, **4**, 75–82.

BURGER, J.E., BOGESWARA, R. & MOHANTY, K.K. 1994. Effect of phase behavior on bypassing in enriched gas floods. *Society of Petroleum Engineers Reservoir Engineering*, **9**, 112–118.

CLOSMANN, P.J. & VINEGAR, H.J. 1993. A technique for measuring steam and water relative permeabilities at residual oil in natural cores: CT scan saturations. *Journal of Canadian Petroleum Technology*, **32**, 55–60.

COLES, M.E., HAZLETT, R.D., SPANNE, P., MUEGGE, E.L. & FURR, M.J. 1996. Characterization of reservoir core using computed microtomography. *Society of Petroleum Engineers Journal*, **1**, 295–301.

COLES, M.E., HAZLETT, R.D., SPANNE, P., SOLL, W.E., MUEGGE, E.L. & JONES, K.W. 1998. Pore-level imaging of fluid transport using synchrotron X-ray microtomography. *Journal of Petroleum Science and Engineering*, **19**, 55–63.

CUTHIEL, D., SEDGWICK, G., KISSEL, G. & WOOLEY, J. 1993. Steam corefloods with concurrent X-ray CT imaging. *Journal of Canadian Petroleum Technology*, **32**, 37–45.

DICARLO, D.A., SAHNI, A. & BLUNT, M.J. 2000*a*. The effect of wettability on three-phase relative permeability. *Society of Petroleum Engineers Journal*, **5**, 82–91.

DICARLO, D.A., SAHNI, A. & BLUNT, M.J. 2000*b*. Three-phase relative permeability of water-wet, oil-wet, and mixed-wet sandpacks. *Transport in Porous Media*, **39**, 347–366.

FRENSHAN, P.B. & JELEN, J. 1986. Displacement of heavy oil visualized by CAT scan. *In*: *Proceedings of the 37th Annual Technical Meeting of the Petroleum Society of CIM, Calgary, 8–11 June*, 605–620.

GANAPATHY, S., WREATH, D.G., LIM, M.T., ROUSE, B.A., POPE, G.A. & SEPEHRNOORI, K. 1991. Simulation of heterogeneous sandstone experiments characterized using CT scanning. *In*: *Proceedings of the Western Regional Meeting, Long Beach, California, 20–22 March*, Paper **SPE 21757**.

HERMAN, G.T. 1980. *Image Reconstruction from Projections. The Fundamentals of Computerized Tomography*. Academic Press, New York.

HICKS, P.J., NARANAYAN, R. & DEANS, H.A. 1990. An experimental study of miscible displacements in heterogeneous carbonate cores using X-ray CT. *In*: *Proceedings of the 65th Annual Technical Conerence and Exhibition of Society of Petroleum Engineers, New Orleans, Louisiana, 23–26 September*, Paper **SPE 20492**.

HICKS, P.J. & DEANS, H.A. 1994. Effect of permeability distribution on miscible displacement in a heterogeneous carbonate core. *Journal of Canadian Petroleum Technology*, **33**, 28–34.

HONARPOUR, M., KOEDERITZ, L. & HARVEY, A.H. 1986. *Relative Permeability of Petroleum Reservoirs*. CRC Press, Boca Raton, Florida.

HOUNSFIELD, G.N. 1972. *A Method of and Apparatus for Examination of a Body by Radiation Such as X or Gamma Radiation*. British Patent No. 1.283.915, London.

HOVE, A.O., RINGEN, J.K. & READ, P.A. 1987. Visualization of laboratory corefloods with the aid of computerized tomography of X-rays. *Society of Petroleum Engineers Reservoir Engineering*, **2**, 148–154.

HOVE, A.O., NILSEN, V. & LEKNES, J. 1990. Visualization of xanthan flood behavior in core samples by means of X-ray tomography. *Society of Petroleum Engineers Reservoir Engineering*, **5**, 475–480.

HUNT, P.K., ENGLER, P. & BAJSCROWICZ, C. 1987. Computed tomography as a core analysis tool: applications and artifact reduction techniques. *In*: *Proceedings of the Society of Petroleum Engineers Annual Technical Conference and Exhibition, Dallas, Texas, 27–30 September*, Paper **SPE 16952**.

JASTI, J.K., JESION, G. & FELDKAMP, L. 1993. Microscopic imaging of porous media with X-ray Computer Tomography. *Society of Petroleum Engineers Formation Evaluation*, **8**, 190–193.

JOHNS, R.A., STEUDE, J.S., CASTANIER, L.M. & ROBERTS, P. 1993. Nondestructive measurements of fracture aperture in crystalline rock cores using X-Ray Computed Tomography. *Journal of Geophysical Research*, **98**, 1889–1900.

KALUKIN, A.R., VAN GEET, M. & SWENNEN, R. 2000. Principal components analysis of multienergy X-ray computed tomography of mineral samples. *IEEE Transactions on Nuclear Science*, **47**, 1729–1736.

KAMATH, J., DEZABALA, E.F. & BOYER, R.E. 1995. Water/oil relative permeability endpoints of intermediate-wet low-permeability rocks, *Society of Petroleum Engineers Formation Evaluation*, **10**, 4–10.

KANTZAS, A., CHATZIS, I., MACDONALD, F. & DULLIEN, F.A.L. 1988. Using a vertical scanner for horizontal scanning in non-medical applications of computer assisted tomography. *CSNDT Journal*, **March/April**, 20–25.

KARACAN, C.O. & OKANDAN, E. 1999. Heterogeneity effects on the storage and production of gas from coal seams. *In*: *Proceedings of the 1999 Society of Petroleum Engineers Annual Technical Conference and Exhibition, Houston, Texas, 3–6 October*, Paper **SPE 56551**.

KLINGENBECK-REGN, K., SCHALLER, S., FLOHR, T., OHNESORGE, B., KOPP, A.F. & BAUM, U. 1999. Subsecond multi-slice computed tomography: basics and applications. *European Journal of Radiology*, **31**, 110–124.

KRILOV, Z., STEINER, I., GORICNIK, B., WOJTANOWICZ, A.J. & CABRAJAC, S. 1991. Quantitative determination of solids invasion and formation damage using CAT scan and barite suspensions. *In*: *Proceedings of the 1991 Offshore Europe Conference, Aberdeen, Scotland, 3–6 September*, Paper **SPE 23102**.

KURU, E., DEMIRAL, B., AKIN, S., KEREM, M. & CAGATAY, B. 1998. An integrated study of drilling-fluid shaly rock interactions: a key to solve wellbore instability problems. *Oil Gas European Magazine*, **24**, 25–29.

LIU, D.B., CASTANIER, L.M. & BRIGHAM, W.E. 1990. Analysis of transient foam flow in 1-D porous media with CT. *In*: *Proceedings of the 60th Regional California Meeting, Ventura, California, 4–6 April*, Paper **SPE 20071**.

LU, X. C., PEPIN, G.P. & MOSS, R.M. 1992. Determination of gas storage in Devonian shales with X-ray computed tomography. *In*: *Proceedings of the 67th Annual Technical Conference and Exhibition of Society of Petroleum Engineers, Washington, DC, 4–7 October*, Paper **SPE 24810**.

MACALLISTER, D.J., MILLER, K.C., GRAHAM, S.K. & YANG, C.T. 1990. Application of X-ray CT scanning to the determination of gas-water relative permeabilities. *In*: *Proceedings of the 65th Annual Technical Conference and Exhibition of Society of Petroleum Engineers, New Orleans, Louisiana, 23–26 September*, Paper **SPE 20494**.

MOHANTY, K.K. & JOHNSON, S.W. 1991. Interpretation of laboratory gasfloods with multidimensional compositional modeling. *In*: *Proceedings of the Society of Petroleum Engineers Symposium on Reservoir Simulation, Anaheim, Califormia, 17–20 February*, Paper **SPE 21204**.

MOHANTY, K.K. & MILLER, A.E. 1988. Factors influencing unsteady relative permeability of a mixed-wet reservoir rock. *In*: *Proceedings of the 63rd Annual Technical Conference and Exhibition of Society of Petroleum Engineers, Houston, Texas, 2–5 October*, Paper **SPE 18292**.

MOSS, R.M., PEPIN, G.P. & DAVIS, L.A. 1990. Direct measurement of the constituent porosities in a dual porosity matrix. *In*: *Society of Core Analysts Conference Proceedings*, Paper **9003**.

NAKASHIMA, Y. 2000. The use of X-ray CT to measure diffusion coefficients of heavy ions in water-saturated porous media. *Engineering Geology*, **56**, 11–17.

OHNESORGE, B., FLOHR, T. & KLINGENBECK-REGN, K. 1999. Efficient object scatter correction algorithm for third and fourth generation CT scanners. *European Radiology*, **9**, 563–569.

PETERS, E.J. & HARDHAM, W.D. 1990. Visualization of fluid displacements in heterogeneities in porous media using computed tomography imaging. *Journal of Petroleum Science and Engineering*, **4**, 155–168.

PETERS, E.J. & AFZAL, N. 1992. Characterization of Heterogeneities in Permeable Media with Computed Tomography Imaging, *Journal of Petroleum Science and Engineering*, **7**, 283–296.

PETERS, E.J. & GHARBI, R. 1993. Numerical modeling of laboratory corefloods. *Journal of Petroleum Science and Engineering*, **9**, 183–205.

PETERS, E.J., GHARBI, R. & AFZAL, N. 1996. A look at dispersion in porous media through computed tomography imaging. *Journal of Petroleum Science and Engineering*, **15**, 23–31.

QADEER, S., AZIZ, K., FAYERS, J., CASTANIER, L.M. & BRIGHAM, W.E. 1994. An error analysis of relative permeabilities calculated using conventional techniques. *In*: *Proceedings of the International Energy Agency Enhanced Oil Recovery Workshop, Bergen, Norway 28–31 August*.

RANGEL-GERMAN, E.R., AKIN, S. & CASTANIER, L.M. 1999. Multiphase-flow properties of fractured porous media. *Proceedings of the 1999 Society of Petroleum Engineers Western Regional Meeting, Anchorage, Alaska, 26–28 May*, Paper **SPE 54591**.

SAHNI, A., BURGER, J.E. & BLUNT, M.J. 1998. Measurement of three phase relative permeability during gravity drainage using CT scanning. *In*: *Proceedings of the Society of Petroleum Engineers and Department of Energy Improved Oil Recovery Conference, Tulsa, Oklahoma 19–22 April*, Paper **SPE 39655**.

SATIK, C. 1998. A study of steam-water relative permeability. *Proceedings of the 68th Annual Society of Petroleum Engineers Western Regional Meeting, Bakersfield, California, 11–15 May*, Paper **SPE 46209**.

SCHEMBRE, J.M., AKIN, S., CASTANIER, L.M. & KOVSCEK, A.R. 1998. Spontaneous water imibition into diatomite. *Proceedings of the 68th Annual Society of Petroleum Engineers Western Regional Meeting, Bakersfield, California, 11–15 May*, Paper **SPE 46211**.

SEDGWICK, G.E. & DIXON, E.M. 1988. Application of X-ray imaging techniques to oil sands experiments. *Journal of Canadian Petroleum Technology*, **27**, 104–110.

SHARMA, B.C., BRIGHAM, W.E. & CASTANIER, L.M. 1997. *CT Imaging Techniques for Two-Phase and Three-Phase In-Situ Saturation Measurements*. US Department of Energy, National Petroleum Technology, Tulsa, Oklahoma, Report **SUPRI TR 107**.

SIDDIQUI, S., HICKS, P.J. & GRADER, A.S. 1996. Verification of Buckley-Leverett three-phase theory using computerized tomography. *Journal of Petroleum Science and Engineering*, **15**, 1–21.

SPANNE, P., THOVERT, J.F., JACQUIN, C.J., LINDQUIST, W.B., JONES, K.W. & ADLER, P.M. 1994. Synchrotron computed microtomography of porous media: topology and transport. *Physical Review Letters*, **73**, 2001–2004.

SPRUNT, E.S., DESAL, K.P., COLES, M.E., DAVIS, R.M. & MUEGGE, E.L. 1991. CT-scan-monitored electrical-resistivity measurements show problems achieving homogeneous saturation. *Society of Petroleum Engineers Formation Evaluation*, **6**, 134–140.

TREMBLAY, B., SEDGWICK, G. & VU, D. 1999. CT imaging of wormhole growth under solution gas drive. *Society of Petroleum Engineers Reservoir Engineering and Evaluation*, **2**, 37–45.

VAN GEET, M. & SWENNEN, R. 2001. Quantitative 3D-fracture analysis by means of microfocus X-ray computer tomography (μCT): an example from coal. *Geophysical Research Letters*, **28**, 3333–3336.

VAN GEET, M., SWENNEN, R. & WEVERS, M. 2001. Towards 3-D petrography: application of microfocus computer tomography in geological science. *Computers & Geosciences*, **27**, 1091–1099.

VAN GOLF-RECHT, T.D. 1982. *Fundamentals of Fractured Reservoir Engineering*. Elsevier, Amsterdam, The Netherlands.

VINEGAR, H.J. & WELLINGTON, S.L. 1987. Tomographic imaging of three-phase flow experiments. *Review of Scientific Instruments*, **58**, 96–107.

VINEGAR, H.J. & KEHL JR, R.P. 1988. *User Guide for Computer Tomography Color Graphic System-CATPIX*. Shell Development Co., Houston, Texas.

WALSH, M.P. & WITHJACK, E.M. 1994. On some remarkable observations of laboratory dispersion using computed tomography [CT]. *Journal of Canadian Petroleum Technology*, **33**, 36–44.

WANG, S.Y., HUANG, Y.B., PEREIRA, V. & GRYTE, C.C. 1985. Application of computed tomography to oil recovery from porous media. *Applied Optics*, **24**, 4021–4027.

WATSON, A.T. & MUDRA, J. 1994. Characterization of Devonian shales with X-ray computed tomography. *Society of Petroleum Engineers Formation Evaluation*, **9**, 209–212.

WELLINGTON, S.L. & VINEGAR, H.J. 1987. X-ray computerized tomography. *Journal of Petroleum Technology*, **39**, 885–898.

WITHJACK, E.M. 1988. Computed tomography for rock property determination and fluid flow visualization. *Society of Petroleum Engineers Formation Evaluation*, **3**, 696–704.

WITHJACK, E.M. & AKERVOLL, I. 1988. Computed tomography studies of 3-D miscible displacement behavior in a laboratory five-spot model. *In*: *Proceedings of the Society of Petroleum Engineers Annual Technical Conference and Exhibition, Houston, Texas, 2–5 October*, Paper **SPE 18095**.

WITHJACK, E.M., GRAHAM, S.K. & YANG, C.T. 1991. CT determination of heterogeneities and miscible displacement characteristics. *Society of Petroleum Engineers Formation Evaluation*, **6**, 447–452.

Study of the microgeometry of porous materials using synchrotron computed microtomography

K. W. JONES[1], H. FENG[2], W. B. LINDQUIST[3], P. M. ADLER[4], J. F. THOVERT[5], B. VEKEMANS[6], L. VINCZE[6], I. SZALOKI[6], R. VAN GRIEKEN[6], F. ADAMS[6] & C. RIEKEL[7]

[1] *Laboratory for Earth and Environmental Sciences, Brookhaven National Laboratory, Upton, New York 11973-5000, USA (e-mail: kwj@bnl.gov)*
[2] *Department of Earth and Environmental Studies, Montclair State University, Upper Montclair, New Jersey 07043, USA*
[3] *Department of Applied Mathematics and Statistics, State University of New York, Stony Brook, New York 11794-3600, USA*
[4] *Institut de Physique du Globe de Paris, F-75252 Paris Cedex 05, France*
[5] *LCD-PTM, F-86960 Futuroscope Cedex, France*
[6] *Department of Chemistry, University of Antwerp, B-2610 Wilrijk, Belgium*
[7] *European Synchrotron Radiation Facility, BP 220, F-38043 Grenoble Cedex, France*

Abstract: A series of measurements of the structure of a variety of porous materials has been made using synchrotron computed microtomography (SCMT). The work was carried out at the Brookhaven National Synchrotron Light Source (NSLS), the Argonne Advanced Photon Source (APS) and the European Synchrotron Radiation Facility (ESRF). The experiments at Brookhaven and Argonne were carried out on bending magnet beam lines using area detectors to obtain CT images based on determination of X-ray absorption coefficients. The work at the ESRF used an undulator beam line, a 13 KeV pencil X-ray beam of 2 μm and an energy dispersive X-ray detector to make tomographic sections of trace element distributions by X-ray fluorescence tomography. Most of the work was done with a pixel/voxel size ranging from 0.002 to 0.010 mm. We examined the structure of unconsolidated estuarine sediments, whose structure is relevant to transport of contaminants in rivers and estuaries. Fluorescent tomography with 2–3 μm resolution was used to ascertain whether or not metals were concentrated on the surface or throughout the volume of a single sediment particle. Sandstone samples were investigated to obtain a set of values describing their microstructures that could be useful in fluid flow calculations relevant to petroleum recovery or transport of environmental contaminants. Measurements were also made on sandstone samples that had been subjected to high-pressure compression to investigate the relation between the microgeometry and the magnitude of the applied pressure. Finally, a Wood's metal-filled sample was scanned for demonstration of resolution enhancement and fluid flow studies.

There are many reasons to study materials of interest in the earth and environmental sciences on a micrometre scale. For instance, measurement of fluid-solid interactions and fluid flow in porous media using experimental data with sufficient spatial resolution can help to give improved understanding of contaminant transport in the vadose zone and help in the assessment of the suitability of possible sites for the long-term storage of nuclear waste. The same type of data is useful in devising improved methods for oil recovery. Data on the structure of rocks as a function of applied pressure are essential for understanding the formation and growth of strain in the rock, with regards to both long-term geological processes and practical applications in petroleum recovery.

Measurement of three-dimensional structures can be approached in a number of ways. These include the destructive approach of cutting serial sections followed by examination with optical or electron microscopes and the non-destructive techniques of confocal microscopy, computed tomography (CT) and magnetic resonance imaging (MRI). The CT approach has been extended over the past decade through the use of

From: MEES, F., SWENNEN, R., VAN GEET, M. & JACOBS, P. (eds) 2003. *Applications of X-ray Computed Tomography in the Geosciences*. Geological Society, London, Special Publications, **215**, 39–49. 0305-8719/03/$15.

high-intensity synchrotron X-ray sources and X-ray focusing optics with X-ray tubes. As a result, it is now possible to carry out synchrotron computed microtomography (SCMT) measurements with spatial resolutions as low as about 0.005 mm with approximately 10^9 voxels.

Our group has used SCMT in the past to investigate structures of several sandstones, in order to provide a foundation for calculations of flow on a macroscopic scale and for applications related to petroleum recovery (Spanne *et al.* 1994; Coles *et al.* 1998). Other investigations include a study of voids and inclusions in micrometeorites (Feng *et al.* 1999) and of voids in volcanic basalts (Song *et al.* 2001). SCMT has also been applied in other fields of scientific research (Kinney *et al.* 1993; Lee *et al.* 1998; Morgan *et al.* 1998).

In this paper, we describe several experiments that extend the previous work. Experiments were carried out on estuarine sediments, since they can be considered a precursor to sandstone formation and are also of importance for understanding the fate and transport of anthropogenic contaminants in fresh water and estuarine environments. Sandstone samples were analysed in their natural state and following the application of stress, to extend the data base for natural sandstones and to delineate changes in structure in the stressed samples that could affect their porosity and permeability. Finally, measurements were made of Wood's metal-filled sandstone to investigate its use for enhancing spatial resolution and for studying fluid flow.

Synchrotron X-ray sources

Three different synchrotron X-ray sources were used for the tomography work described here. They were the National Synchrotron Light Source (NSLS) at Brookhaven, the Advanced Photon Source (APS) at Argonne and the European Synchrotron Radiation Facility (ESRF) at Grenoble, France. The three X-ray beams had very different parameters. The NSLS operates with stored electron beams at an energy of 2.7 GeV, compared to the positron energies of 7 and 6 GeV used at the APS and ESRF, respectively. Furthermore, both the APS and ESRF facilities have undulator beam lines that produce much higher X-ray fluxes than the bending magnet beam lines.

At the NSLS, a filtered white beam from a bending magnet was used to obtain adequate intensities of X-rays in the region of 50 KeV. Typical filters were 1.59 cm of aluminium and 0.125 mm of zirconium. The bending magnet X-ray spectrum at the APS extends to higher energies and makes possible the use of monoenergetic beams at even higher energies. At the ESRF, we used an undulator beam line to obtain very high flux beams at 13 KeV for fluorescent X-ray, rather than absorption X-ray tomography.

The energy spectra for the X-ray beams produced at the three facilities are shown in Figure 1. It can be seen that the brilliance of the third-generation bending magnet beam lines and undulator beam lines is higher than the brilliance produced at the second-generation NSLS facility. However, the NSLS beams are sufficiently intense to make exposure times for absorption tomography small compared to exposure times involved in data acquisition when monoenergetic low-energy beams or high-energy filtered white beams are used. Undulator beam lines are

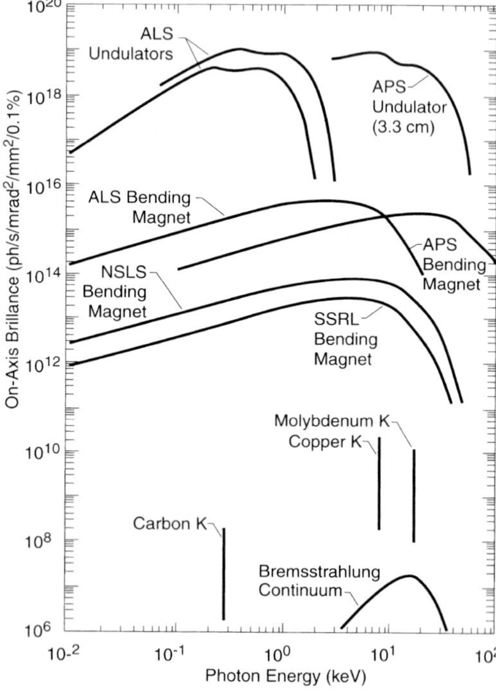

Fig. 1. X-ray brilliance produced at the Brookhaven National Synchrotron Light Source (NSLS), Argonne Advanced Photon Source (APS), Berkeley Advanced Light Source (ALS) and the Stanford Synchrotron Radiation Laboratory (SSRL). The brilliance for the European Synchrotron Radiation Facility X-ray beams are similar to those produced at the APS. Comparison is also made to conventional X-ray sources. The advantages of synchrotron sources for use in SCMT experiments are clear from examination of the figure.

necessary to provide the high intensities needed for fluorescent tomography. Note that the data for the APS are included as representative of the characteristics of third-generation synchrotrons including the ESRF. The characteristics of other conventional X-ray sources are included to show why the use of a high-intensity synchrotron source is advantageous for rapid data accumulation with high spatial resolution where high X-ray intensity is a necessity.

SCMT apparatus

Hardware

The NSLS SCMT experimental hardware is typical of apparatus used at several synchrotron facilities. The X-ray beam is produced at a bending magnet and can be used as either a filtered white beam or made monoenergetic with a multilayer spectrometer. The experiments reported here all used the filtered white beam mode, since it gives the most photons at the higher energies used in these experiments. A beam with a horizontal dimension of about 5 mm and a vertical dimension of about 1 mm passes through the sample and impinges on a thin yttrium-aluminium-garnet (YAG) scintillation X-ray detector. Light from the scintillator is imaged on a CCD camera after a 90° reflection from a mirror placed at 45° to the beam and passage through a focusing lens. Most of the data were recorded using a camera with a Kodak CCD with a 0.0067 mm pixel size and an area of 1317 × 1035 pixels. Data are acquired at a number of discrete angles as the sample is rotated from 0 to 180°. The number of views is generally about 1800 for each volume. The camera provides a 12-bit digitization of the scintillator light intensity. A schematic diagram of the apparatus is given in Figure 2. A detailed description of the equipment is given by Dowd *et al.* (1999*a*, *b*).

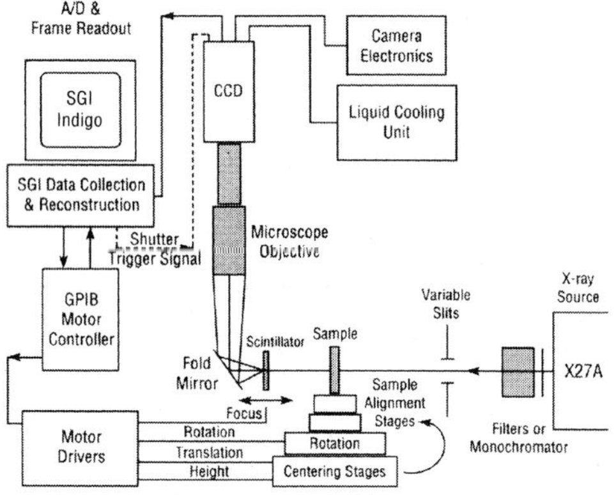

Fig. 2. Schematic diagram of the major components in the SCMT apparatus used at the BNL X27A bending magnet beam line.

The apparatus at the APS is conceptually the same as that used at the NSLS. A crystal monochromator is used to provide monoenergetic high-energy X-ray beams up to about 100 KeV. This is made possible by the high brilliance of the beams produced at the APS bending magnets. The SCMT apparatus was located at the GSECARS BM13 beam line.

Making SCMT measurements based on detection of characteristic X-rays demands the high X-ray intensities provided by an undulator source. The fluorescent SCMT experiment described below was carried out at the ESRF ID13 undulator beam line. The beam line used a Si(111) liquid-nitrogen-cooled monochromator to produce a 13 KeV beam. The beam was then focused with a rhodium-coated ellipsoidal mirror and a tapered glass capillary to a 0.002 mm size. A silicon drift chamber was used for X-ray detection because it has the advantage of excellent energy resolution and high counting rate capacity. The use of the pencil beam required that data be taken in a translation-rotation mode. In this case the beam was translated across the sample in steps of 0.002 mm. The sample was rotated from 0 to 180° with approximately 50 frames taken at each angle and, typically, a 3° angular step. The minimum detection limits for the elements of interest ranged from 0.04 to 1 fg.

Reconstruction software

At Brookhaven, volumetric data were obtained from the recorded views by using a software program based on a Fast Filtered Back-Transform algorithm that implements a gridding technique for fitting to Cartesian co-ordinates (R. B. Marr, pers. comm.). The data reconstruction proceeds in three phases. In Phase one, a white field normalization and any filters needed are used and files containing the data for all views for a single slice are created. In Phase two, each slice is processed independently. It applies the view-by-view air value normalization, optionally applies a filter to reduce the ring artefacts, computes the location in the images of the centre of rotation and converts the data to a sinogram. Phase three is the actual reconstruction. It generates a square array with dimensions of the horizontal row size. The visualization process following this reconstruction is a much more varied process and depends strongly on the particular sample being analysed.

Software for visualization and data analysis

Visualization of the reconstructed data can be satisfactorily accomplished using open or proprietary software. The large data sets present a challenge if a rapid visualization of the data is needed in order to decide on the course of the experiment, or for rapid examination of different portions of the volume. A promising approach is the use of parallel computing techniques (Feng *et al.* 2001).

Analysis of the data reported here has benefited from the application of specialized software for extracting specific parameters. Some of the software is described by Lindquist *et al.* (1996, 2000) and Thovert *et al.* (1990, 1993). Extraction of values for porosity, permeability, connectivity, tortuosity, specific surface area and other parameters is possible. Detailed calculations of fluid flow in sandstones have been reported by Zhang *et al.* (2000), using a lattice Boltzmann model.

If this method can be applied across the entire spectrum of steps in the creation of the final data, rendered volumes with theoretical results for flow or other parameters could be provided on a quasi-real time basis, which would greatly increase the power of CT techniques.

Experimental results

Investigation of microstructure and metals in sediments from New York/New Jersey Harbour and the North Sea

Sediments found in rivers, lakes, estuaries and marine environments are associated with both organic and inorganic compounds, which may be of both natural and anthropogenic origin. The effect of the anthropogenic compounds on the environment can be substantial, thus necessitating steps for isolation of the contaminants through containment or *in-situ* or *ex-situ* treatment for their immobilization or destruction.

The sediments found in the New York/New Jersey Harbour serve as an excellent example of the problem. The sediments are rich in organic materials with a major fraction of anthropogenic origin including polychlorinated biphenyls (PCBs), polynuclear aromatic hydrocarbons (PAHs), dioxins, furans, insecticides and pesticides. Metals include Ag, Cd, Cr, Pb, Cu, Zn and Hg. These compounds exist in a complex stew of materials, with the sediment solids including clays and silts, sand and gravel in varying concentrations. Much of the Harbour sediment is very fine-grained, with perhaps 50% by weight having a size less than 0.015 mm. The behaviour of the contaminant compounds is affected by the chemistry of the sediment environment and is also dependent on interactions with bacteria and other benthic organisms (Rittmann & McCarty 2001).

Determination of the microstructure of the sediments and the ways in which the contaminants interact with each other and with the sediments is needed to provide the microscopic basis for developing detailed macroscopic models for their fate and transport. These models are essential for developing programs that minimize the environmental impact of the contaminants and for developing treatment methods.

An exploratory investigation of the microstructure was carried out at BNL. Sandy sediments from the New York/New Jersey Harbour were allowed to settle in a polyethylene tube and then measured using SCMT with a voxel resolution of 0.0068 mm. A portion of the data is shown in Figure 3. The results show that it is possible to successfully apply CT to wet sediment samples and thus measure sediment structures formed by settling of the sediment particles from the water column under gravitational forces without any manipulation of the material through freezing or addition of polymers. This type of data can be used to verify theoretical predictions of the structure (Quintanilla & Torquato 1997; Coelho *et al.* 1997).

The location of the metals in the sediment is also of prime importance. How do the metals fractionate between binding to organic materials

Fig. 3. A SCMT volume of sandy sediments from the New York/New Jersey Harbour. The sediments were sieved and grain sizes greater than 0.063 mm were investigated. The pore space is black whereas sediment grains are shown in greys and whites. The white voxels represent the highest X-ray attenuation coefficients. The voxel size used for the measurements was 0.0068 mm. The volume displayed is 1.36 mm × 1.36 mm × 1.02 mm. The volume dimensions in voxels are shown on the figure axes.

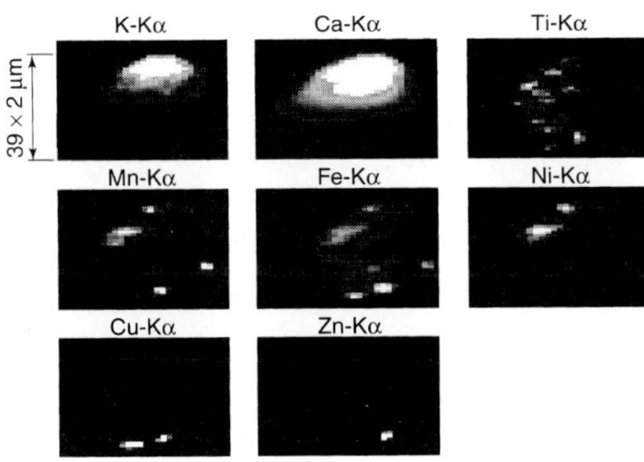

Fig. 4. Distribution of elements found in a single grain of sediment from the New York/New Jersey Harbour. These results were obtained at the ESRF using fluorescent tomography. The voxel size is 0.002 mm. The scale is shown at the upper left of the figure.

and to the sediment particles? Are they preferentially found on smaller particles or are they associated with organic compounds? These questions were addressed using fluorescent tomography at the ESRF. Single sediment particles were mounted on the end of a glass capillary using a micromanipulator. The size of the particles was on the order of 0.005 mm. Measurements were made with a spatial resolution of 0.002 mm. Particles from Newark Bay in the New York/New Jersey Harbour and from the North Sea (expected to be relatively uncontaminated) were investigated. Results for a single section through a particle from the New York/New Jersey Harbour are shown in Figure 4. We were not able to detect any evidence indicating that the metals were preferentially located in a narrow biofilm coating the particle. There was, however, evidence showing high concentrations both in small regions around the periphery of the particles and within the particles themselves. The first observation could be evidence for smaller sediment particles or organic aggregates containing metals adhering to the single grain. Similar results were found for measurements made on sediment particles from the North Sea. On the other hand, the distributions found for a sediment grain subjected to a proprietary soil washing process (BioGenesis Enterprises) revealed metals throughout the particle, but did not exhibit points of high concentrations at the surface. This could indicate that the metals are associated with smaller surface particles that can be removed by the mechanical washing process. More refined experimental measurements are necessary to clarify these points. It may be that fluorescent SCMT lacks the spatial resolution and detection sensitivity to fully resolve the questions posed.

Sandstone studies

The microgeometry of sandstones is of particular interest since sandstone structures are of importance in understanding the behaviour of petroleum reservoirs and for providing a rational basis for development of improved petroleum recovery methods. The thrust of the experiments includes measuring structures of unperturbed sandstones and extracting information that can be used for flow calculations. Measurements of a variety of sandstones are needed in this regard in order to establish the range of variations in the materials that may occur in different oil fields. In addition, the mechanical properties of the sandstones are of interest in order to facilitate understanding the changes in parameters such as pore size, porosity and permeability. A number of experiments that focused on these topics have been performed at the NSLS. They are summarized below.

Investigation of red sandstone from the Vosges Mountains, France

In this paragraph, we briefly illustrate how a SCMT image can be used to measure and compute many geometrical and transport properties of a porous material. A more detailed description of the conceptual and numerical tools and a more extensive set of results for another sample are provided by Thovert *et al.* (2001). The sandstone sample considered in the present work was among those samples investigated by Lucet (1989).

The SCMT volume measured at the NSLS was reduced to a binary form containing only pore or rock voxels using the segmentation procedure described by Lindquist *et al.* (1996). Representative slices in this binary format are shown in Figure 5. Porosities determined using the segmentation procedure are in good agreement with those found by other methods.

The digitized data can be represented by the binary phase function Z

$$Z(\mathbf{x}) = \begin{cases} 1 & \text{if } \mathbf{x} \text{ belongs to the pore space} \\ 0 & \text{otherwise.} \end{cases} \quad (1)$$

The porosity, ε (i.e. the volume fraction of the pores) and the correlation, $R_z(\mathbf{u})$, can be defined by the statistical averages (denoted by brackets $\langle\ \rangle$).

$$\varepsilon = \langle Z(\mathbf{x}) \rangle \quad (2)$$

$$R_Z(\mathbf{u}) = \frac{\langle (Z(\mathbf{x}) - \langle (Z \rangle) \cdot (Z(\mathbf{x}+\mathbf{u}) - \langle Z \rangle) \rangle}{\varepsilon(1-\varepsilon)}. \quad (3)$$

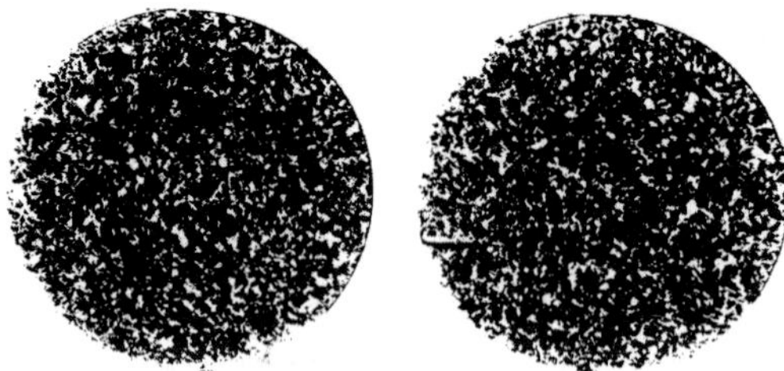

Fig. 5. Typical sections through a sample of red Vosges sandstone measured at the BNL X27A beam line. The data have been segmented into pore space (white) and solids (black). The sample diameter is 5.7 mm. The sections displayed have a pixel size of 0.006 mm and a diameter of 4.2 mm.

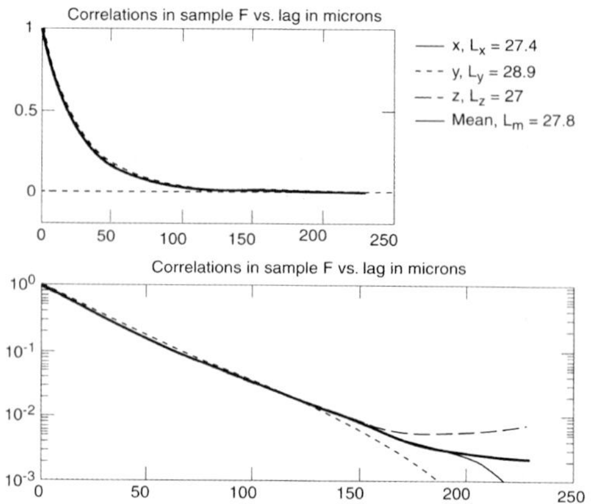

Fig. 6. Two-dimensional correlation functions for red Vosges sandstone. They were calculated from the type of data shown in Figure 5.

The correlation function is directly related to the probability that two points $\mathbf{x}$ and $\mathbf{x}+\mathbf{u}$, separated by a vector (or lag) $\mathbf{u}$, are in the same phase (solid or pore). It is equal to one for $\mathbf{u}=0$ and generally vanishes for large distances. Its initial derivative is also related to the volumetric pore surface area. For isotropic materials, R_Z is a function of the modulus of the lag $u=\{\mathbf{u}\}$ only, $R_Z(\mathbf{u})=R_Z(u)$. Otherwise, the correlations for $\mathbf{u}$ parallel to the x-, y- and z-axes are denoted R_{Zx}, R_{Zy}, and R_{Zz}, respectively. Note that $\varepsilon(1-\varepsilon)$ in equation (3) equals var(Z) since $Z^2(\mathbf{x})=Z(\mathbf{x})$. The correlation length L is defined as the integral of the correlation function:

$$L=\int_0^\infty R_Z(u)\,\mathrm{d}u. \tag{4}$$

For anisotropic materials, L_x, L_y and L_z are defined accordingly from R_{Zx}, R_{Zy} and R_{Zz}. The correlation length is a characteristic of the material texture. For granular porous media, such as the present sandstones, it is of the order of the grain size. Obviously, any mean property of the material, such as the conductivity considered below, should be measured on samples much larger than L.

The correlations $\overline{R_{Zx}}$, $\overline{R_{Zy}}$, $\overline{R_{Zz}}$, measured for the sample of red Vosges sandstone are shown in Figure 6. The correlation functions appear to be isotropic. Differences between the three axes can be distinguished only in the semi-logarithmic plots for relatively large lags. They are probably merely random statistical fluctuations since the correlation, which is the ratio of a covariance to a variance i.e. a signal-to-noise ratio, is smaller than 10^{-2} in this range. The correlation lengths and their average are given in Table 1. Fits to the semi-logarithmic plots show that the correlation functions are well described by a negative exponential function:

$$R_Z=(u)=\mathrm{e}^{-(u/\lambda)}. \tag{5}$$

The decay length λ, evaluated from a least square fit over a 100 µm lag range, is also listed in Table 1. It is in good agreement with the correlation length L_m.

An important transport property to be determined is whether the pore space in the material percolates or not, i.e. whether a continuous path through the pore space exists between two opposite faces of the sample. In the absence of percolation, all the macroscopic coefficients for transport processes in the pore space are trivially zero.

A calculation of the electrical conductivity of the sample relative to a sample totally filled with a conducting liquid was made with the assumption that the solid phase is an insulator. The analytical approach followed that described by Thovert *et al.* (1990, 1993). The present calculations were performed for blocks of voxels cut from the overall volume, ranging in size from 32^3 to 170^3. Results of the calculations for porosity and conductivity are shown in Figure 7.

The results of the calculations showed that for this sample the probability of percolation along one axis depends on the block size. Approximately 30% of the 32^3 voxel size blocks did not percolate. The results for this sandstone can be compared to the results obtained in a study of

Table 1. *Average porosities* ε *and correlation lengths* L_x, L_y *and* L_z *for the red Vosges sandstone sample*

$\bar{\varepsilon}$	L_x (µm)	L_y (µm)	L_z (µm)	L_m (µm)	λ (µm)
0.1873	27.4	28.9	27.0	27.8	29.6

L_m is the average of L_x, L_y and L_z; λ is the decay length in equation (5).

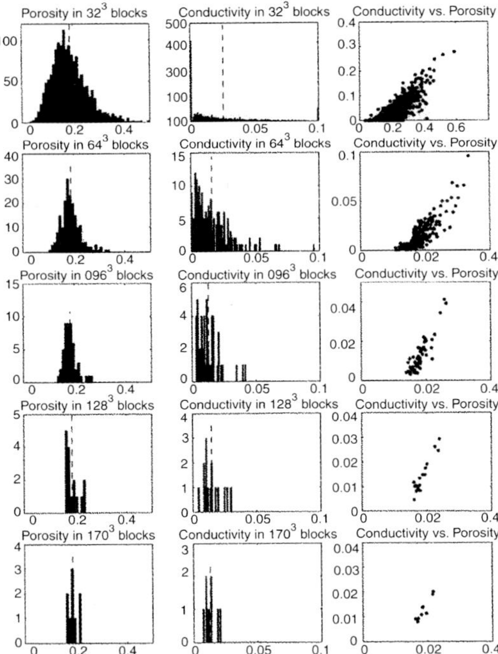

Fig. 7. Results of calculations of porosity, conductivity and porosity versus conductivity for the red Vosges sandstone sample for different size data blocks extracted from the complete volume.

Fontainebleau sandstone (Spanne *et al.* 1994). They found values for the porosity of 17.9% and a correlation length of 30 μm. However, (dimensionless) conductivity was found to be 0.037, which is approximately three times the value determined for this red Vosges sandstone. The physical reasons for the difference are not clear at this time.

Effect of compaction on sandstone grain structure

Two investigations of laboratory induced compaction of sandstone have been carried out to determine stress-induced changes in the sandstone microstructure. Knowledge of these changes is useful in considering transport in petroleum reservoirs and for developing methods for increasing flow from the reservoir to the well casing.

Laboratory study of compaction of Darley Dale sandstone. A laboratory study of stress-induced changes was carried out in collaboration with P. Baud and T. F. Wong at the State University of New York at Stony Brook.

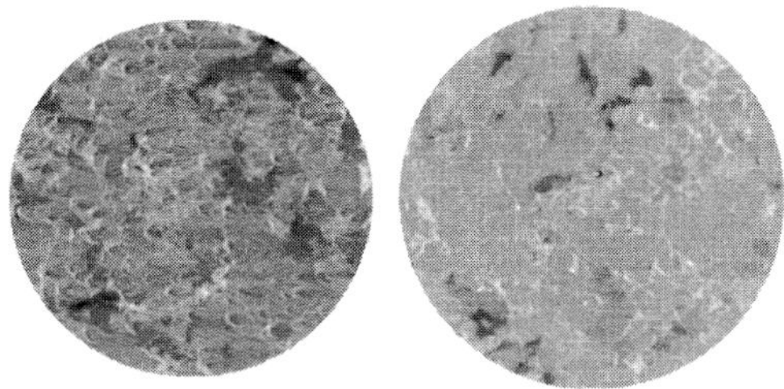

Fig. 8. SCMT sections taken through samples of Darley Dale sandstone in its natural state (left) and following compaction at 4.5% axial strain (right). Changes in grain structure are clearly revealed. The sections displayed have a diameter of 4.2 mm and a pixel size of 0.006 mm.

The samples were cored from a block of Darley Dale sandstone, a well-indurated feldspathic sandstone with a siliceous cement from the north of England. This sandstone has an average grain size (mean intercept length) of 0.22 mm and a model composition of 67% quartz, 14% feldspar, 2% mica and 6% clay. The grains are sub-angular and poorly sorted, with sizes ranging from 0.08 to 0.8 mm. The mean porosity of the samples was 13.4%.

A series of experiments were performed in the cataclastic flow regime. The samples were stressed into different strain stages, then unloaded and retrieved from the pressure vessel for microscope observations and quantitative microstructural analysis or X-ray μCT. At an effective pressure of 200 MPa, shear localization was inhibited and the Darley Dale sandstone failed by cataclastic flow.

Tomographic volumes were obtained for sandstone samples in the natural and deformed states. Results obtained for a sample deformed with 4.5% axial strain are shown in Figure 8. Porosity calculated from this measurement was in agreement with values found by conventional means. In the representation shown in the figure, the x, y plane is perpendicular to the major stress (vertical). Examination of the data showed no significant anisotropy, in agreement with two-dimensional analysis based on optical microscopy.

The three-dimensional tomographic volumes reveal the geometric complexity of the damage associated with stress-induced grain crushing and pore collapse. While the connection of pore space is enhanced by damage accumulation, the pore sizes are reduced; this results in an overall reduction in permeability. Thus, the data from microtomography can be used to simulate the coupling of mechanical compaction and transport properties in reservoir rocks.

Dynamic compaction of sandstone. A second compaction study was aimed at obtaining a better understanding of heterogeneous grain-to-grain interactions under the dynamic loading that occurs when a shaped-charge is used to perforate an oil well casing, to provide connectivity to the reservoir rock (Hiltl *et al.* 1999). A single-stage light-gas gun at Lawrence Livermore National Laboratory was used for production of a short high-pressure pulse at the front surface of a sandstone sample. The specifications for the gun are as follows: 35 mm bore diameter, 5–15 mm sandstone thickness, 22.4 mm sandstone diameter, 1.3 to 9.8 GPa stress levels at front of sample, interior pressures 35–40% lower than at the front of the capsule and 1 or 2 μs shock pulse durations. Dry or wet samples can be used.

SCMT measurements were made on 4 mm large samples of Berea sandstone prepared from unshocked and dry shocked materials. The shock pulse was 1 μs long with a maximum pressure of 61 Kbar, corresponding to 6.1 GPa. The porosity found for the undamaged material was 21.0% compared to 13.3% found for the shocked material. The value for the undamaged region is in agreement with the accepted value of 21.9%. Tomographic sections through the two specimens are presented in Figure 9. The effects produced by the compression pulse are clearly displayed.

The data obtained can be used in simulations of flow through the material and hence lead to prediction of the usefulness of shaped charges for improving connectivity to the reservoir rock.

Wood's metal-filled sandstone

There are several reasons for using the Wood's metal intrusion technique in preparing samples for investigations of porous media using SCMT: (i) its X-ray absorption coefficient is much higher than the coefficient for sandstone or other rocks, allowing identification of filled volumes that are smaller than the volume of a single instrumental voxel; (ii) the ability to freeze the flow of the Wood's metal in the media makes possible the detailed study of the flow of a liquid through a porous medium, which would be difficult or impossible with a fluid that is liquid at room temperature; (iii) the study of microcracks as a function of pressure is facilitated; and (iv) pore-scale SCMT results may be helpful in the understanding and interpretation of mercury intrusion porosimetry. An example of the use of Wood's metal porosimetry for study of cement pore structure is given by Willis *et al.* (1998). Porosimetry using Wood's metal was compared to mercury porosimetry by Dullien (1981).

We used synchrotron SCMT to investigate the pore structure of Fontainebleau sandstone filled with Wood's metal. The SCMT experiments were carried out at the Brookhaven NSLS and at the Argonne APS. A third-generation apparatus was used in both cases at voxel sizes of approximately 0.020 mm. A filtered white beam with a mean energy of approximately 55 keV was used at the NSLS. The work at the APS was done with a monoenergetic beam at several energies between approximately 68 and 80 keV. The same sample of Fontainebleau sandstone with a nominal porosity of 5% (prepared by Michel Darot, Strasbourg, France) was studied in both experiments.

The three-dimensional distribution of Wood's metal in the sandstone that was measured at the NSLS is shown in Figure 10. The pathways of the Wood's metal between filled pores was not revealed in this display. The percolation front between the Wood's metal filled space and native sandstone measured at the APS is shown in Figure 11.

The observed distribution of X-ray attenuation coefficients is dependent on the grain size distribution and the spatial resolution of the apparatus. This was seen in the histograms found for measurements at the NSLS (20 μm) and APS (100 μm). A peak corresponding to filled voxels is seen for the higher resolution data. The measurements did not show connectivity between the pores filled with Wood's metal, which is attributed to a spatial resolution that is large relative to the pore structures. They do show that Wood's metal impregnation can improve the volume resolution by a factor of three or more. The type of data presented here should be useful in theoretical modelling of the flow of Wood's metal in sandstone.

Conclusions

Our intent in this paper was to discuss recent experiments applying synchrotron CT techniques to the study of sediments (as a precursor to sandstone formation) and to examinations of sandstones in their natural and compacted states. Attention was given to experiments at three different synchrotron facilities, to demonstrate the availability of facilities for making these examinations and to document the hardware and software that were used in the experiments. These experiments, taken together with results from other groups, show that SCMT has achieved a solid position as a tool for examining microstructures found in samples that are relevant to

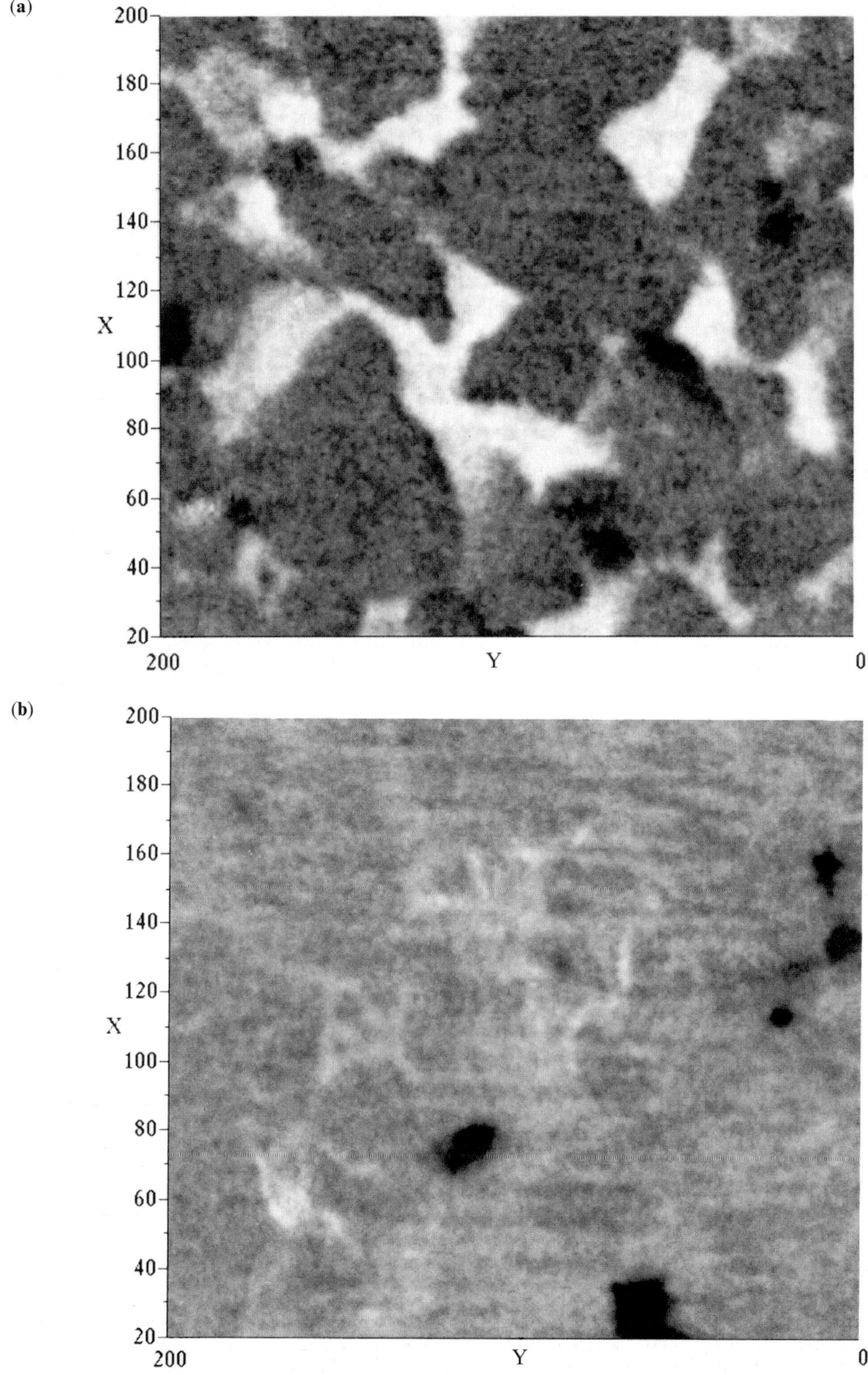

Fig. 9. Sections through Berea sandstone in its natural state (top) and in a compacted state following shock compaction (bottom). The pixel size is 0.0036 mm and the size of the area displayed is 0.72 mm × 0.72 mm.

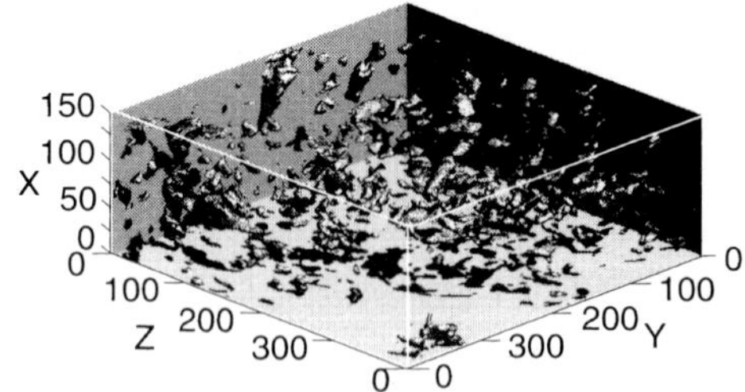

Fig. 10. CT image obtained at the NSLS. It shows the three-dimensional distribution of Wood's metal in the sample. The spatial resolution is not sufficient to reveal the pathways between the observed filled pores. The volume shown has dimensions of 1.8 mm × 1.8 mm × 0.9 mm with a pixel size of 0.006 mm.

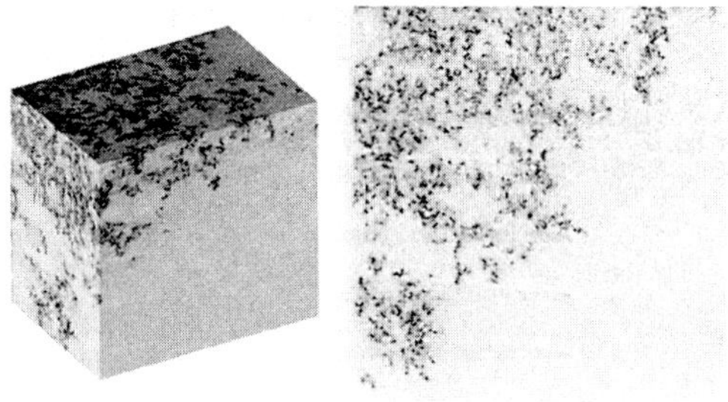

Fig. 11. CT image obtained at the APS showing the Wood's metal percolation front in a sample of Fontainebleau sandstone. The right-hand picture shows a vertical section through the block shown on the left. The sandstone grains are not shown in this display. The light grey voxels and pixels represent the sandstone. The dark grey and black voxels represent the Wood's metal. The voxel size for the measurements was 0.045 mm and the size of the volume displayed is 20.25 mm × 10.53 mm × 19.2 mm.

the earth and environmental sciences. It is also apparent that technical improvements, such as the use of parallel computing techniques, are being brought into practice at a rapid rate and will lead to more sophisticated experiments in the near future.

Work supported in part by the US Department of Energy under Contract No. DE-AC02–98CH10886 (KWJ, HF) and use of the Advanced Photon Source by the US Department of Energy, Office of Science, Office of Basic Energy Sciences, under Contract No. W-31-109-ENG-38. This work was performed at the GeoSoilEnviroCARS (GSECARS) Sector 13, beam line BM13. GSECARS is supported by the National Science Foundation (Earth Sciences), Department of Energy (Geosciences), W. M. Keck Foundation, and the US Department of Agriculture.

References

COELHO, D., THOVERT, J.F. & ADLER, P.M. 1997. Geometrical and transport properties of random packings of spheres and aspherical particles. *Physical Review E*, **55**, 1959–1978.

COLES, M.E., HAZLETT, R.D., MUEGGE, E.L., JONES, K.W., ANDREWS, B., DOWD, B. SIDDONS, P., PESKIN, A., SPANNE, P. & SOLL, W.E. 1998. Developments in synchrotron X-ray microtomography with applications to flow in porous media. *Society of Petroleum Engineering Reservoir Evaluation & Engineering*, 288–296.

DOWD, B.A., ANDREWS, A.B., MARR, R.B., SIDDONS, D.P., JONES, K.W. & PESKIN, A.M. 1999*a*. Advances in X-ray computed microtomography at the NSLS. *Advances in X-Ray Analysis*, **42** (CD-ROM). ICDD, Newtown Square, Pennsylvania.

DOWD, B.A., CAMPBELL, G.H., MARR, R.B., NAGARKAR, V., TIPNIS, S., AXE, L. & SIDDONS, D.P. 1999*b*. Developments in synchrotron X-ray computed microtomography at the National Synchrotron Light Source. *Developments in X-ray Tomography II, Proceedings of SPIE*, **3772**, 224.

DULLIEN, F.A.L. 1981. Wood's metal porosimetry and its relation to mercury porosimetry. *Powder Technology*, **29**, 109–116.

FENG, H., JONES, K., STEWART, B., HERZOG, G.F., SCHNABEL, C. & BROWNLEE, D.E. 1999. Internal Structure of two Type-I deep-sea spherules by X-ray computed microtomography. *In*: *Proceedings of the 30th Lunar and Planetary Science Conference, Johnson Space Center, Houston, Texas, March 15–19 (CD-ROM)*. Lunar and Planetary Institute, Houston, Texas.

FENG, H., JONES, K.W., MCGUIGAN, M., SMITH, G.J. & SPILETIC, J. 2001. High-performance computing for the study of earth and environmental science materials using synchrotron X-ray computed microtomography. *In*: ANTONIOU, G. & DEREMER, D. (eds) *Proceedings of the International Conference on Computing and Information Technologies (ICCIT 2001) – Exploring Emerging Technologies*. World Scientific, New Jersey, 471–481.

HILTL, M., HAGELBERG, C.R., SWIFT, R.P., CARNEY, T.C. & NELLIS, W.J. 1999. Dynamic response of Berea Sandstone shock-loaded under dry, wet and water-pressurized conditions. *In*: *Proceedings of the 17th AIRAPT International Conference on High Pressure Science and Technology (AIRAPT-17), July 25–29, University of Hawaii, Honolulu, Hawaii.*

KINNEY, J.H., BREUNIG, T.M., STARR, T.L., HAUPT, D., NICHOLS, M.C., STOCK, S.R., BUTTS, M.D. & SAROYAN, R.A. 1993. X-ray tomographic study of chemical vapor infiltration processing of ceramic composites. *Science*, **260**, 789–792.

LEE, S., STOCK, S.R., BUTTS, M.D., STARR, T.L., BREUNIG, T.M. & KINNEY, J.H. 1998. Pore geometry in woven fiber structures: 0°/90° plain-weave cloth lay-up preform. *Journal of Materials Research*, **13**, 1209–1217.

LINDQUIST, W.B., LEE, S.M., COKER, D.A., JONES, K.W. & SPANNE, P. 1996. Medial axis analysis of void structure in three-dimensional tomographic images of porous media. *Journal of Geophysical Research*, **101**, 8297–8310.

LINDQUIST, W.B., VENKATARANGAN, A., DUNSMUIR, J. & WONG, T.-F. 2000. Pore and throat size distributions measured from synchrotron X-ray tomographic images of Fontainebleau sandstones. *Journal of Geophysical Research*, **105**, 21 509–21 527.

LUCET, N. 1989. *Vitesse et Atténuation des Ondes Élastiques Soniques et Ultrasoniques dans les Roches sous Pression de Confinement*. PhD Thesis, University of Paris VI, Paris.

MORGAN, H., WILSON, R.M., ELLIOTT, J.C., DOWKER, S.E.P. & ANDERSON, P. 1998. Cells for the study of acidic dissolution in packed apatite powders as model systems for dental caries. *Caries Research*, **32**, 428–434.

QUINTANILLA, J. & TORQUATO, S. 1997. Microstructure functions for a model of statistically inhomogeneous media. *Physical Review E*, **55**, 1558–1565.

RITTMANN, B.E. & MCCARTY, P.L. 2001. *Environmental Biotechnology: Principles and Applications*. McGraw Hill, New York.

SONG, S.R., JONES, K.W., LINDQUIST, W.B., DOWD, B.A. & SAHAGIAN, D.L. 2001. Synchrotron X-ray computed microtomography: studies on vesiculated basaltic rocks. *Bulletin of Volcanology*, **63**, 252–263.

SPANNE, P., THOVERT, J.-F., JACQUIN, C.G., LINDQUIST, W.B., JONES, K.W. & ADLER, P.M. 1994. Synchrotron computed microtomography of porous media: topology and transports. *Physical Review Letters*, **73**, 2001–2004.

THOVERT, J.F., WARY, F. & ADLER, P.M. 1990. Thermal conductivity of random media and regular fractals. *Journal of Applied Physics*, **68**, 3872–3883.

THOVERT, J.F., SALLES, J. & ADLER, P.M. 1993. Computerized characterization of the geometry of real porous media: their discretization, analysis and interpretation. *Journal of Microscopy*, **170**, 65–79.

THOVERT, J.F., YOUSEFIAN, F., SPANNE, P., JACQUIN, C.G. & ADLER, P.M. 2001. Grain Reconstruction of porous media: application to a low-porosity Fontainebleau Sandstone. *Physical Review E*, **63**, 61 307–61 323.

WILLIS, K.L., ABELL, A.B. & LANGE, D.A. 1998. Image-based characterization of cement pore structure using Wood's metal intrusion. *Cement and Concrete Research*, **28**, 1695–1705.

ZHANG, D.X., ZHANG, R.Y., CHEN, S.Y. & SOLL, V.E. 2000. Pore scale study of flow in porous media: scale dependency, REV, and statistical REV. *Geophysical Research Letters*, **27**, 1195–1198.

Porosity measurements of sedimentary rocks by means of microfocus X-ray computed tomography (μCT)

M. VAN GEET[1,3], D. LAGROU[2] & R. SWENNEN[1]

[1] *Katholieke Universiteit Leuven, Fysico-chemische Geologie, Celestijnenlaan 200C, B-3001 Heverlee, Belgium (e-mail: mvgeet@sckcen.be)*

[2] *Vito, Energy Technology, Boeretang 200, B-2400 Mol, Belgium*

[3] *SCK-CEM, Waste and Disposal Department, Boeretang 200, B-2400 Mol, Belgium*

Abstract: Porosity of reservoir rocks is an important petrophysical characteristic, used as a basic parameter in simulation studies for predicting reservoir quality. An extensive debate continues about the techniques that are available for porosity measurements and visualization. One aspect is the fact that petrophysical measurements are performed on volumetric samples, whereas classical geological petrography using a petrographical microscope is restricted to 2D analysis. This leads to a discrepancy between petrographical and petrophysical studies. This paper aims to evaluate microfocus X-ray computed tomography (μCT) as a technique that can link petrography and petrophysics. A short overview of the μCT technique is given, together with a discussion of its limitations, mainly due to artefacts. Optimization of image quality and procedures for quantification are outlined. μCT results for porosity measurements of a limestone and a sandstone are compared with results obtained by other techniques.

The most common porosity measurement techniques, such as mercury porosimetry, are indirect (Lindquist *et al.* 2000), because a geometric/fluid model is used to translate measured pressure jumps into effective pore radius. Moreover, the distribution of effective pore radii that is otained is often interpreted as a pore size distribution, although in reality represents a distribution of the effective radii of pore throats. Until the 1990s, direct measurement of pore space characteristics was largely restricted to the stereological study of thin sections and analysis of serial stacks of thin sections. The disadvantages of this procedure are the long working hours required for polishing and slicing and the destructive nature of this approach. Direct measurements can now be performed by means of synchrotron X-ray computed tomography (Flannery *et al.* 1987; Spanne & Rivers 1987; Damico *et al.* 1989; Kinney & Nichols 1992; Lindquist *et al.* 2000). These facilities can produce high resolution (micron-scale) 3D images of small rock samples, but access to these instruments is limited and expensive. Laser scanning confocal microscopy has also been used for direct porosity characterization (Frederich 1999), but this technique is limited to small samples (about 1 mm^2) and the depth of analysis is only about 170 μm. Another tool is microfocus X-ray computed tomography (μCT), which is a non-destructive 3D visualization and quantification tool. Compared to medical CT instruments, a much higher resolution can be reached (presently up to 5 μm in three dimensions). Compared to synchrotron facilities, these instruments are obviously less expensive and access is much easier.

Instrumentation

μCT is based on recording X-ray projections of the studied object at different angles. A back-projection algorithm is used to reconstruct a virtual slice through the object, which represents a visualization of variations in linear attenuation coefficient. The attenuation coefficient depends on the X-ray energy used and on the atomic number and density of the studied object. Stacking several sequential slices enables a reconstruction of 3D distributions. Descriptions of μCT instruments and reconstruction algorithms can be found in the literature (Brooks & DiChiro 1976; Sasov 1987; Kak & Slaney 1988). The instrument used for this study is a Skyscan 1072 Microtomograph. Scans were made at 130 kV and/or 100 kV, and 300 μA. Scanning time was about 3 hours for one object. Voxel size was about 22 μm in three dimensions for samples of 8 mm in diameter.

From: MEES, F., SWENNEN, R., VAN GEET, M. & JACOBS, P. (eds) 2003. *Applications of X-ray Computed Tomography in the Geosciences*. Geological Society, London, Special Publications, **215**, 51–60. 0305-8719/03/$15.

For verification of the μCT results, image analysis was carried out on polished surfaces to measure porosity along a trace with reflected light microscopy. A Foster Findlay PC_Image for Windows system was used for this, equipped with a Sony DXC-930P video camera and a Zeiss Axioskop microscope. The scanned image is transmitted to the hardware, enabling digital storage of the images in a tiff format (576 × 768 pixels). Two different magnifications were used, resulting in pixel sizes of 3.1 × 3.1 μm and 0.82 × 0.82 μm respectively.

Artefacts

Although μCT is a promising technique for visualizing and quantifying internal features of rocks, μCT images are not free of artefacts. Minimization or exclusion of these artefacts is necessary before quantification can be performed (Joseph 1981; Van Geet *et al.* 2000). Beam hardening is the most cumbersome artefact with regard to quantification. Beam hardening originates from the fact that a polychromatic X-ray source is used in (μ)CT instruments. In Figure 1 the theoretical attenuation of X-rays for calcite ($CaCO_3$) and quartz (SiO_2) are given. From these plots it is clear that low X-ray energies are preferentially absorbed. Consequently, the X-ray energy spectrum of polychromatic X-rays changes while travelling through the object, hence the attenuation will change also. It should be noted that earlier CT studies of rocks, mainly using medical CT instruments, neglected or did not even mention this artefact.

A first possible method to minimize beam hardening artefacts is using hardware filters (Jennings 1988; Van Geet *et al.* 2000). Metal foils placed between the X-ray source and the object will absorb the low energy X-rays before they reach the object. The disadvantage of this procedure is that the X-ray beam is less effectively used and that a great increase in scanning time might be needed to obtain good signal-to-noise ratios. Moreover, the artefacts are only minimized. For highly attenuating materials, considerable filter thicknesses would be needed, which would greatly decrease the signal-to-noise ratio. The major advantages of introducing hardware filters is their ease of use and that they can be applied to heterogeneous materials.

A second beam hardening correction method is linearisation (Hammersberg & Mangard 1998;

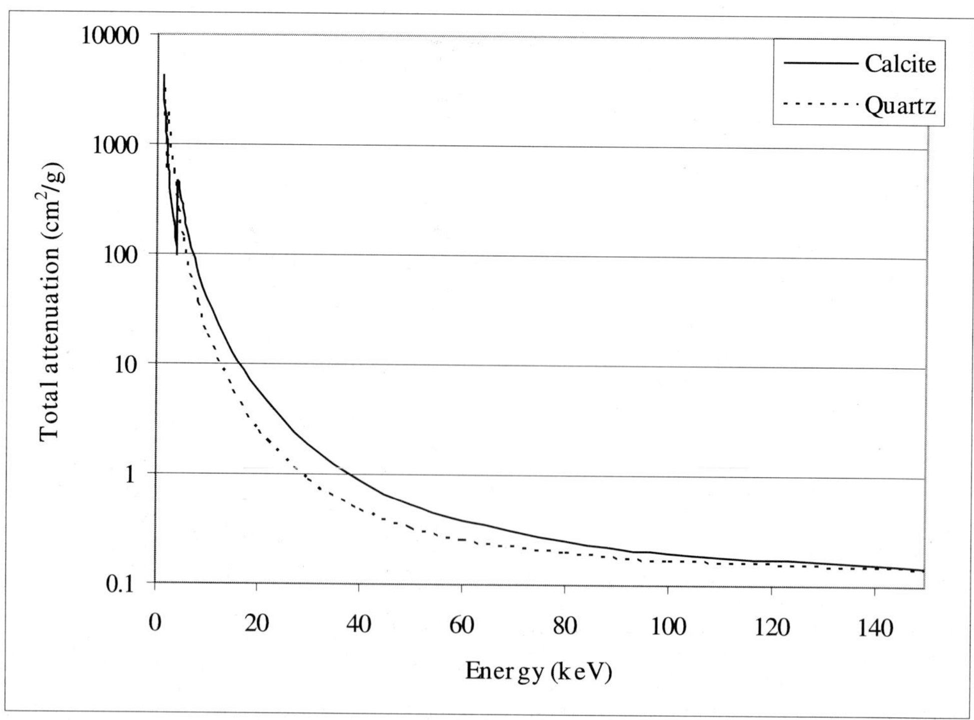

Fig. 1. Plots of the total linear attenuation coefficient of calcite and quartz at different X-ray energies.

Van Geet 2001). From Beer's law, the basis of CT, it follows that:

$$-\ln\left(\frac{I}{I_0}\right) = \mu h$$

where I is the measured intensity of the X-rays, I_0 is the emitted X-ray intensity, μ is the linear attenuation coefficient and h is the thickness of the object. The left part of this equation is measured with CT instruments. It is clear that for monochromatic X-rays the linear attenuation coefficient μ is energy independent. Consequently, for one material (constant μ) a linear relationship should be found between measured intensity and object thickness. However, due to polychromaticity, a deviation from linearity is found. These measured data points can be fitted with a low order polynomial. Once this calibration is performed a correction of every measured intensity value can be achieved, resulting in images that are corrected of beam hardening. The disadvantage of this approach is that the calibration is material-dependent and can therefore only be used for objects consisting of a single material.

Quantification

Once the quality of CT images is optimized and artefacts are excluded or minimized, quantification can be performed. The parameter that is visualized is the linear attenuation coefficient, which is a parameter that is difficult to interpret in terms of rock properties. For homogeneous monomineralic samples, porosity measurements can be performed by means of calibration with a pure non-porous sample. This is not the case for heterogeneous samples. A dual energy technique can be used to solve this problem (Alvarez & Macovski 1976; Coenen & Maas 1994; Van Geet *et al.* 2000). This procedure is based on the fact that the attenuation coefficient depends on the density and atomic number of the object. A generalized equation for the attenuation coefficient is given by Wellington & Vinegar (1987):

$$\mu = \rho\left(a + b\left(\frac{Z^{3.8}}{E^{3.2}}\right)\right)$$

where μ is the linear attenuation coefficient, ρ is the density of the object, Z is the atomic number of the object, E is the used X-ray energy and a and b are two energy-dependent constants.

From Eq. 1 it can be deduced that scanning the object at two different energies enables a re-arrangement of the attenuation values to obtain a density and atomic number image. Van Geet *et al.* (2000) have shown that this procedure can also be used for mineralogical characterization. The major drawbacks of the technique are that anomalous values are measured for strongly attenuating particles and that noise needs to be minimized.

Applications

Porosity measurements in limestone

Many carbonate reservoirs consist predominantly of pure limestone, i.e. calcite, enabling the use of a linearization procedure to exclude beam hardening artefacts. A homogeneous monocrystalline calcite crystal was used as a calibration sample. The deviation from a linear correlation between measured intensity and thickness of the calcite was fitted with a fifth order polynomial, which was then used for correcting beam hardening. Visual inspection showed a complete elimination of beam hardening artefacts in calcite crystals and calcareous mudstones.

Once beam hardening is corrected, the measured linear attenuation coefficient can be correlated with density or porosity of the object. Because the linear attenuation coefficient reflects the mean attenuation of the whole voxel the measured data include total porosity, including

Fig. 2. Macroscopic view of a turbiditic carbonate sample used to compare the porosity measurements by μCT and classical optical microscopy. The polished half cylinder (8 mm diameter) that was analysed with classical microscopy using image analysis is shown. The black coloured part is oil-impregnated.

microporosity related to voids that are not visible as individual pores. A calcite crystal was used for calibration (0% porosity).

To determine the possibilities of quantifying porosity, a sample containing porous and non-porous zones was used (Fig. 2). The limestone sample is 8 mm in diameter, derived from a turbiditic oil reservoir formation of the Ionian Zone, Albania (Van Geet *et al.* 2002). It contains a porous zone at the top, with a dark colour due to oil impregnation. The bottom part is non-porous, due to early diagenetic calcite cementation, and lacks any oil impregnation. About 300 μCT-slices were scanned for this sample, perpendicular to the longitudinal axis of the cylindrical object. The mean attenuation coefficient was measured in every slice to calculate the mean density, which was subsequently converted to porosity.

Petrographical research, including cathodoluminescence microscopy, reveals that the studied turbidite beds consist of non-porous pack- to grainstones, laminated bioclastic to pelletoidal wackestones and compacted mudstones. The reservoir rocks are represented by the medium-grained bioclastic to pelletoidal wackstones. The most prominent cement phase is a beige to orange luminescent syntaxial cement, developed on rudist fragments and other monocrystalline bioclasts. Together with equant calcite that has similar luminescence characteristics, this syntaxial cement is framework-stabilizing and preserves some interparticle porosity. This contrasts with the bioclastic pack- and grainstones of the same formation, which were so pervasively cemented during early diagenesis that all porosity became occluded. The mudstones are not cemented, but they are so intensely compacted that all porosity was lost. The layering shown in Figure 2 is caused by the alternation of wackestone and pack/grainstone.

To verify the μCT results, image analysis was carried out with reflected light microscopy on a polished surface to measure porosity along

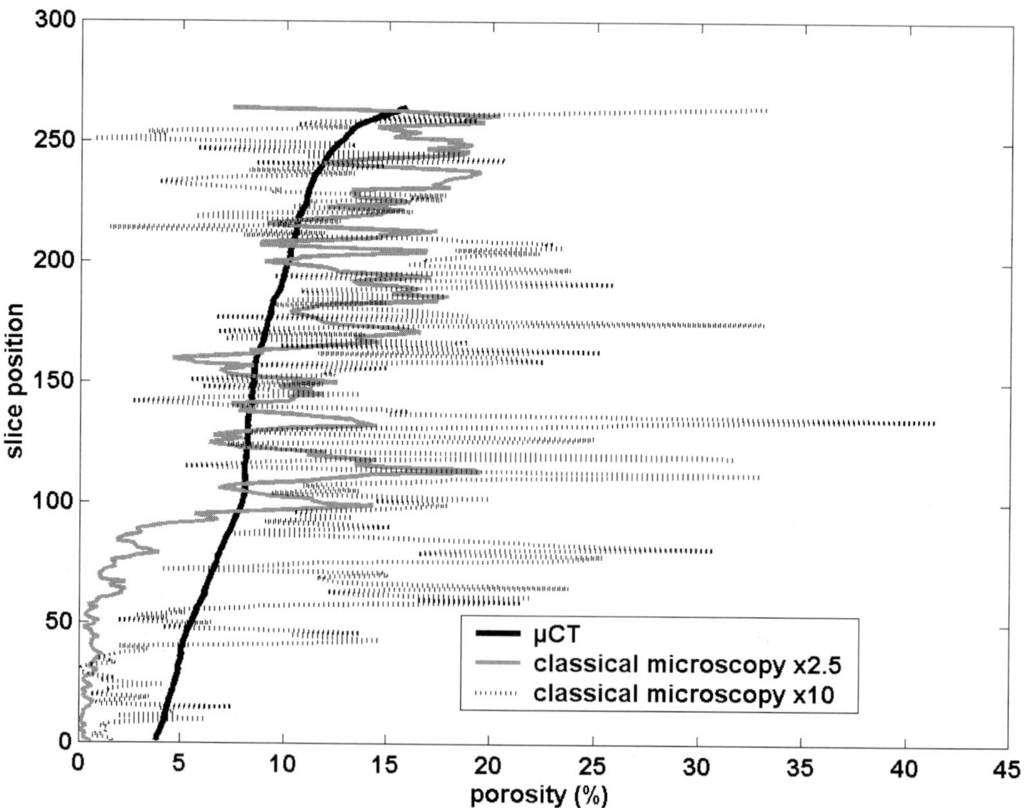

Fig. 3. Evaluation of porosity measurements by μCT and classical reflected light microscopy carried out on a polished surface (Cretaceous turbidite, Ionian Zone, Albania). This sample comprises an originally oil-impregnated zone (slice numbers above 120) and a non-impregnated section (slice numbers below 120) (see Fig. 2).

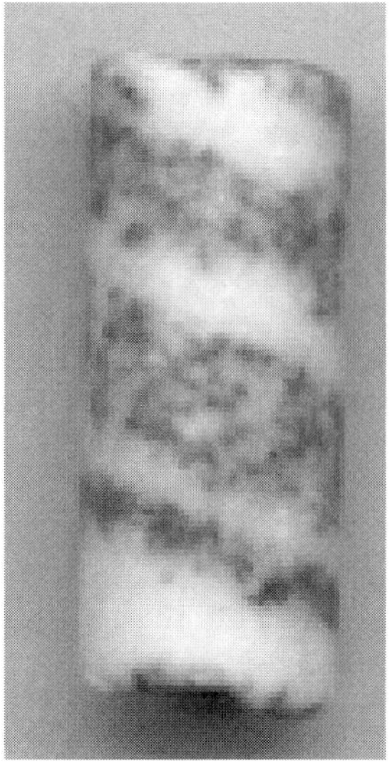

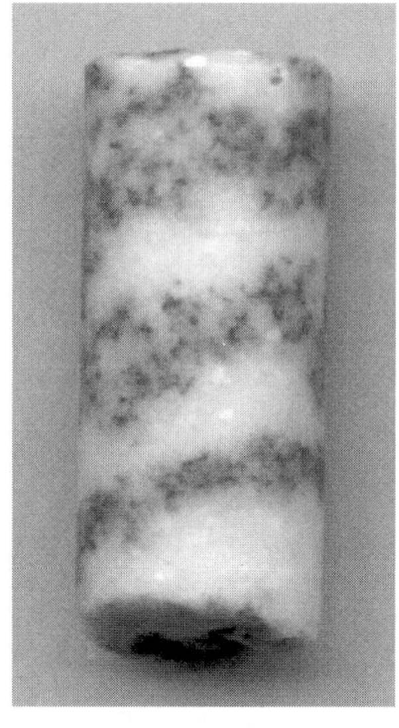

Fig. 4. Macroscopic views of the sample illustrated in Figure 5 imaged from two different angles (0° and 180°). (**a**) Representation of the sample from the same side as illustrated in Figure 5; the impregnated zones (dark) and non-impregnated zones (light grey) can be clearly distinguished and correspond to the zones outlined in Figure 5. (**b**) Representation showing that the impregnated porosity is more diffuse at the top of the sample, which explains the lack of lamination in the upper part of the sample in Figure 5.

a trace parallel to the longitudinal axis of the object. Note that with this approach it was not possible to differentiate isolated pores from connected pores, which make up the network porosity that can be accessed by fluids. However, this differentiation is also not possible for μCT-derived porosity measurements.

Discrimination between porosity and limestone was evaluated by thresholding. The measured porosity profiles obtained by both techniques are plotted in Figure 3. Although different types of information are compared (two dimensional versus three dimensional), a similar trend is found in all measurements, with a particularly good correlation between the μCT results and optical microscopy measurements at the highest of the two magnifications that were used. At this magnification micro-porosity is incorporated as well, which is also included in the μCT data. This shows that μCT can clearly be used for total porosity measurements in limestone samples after calibration.

μCT has the additional advantage of allowing visualization of the largest pores in three dimensions, providing information about their interconnectivity. To evaluate this, a turbidite limestone reservoir sample composed of several bands of porous (oil-impregnated) and non-porous zones was scanned (Fig. 4). Figure 5

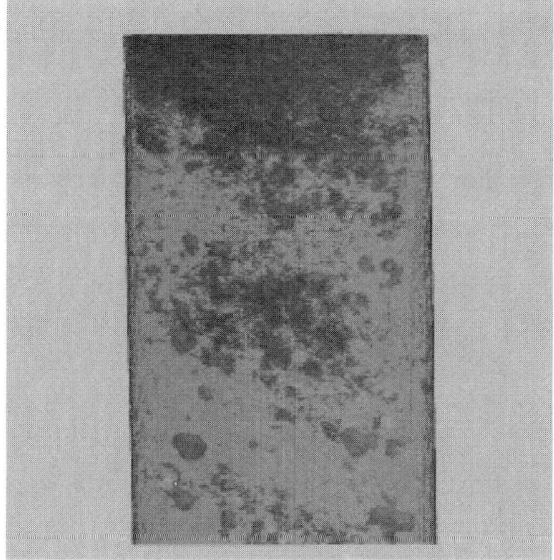

Fig. 5. 3D visualization of the distribution of macroporosity for a turbiditic carbonate sample of the Ionian Zone (Albania), based on μCT-analysis. The carbonate is made 'transparent' (i.e. light grey) and the macroporosity is visualized as darker grey sections by means of thresholding, within a fuzzy outline of a cylindrical sample. Inclined laminae with higher porosities alternate with non-porous laminae.

illustrates the μCT results for this sample as a kind of radiograph in which limestone is made transparent (here light grey) and porosity is shown as dark grey features. It is clear that a banded structure and bedding can be seen, illustrating that μCT data provide information about macroporosity distributions, in addition to total porosity measurements.

Porosity measurements in sandstone

Sandstone samples can be quite heterogeneous and may contain different minerals. Consequently, correction of beam hardening by linearisation cannot be used. However, the use of a dual-energy technique allows the density of the samples to be defined. This information might allow extracting information about porosity.

Two samples of compact Westphalian C fluvial channel sandstones from the coal measures of the Campine Basin (NE Belgium) were studied. From the 3D μCT image reconstruction (Fig. 6) it can be deduced that different minerals are present (different grey values). Classical petrography shows that both sandstone samples are medium grained. The detrital minerals mainly consist of mono- and polycrystalline quartz, clay minerals and small amounts of chert, rock and coal fragments and mica. Nearly all feldspar is altered to clay minerals (mainly fine-grained kaolinite booklets). The authigenic mineral assemblage of the sandstone consists of carbonates (ferroan calcite, siderite and dolomite), quartz overgrowths, kaolinite and illite. The precipitation of these authigenic minerals had a severely deteriorating effect on the porosity. The presence of authigenic kaolinite seems to be the main occluder of porosity (Van Keer 1999). According to the sandstone classification of Pettijohn *et al.* (1987), these sandstones are classified as sublitharenites.

Strongly attenuating phases are difficult to characterize with dual-energy methods. Moreover, noise can cause anomalous measurements. Therefore, several processing steps need to be performed on all slices:

- application of a threshold to eliminate high attenuation voxels;
- calculation of the mean attenuation value of the remaining material;
- deduction of density by means of dual-energy measurements;
- calculation of porosity in each slice, assuming a density of 2.65 g/cm^3 for pure quartz.

These processing steps allow the removal of strongly attenuating particles that are not relevant for porosity studies. Moreover, the calculation of density and porosity over the whole volume of a slice minimizes the effect of noise. However, a threshold needs to be defined, which introduces some subjectivity. A representative histogram of the attenuation values through one slice is given in Figure 7. This histogram shows a strong peak with a maximum around the attenuation value of 0.083, which corresponds to voxels composed of quartz containing some porosity. A second peak, for strongly attenuating particles, seems to be absent. However, a fit with a theoretical normal distribution

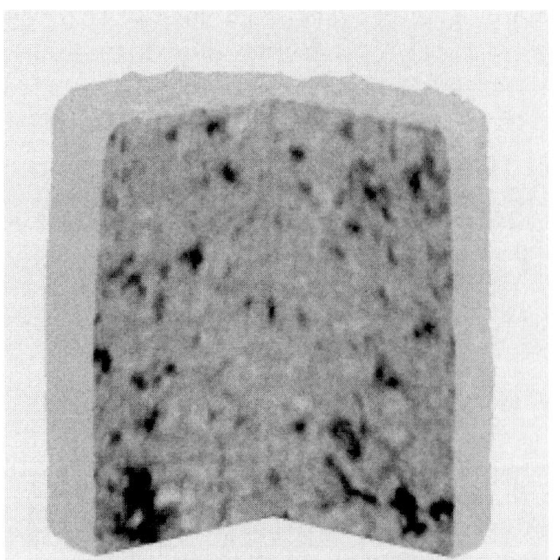

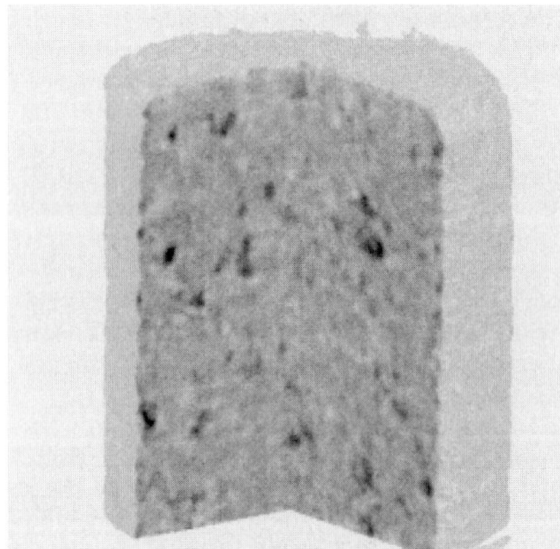

Fig. 6. 3D visualization of heterogeneous Westphalian sandstone samples (8 mm in diameter). Darker colours correspond with high attenuation values. (**a**) Sample of core KB172. (**b**) Sample of core KB146.

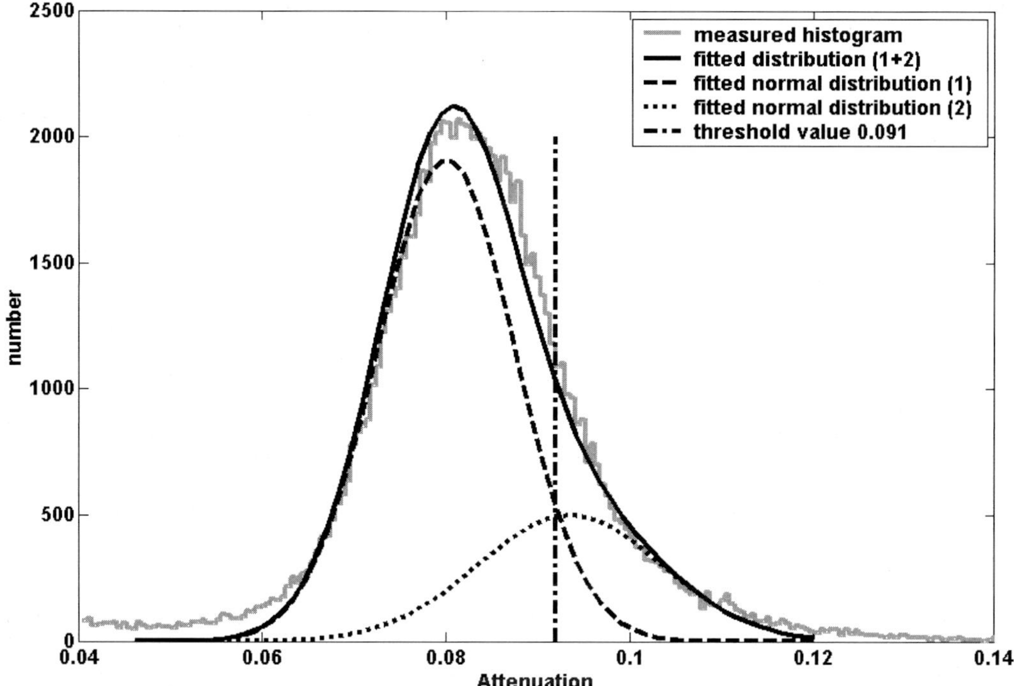

Fig. 7. Representative measured histogram of a slice through sample KB172. A good fit can be obtained by assuming two populations with normal distributions ($\mu_1 = 0.08$, $\sigma_1 = 0.0075$; $\mu_2 = 0.0935$, $\sigma_2 = 0.01$). The crossing of these fitted distributions at an attenuation value of 0.091 can be used as threshold value.

shows that a second population with higher attenuation values is present as well. Due to the strong overlap of both populations, a distinction between the two peaks is not possible. The choice of the threshold value has been based on the crossing of the normal distributions of both populations at an attenuation value of 0.091. This choice is supported by the fact that the attenuation of a pure glass sample has a mean attenuation value of about 0.09.

The deduced porosity values are shown in Figure 8. Here, porosity variation within every slice is given. The mean porosity of both samples is given in Table 1 and compared with porosity

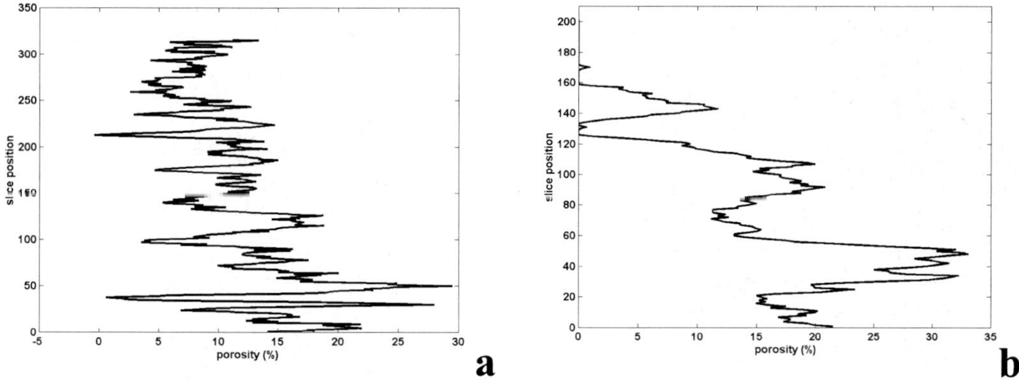

Fig. 8. Results of mean porosity measurements of sequential slices by μCT for two Westphalian sandstones. (**a**) Sample of core KB172, with a mean porosity of 11.4%. (**b**) Sample of core KB146, with a mean porosity of 12.1%. Note the cyclic banding of porosity in this sample.

Table 1. *Comparison of the results of different porosity measurements (in %) for two samples of Westphalian sandstone from the Campine Basin (NE Belgium)*

	KB 172	KB 146
Image analysis on thin section	(11.0–12.0)–(16.8–23.0)	(1.1–2)–(10.0–23.0)
Hg-porosimetry	10.5	11
Vacuum saturation	10.45	11.12
μCT	11.4	12.1

A range is given for the image analysis results, whereby the first numbers refer to intergranular porosity and the second numbers refers to the sum of inter- and intragranular porosity.

measurements by image analysis of thin sections, Hg-porosimetry (both performed on adjacent samples) and vacuum saturation test (performed on the same sample). It is concluded that quite similar porosity values are obtained by μCT if compared with vacuum saturation and Hg-porosimetry. Porosity measurements with μCT do seem to give slightly higher values. This is partly related to the fact that μCT gives a total porosity value, whereas the other techniques are based on a connectivity of the pores. The presence of isolated pores might explain the slightly higher values obtained by μCT analysis. For petrographical analysis, the macroporosity of the sandstones was accentuated by adding a fluorescent dye in the epoxy of the vacuum-impregnated samples. This allows automatic measurement of intergranular and intragranular porosities by means of image analysis (Zeiss AxioVision 3.0). The measurements show very large variations, depending on whether only inter- or both inter- and intragranular porosity is selected.

The use of μCT for porosity measurements has the advantage of being able to visualize directional variations. These variations might be correlated with grain size distribution, microstructure, etc. This is illustrated by a kind of cyclicity that can be observed in the porosity measurements for one of the samples (Fig. 8). Further research is needed to check whether this cyclicity is related to natural layering in the sample.

Apart from obtaining total porosity values for each slice, thresholding can be used to visualize and quantify the distribution of large pores in 3D. As an example, the distribution of large pores (about 300 μm in diameter) in both sandstone samples is given in Figure 9. These pores make up 0.53% of the sample volume in sample A and 0.57% in sample B.

Conclusions and perspectives

μCT allows the internal features of objects to be visualized. Some artefacts might, however,

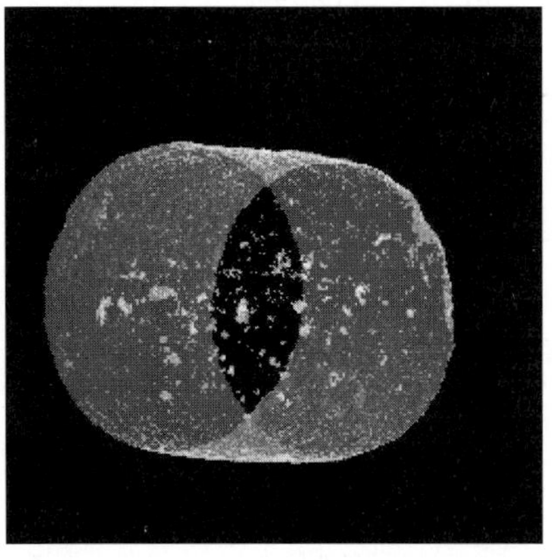

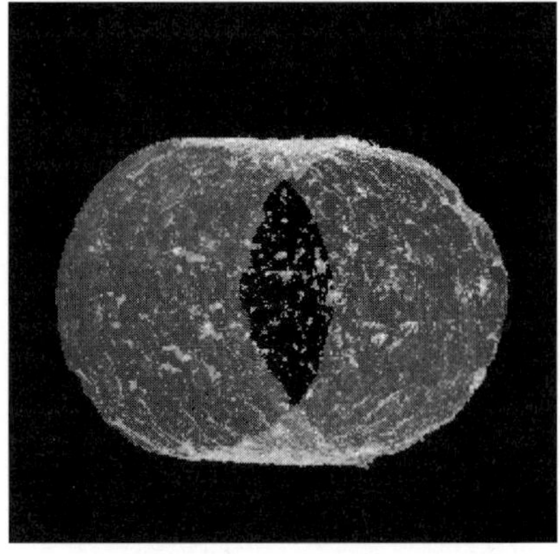

Fig. 9. 3D visualization of the largest pores in the studied sandstone samples. These images are generated by thresholding on the three-dimensional μCT data volume, based on a visual inspection of the grey value histograms. (**a**) Sample of core KB172, in which the pores correspond with 0.53% of the sample volume. (**b**) Sample of core KB146, in which the pores correspond with 0.57% of the sample volume.

hinder quantitative measurements. The most important artefact with regards to this is related to beam hardening. The use of a linearization technique can eliminate this artefact for porous monomineralic materials, whereas the use of hardware filters can minimize the artefact in heterogeneous materials. Quantification of features can be straightforward by measuring a calibration sample, as illustrated by measuring the porosity of a limestone sample. For heterogeneous materials such as sandstones, this is much more difficult. For these materials, the use of metal foils can strongly reduce beam hardening artefacts and a dual-energy procedure allows quantification of density and thus of porosity. The use of a threshold value is the most critical step. Calibration with other porosity analysis techniques at an initial stage is recommended. It should be noted that total porosity is measured by μCT, regardless of connectivity. This is a potential drawback for studies that are focused on permeability measurements. μCT is, therefore, mainly a useful additional technique for porosity measurements.

Apart from measuring total porosity, μCT also allows the study of macroporosity. Using a threshold value allows visualization of the 3D distribution of the pores within a volume. Changing this threshold value might allow obtaining a kind of porosity distribution variation, comparable with the results of Hg-porosimetry. Future research should be focused on this possibility, as this would largely increase the possible use of μCT in porosity studies.

This study shows that μCT can bridge the gap between the petrographical and petrophysical measurements. Moreover, because the technique is entirely non-destructive, as no sample preparation or contrast agents are needed, the same sample may be used for further petrographical and petrophysical research.

The work of M. Van Geet was financially supported by a grant of the 'Flemish Institute for the Promotion of Scientific-Technological Research in Industry' (IWT). The purchase of the Skyscan instrument was financially supported by the IWT, project INM/950330.

References

Alvarez, R.E. & Macovski, A. 1976. Energy-selective reconstructions in X-ray computerized tomography. *Physics in Medicine and Biology*, **21**, 733–744.

Brooks, R.A. & DiChiro, G. 1976. Principles of computer assisted tomography (CAT) in radiographic and radioisotopic imaging. *Physics in Medicine and Biology*, **21**, 689–732.

Coenen, J.G.C. & Maas, J.G. 1994. Material classification by dual-energy computerized x-ray tomography. *In*: *Proceedings of the International Symposium on Computerized Tomography for Industrial Applications, June 8–10, Berlin*, pp. 120–127.

Damico, K.L., Deckman, H.W., Dunsmuir, J.H., Flannery, B.P. & Roberge, W.G. 1989. X-ray microtomography with monochromatic synchrotron radiation. *Review of Scientific Instruments*, **60**, 1524–1526.

Flannery, B.P., Deckman, H.W., Roberge, W.G. & Damico, K.L. 1987. 3-dimensional X-ray microtomography. *Science*, **237**, 1439–1444.

Frederich, J.T. 1999. 3D imaging of porous media using Laser Scanning Confocal Microscopy with application to microscale transport. *Physics and Chemistry of the Earth (A)*, **24**, 551–561.

Hammersberg, P. & Mangard, M. 1998. Correction of beam hardening artefacts in computerized tomography. *Journal of X-ray Science and Technology*, **8**, 75–93.

Jennings, R.J. 1988. A method for comparing beam-hardening filter materials for diagnostic radiology. *Medical Physics*, **15**, 588–599.

Joseph, P.M. 1981. Artifacts in computed tomography. *In*: Newton, T.H. & Potts, D.G. (eds) *Radiology of the Skull and Brain: Technical Aspects of Computed Tomography, Volume 5*. The, C.V. Mosby Company, St Louis, Missouri, 4058–4095.

Kak, A.C. & Slaney, M. 1988. *Principles of Computerized Tomographic Imaging*. IEEE Press, New York, USA.

Kinney, J.H. & Nichols, M.C. 1992. X-ray tomographic microscopy (XTM) using synchrotron radiation. *Annual Review of Materials Science*, **22**, 121–152.

Lindquist, W.B., Venkataragan, A., Dunsmuir, J. & Wong, T.F. 2000. Pore and throat size distributions measured from synchrotron X-ray tomographic images of Fontainebleau sandstones. *Journal of Geophysical Research (B)*, **105**, 21 509–21 527.

Pettijohn, F.J., Potter, P.E. & Siever, R. 1987. *Sand and Sandstone. 2nd Edition*. Springer-Verlag, New York, USA.

Sasov, A.Y. 1987. Microtomography I. Methods and equipment. *Journal of Microscopy*, **147**, 169–178.

Spanne, P. & Rivers, M.L. 1987. Computerized microtomography using synchrotron radiation from the NSLS. *Nuclear Instruments & Methods in Physics Research*, **B24**, 1063–1067.

Van Geet, M., Swennen, R. & Wevers, M. 2000. Quantitative analysis of reservoir rocks by means of microfocus X-ray computer tomography. *Sedimentary Geology*, **132**, 25–36.

Van Geet, M. 2001. *Optimisation of microfocus x-ray computer tomography for geological research with special emphasis on coal components (macerals) and fractures (cleats) characterisation*. PhD thesis, K.U. Leuven, Belgium.

Van Geet, M., Swennen, R., Durmishi, C., Roure, F. & Muchez, Ph. 2002. Paragenesis of Cretaceous to Eocene carbonate reservoirs in the Ionian fold and thrust belt (Albania): relation between tectonism and fluid flow. *Sedimentology*, **49**, 697–718.

VAN KEER, I. 1999. *Mineralogical variations in sandstone sequences near coal seams, shales and mudstones in the westphalian of the Campine Basin (NE-Belgium): its relation to organic matter maturation*. PhD thesis, K.U. Leuven, Belgium.

WELLINGTON, S.L. & VINEGAR, H.J. 1987. X-ray computerized tomography. *Journal of Petroleum Technology*, 885–898.

Quantitative characterization of fracture apertures using microfocus computed tomography

K. VANDERSTEEN[1], B. BUSSELEN[1], K. VAN DEN ABEELE[2] & J. CARMELIET[1]

[1] *Laboratory of Building Physics, Department of Building Engineering, Catholic University of Leuven, Kasteelpark Arenberg 51, B-3001 Heverlee, Belgium (e-mail: jan.carmeliet@bwk.kuleuven.ac.be)*

[2] *Faculty of Sciences, Catholic University of Leuven at Kortrijk, Sabbelaan 53, B-8500 Kortrijk, Belgium*

Abstract: Microfocus X-ray computed tomography (μCT) was used as a tool to determine the apertures of a fracture in a cylindrical sample of crinoidal limestone. After scanning, artefacts were removed from the images. Phantom objects were used to establish a calibration relationship between real fracture apertures and fracture aperture measurements on the μCT images. The performance of different procedures for quantitative fracture determination was examined. The calibration relationship was then used to determine the fracture apertures in a naturally fractured sample. A comparison of the μCT technique and a microscope technique shows a good agreement between their results.

Research on fluid flow in fractured media is of growing importance in many disciplines, e.g. hydrology, petroleum engineering and civil engineering. Until some decades ago, fractures were described as parallel plates with no variation in aperture. However, it is well known that the variability of apertures in naturally fractured materials can considerably influence the flow in such fractures (e.g. Abelin *et al.* 1985). As a result, detailed knowledge of fracture apertures is needed. Different techniques have been used in the laboratory to accurately determine the variable aperture of a fracture, such as scanning of the two fractures surfaces (Brown *et al.* 1986), injecting of hardening resins (Gentier *et al.* 1989) or metals (Pyrak-Nolte *et al.* 1987). All these methods render the fracture unusable for further flow experiments. Recently, the use of various non-destructive techniques has been explored to determine fracture apertures, e.g. X-ray tomography (Van Geet 2001; Verhelst *et al.* 1995; Johns *et al.* 1993; Keller 1998; Timmerman *et al.* 1999; Bertels *et al.* 2001), transmitted light (transparent systems) (Detwiler *et al.* 1999; Renshaw *et al.* 2000) and nuclear magnetic resonance (Kumar *et al.* 1995; Renshaw *et al.* 2000).

For the study described in this paper, we used microfocus X-ray computed tomography (μCT) as a tool to determine the apertures of a vertical fracture in a cylindrical sample of crinoidal limestone (diameter 22 mm). Microfocus CT was used because of the high quality of the images and the high resolution of the apparatus. The results were compared to those of a microscope technique.

The scanner used in the experiments is a Skyscan 1072 desk-top X-ray microtomograph. It contains an X-ray source with a focal spot size of 10–40 μm a CCD camera as detector, an image intensifying screen, a lens and an object manipulator. The data are gathered in a cone beam configuration, but they are automatically transferred into a parallel beam configuration during measurement. Scanning of the samples was done using a tube voltage of 117.8 kV and a current of 300 μA. One projection was taken every 0.9° over 180°. Other parameters were chosen so as to minimize noise in the images. At the magnification level that was used, pixel size was 55.804 μm. The images were reconstructed using a filtered back-projection algorithm (Van Geet 2001). Beam hardening artefacts were corrected using a linearization procedure (Hammersberg & Mangard 1998). The images are also corrected for other artefacts, such as outlining artefacts, ring artefacts and line artefacts (for a detailed description, see Van Geet 2001).

From: MEES, F., SWENNEN, R., VAN GEET, M. & JACOBS, P. (eds) 2003. *Applications of X-ray Computed Tomography in the Geosciences*. Geological Society, London, Special Publications, **215**, 61–68. 0305-8719/03/$15.

Procedure to quantify fracture apertures

The image of a fractured limestone sample in a horizontal plane of the object is represented in Figure 1. A dip in the profile is noticed where the fracture is present. In theory, the image of a fracture can be seen as a convolution of a rectangular fracture profile with a point spread function (PSF), which is Gaussian (Fig. 2).

Over the years, several methods have been described to quantitatively determine fracture apertures (Fig. 1). Johns *et al.* (1993) used a missing attenuation value (MA), which is the integrated loss of the attenuation coefficient due to the presence of the fracture. Peyton (1992) used the value of the full width at half maximum (FWHM) of the attenuation profile in the fracture. Verhelst *et al.* (1995) used the peak height value (PH) in the fracture attenuation profile. The drawback of the PH value is that it can only be used when the attenuation coefficient in the fracture is higher than the attenuation of air. The advantage of PH in comparison to MA and FWHM is its independence on the direction of measurement in the fracture, whereas MA and FWHM need to be measured perpendicular to the direction of the fracture plane. Van Geet (2001) examined the performance of MA and FWHM and concluded that MA is the better parameter. In the following presentation, we will focus on the performance of MA and PH.

Generally, the values of MA, PH and FWHM are calculated directly from the CT images (crude method). This can be considered as a good method when the signal-to-noise ratio of the image is high. However, heterogeneities and artefacts may be present in CT images, which could negatively influence the signal-to-noise ratio. Also, for small fractures the signal-to-noise ratio is expected to be quite low. Therefore, a new method is proposed that consists of fitting the attenuation profiles prior to calculating MA and PH values.

The dip in the attenuation profile, due to the presence of the fracture, is fitted by a Gaussian function. A Gaussian curve is justified because the μCT image of a fracture can be considered as a convolution of a rectangular fracture profile with a Gaussian point spread function. The result is an approximation of a Gaussian function (Fig. 2). In a first approximation, the noise in the μCT attenuation profiles of the host rock is described by a sine function. In the images used in this study, a certain pattern in the noise was observed, which could indicate the presence of an artefact, as discussed below. A sine function described the noise in

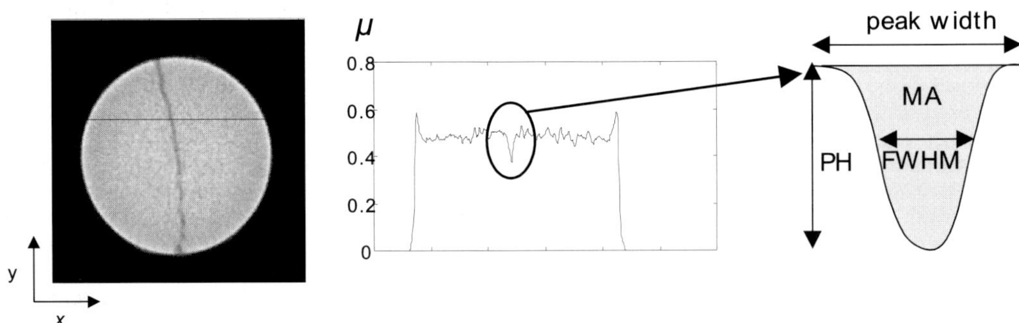

Fig. 1. CT image of a limestone sample with variable aperture fracture, corrected for artefacts (left and centre). Different methods to quantify fracture apertures (right).

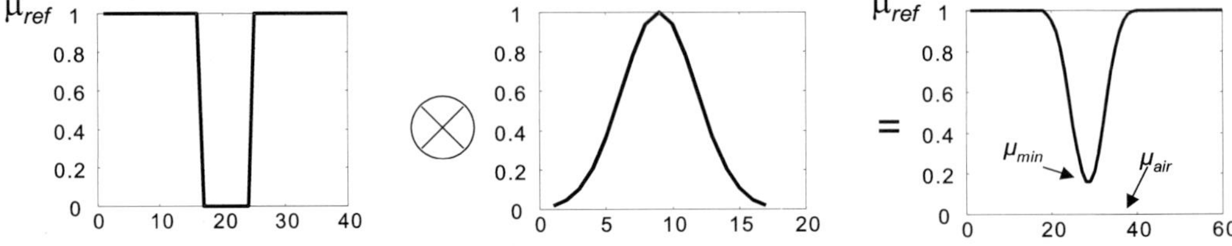

Fig. 2. Convolution of a rectangular profile (fracture) with a PSF, resulting in the image of a fracture. $\mu_{\mathrm{ref}} = (\mu - \mu_{\mathrm{air}})/(\mu_{\mathrm{mat}} - \mu_{\mathrm{air}})$; it is supposed that $\mu_{\mathrm{air}} = 0$.

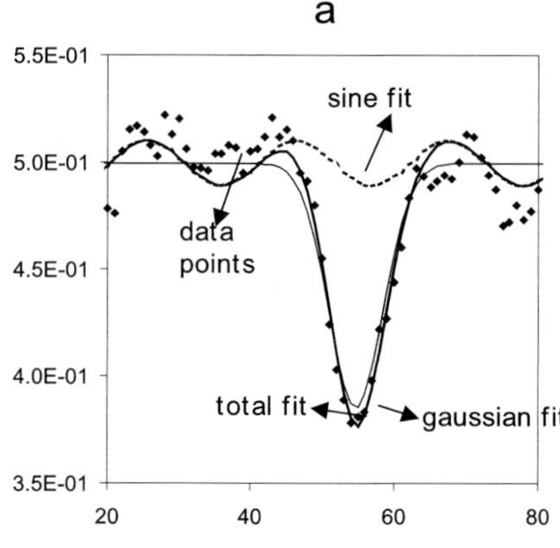

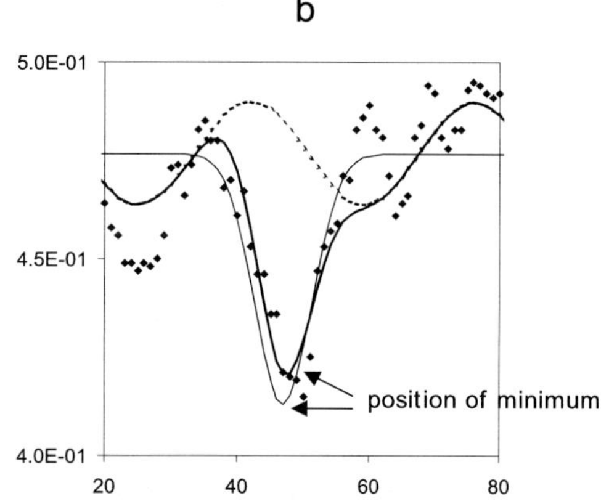

Fig. 3. Fit of Gaussian and sine function through the fracture attenuation profile determined for images with a high signal-to-noise ratio, (a) and a low signal-to-noise ratio, (b).

the images quite well; an attenuation profile Y containing a fracture, can then be described as follows:

$$Y = PH \exp\left[-\frac{1}{2}\left(\frac{X - x_p}{\Delta x_p}\right)^2\right] + \mu_{\mathrm{hr}} + A \sin\left(\frac{2\pi}{\lambda} X + \phi\right) \quad (1)$$

where x_{p} is the position of the peak, Δx_{p} is the width of the peak within a 68% confidence interval, μ_{hr} is the mean attenuation value in the host rock, A is its amplitude, λ is its wavelength and ϕ is the phase angle. Missing attenuation (MA) can be derived as follows:

$$\mathrm{MA} = \sqrt{2\pi \cdot PH \cdot \Delta x_p} \quad (2)$$

As shown in Figure 3, this fitting combination approximates the measurement data quite well. Figure 3a represents a fracture with a high signal-to-noise ratio, while in Figure 3b, the signal-to-noise ratio is poor. The noise has an effect on peak height, on the position of the minimum in the fracture and on the width of the dip.

Calibration measurements

In this section, the performance of the MA and PH values is examined and the fitting procedure for quantification purposes of fracture apertures with variable width is evaluated.

Calibration measurements were performed on phantom objects of crinoidal limestone consisting of two polished halves of a cylinder. These were fixed at certain distances from each other using metal foils of constant thickness, thus reproducing different known fracture apertures. Fracture apertures ranging from 0.05 mm to 1.0 mm were examined. Even at aperture values of 1.0 mm, the peak height measure (PH) has not reached its limit of $\mu_{\mathrm{min}} - \mu_{\mathrm{air}} > 0$. Our research mainly focused on small fractures exhibiting sufficient capillary behaviour (with apertures <0.3 mm). However, larger apertures can occasionally occur locally in naturally fractured materials. The measured μCT fracture attenuation profiles were fitted using the combination of a Gaussian and a sine function, as described in the previous section. From the fitted profiles, the parameters MA and PH are obtained (Fig. 4). For each fracture aperture, one slice was chosen at random in the sample and 280 fracture attenuation profiles per slice were used for the calculation. Because results were not significantly different for different slices, a one-dimensional analysis of the problem could suffice. MA as function of fracture aperture results in a linear relationship, whereas PH yields a curvilinear relationship (results are from the fitted profiles). The slope for PH initially has a high value and decreases for larger apertures. This indicates that for smaller apertures, PH is a more sensitive parameter than MA. The reverse is true for larger apertures. The standard deviation of both curves stays rather constant.

To study the influence of noise, we compared these fits with data fitted by a Gaussian curve only. In Figure 5 real apertures are compared with measured apertures determined using MA and PH. This figure shows that the mean values

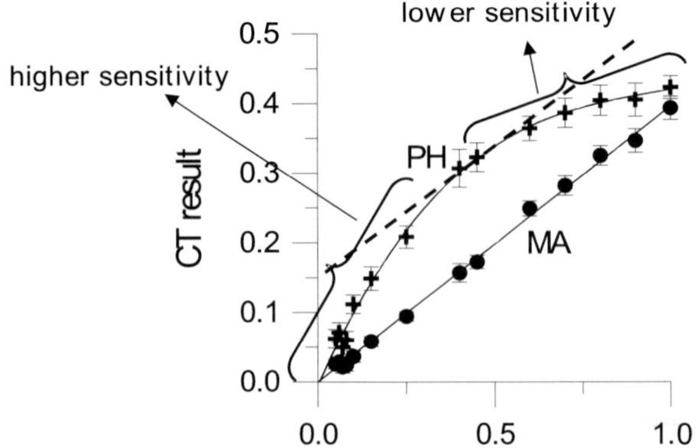

Fig. 4. PH and MA calculated from the fitting procedure (Gaussian fit and sine) for apertures ranging from 0.05 mm to 1.0 mm.

of PH and MA coincide very well with the theoretical values given by the linear curve both for the fitting procedures and for the crude method. The standard deviation of MA is relatively constant for different fracture apertures, while the standard deviation of PH increases dramatically at larger fracture apertures.

In Figure 6 a detailed plot of the coefficient of variation (CV = 100 * standard deviation/mean aperture) for PH and MA is given as a function of aperture for the three calculation methods. For fractures equal to 0.15 mm, the CV of PH is approximately equal to the CV of MA. For lower fracture apertures, the difference between

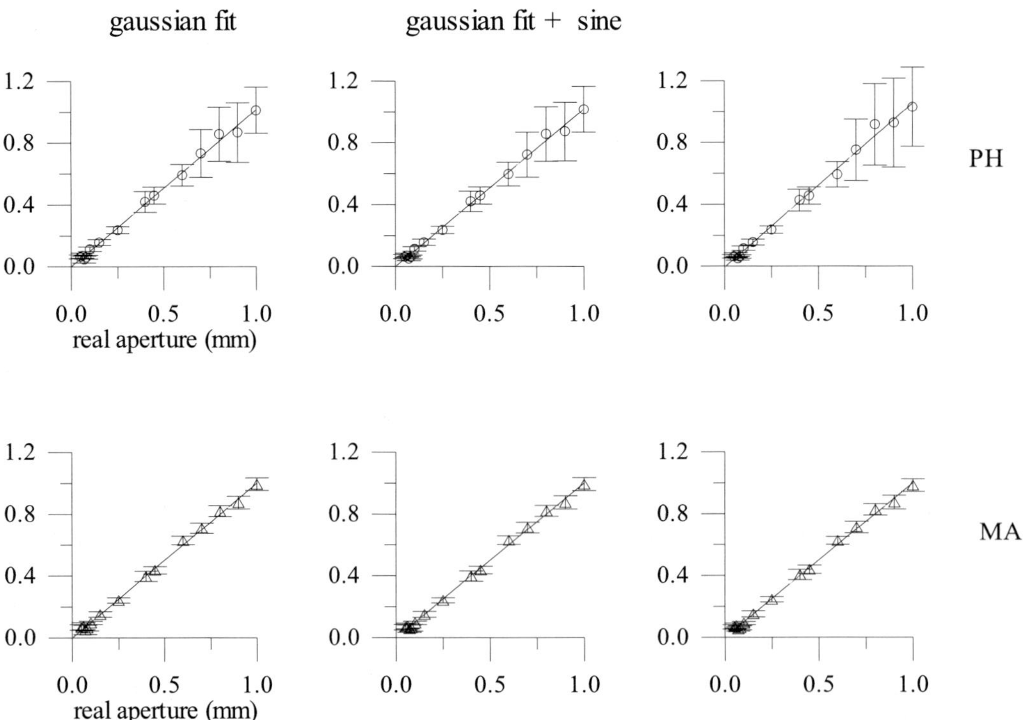

Fig. 5. Comparison between the performance of PH and MA calculated from the fitting procedures and the crude method. Apertures range from 0.05 mm to 1.0 mm (○ = PH, Δ = MA).

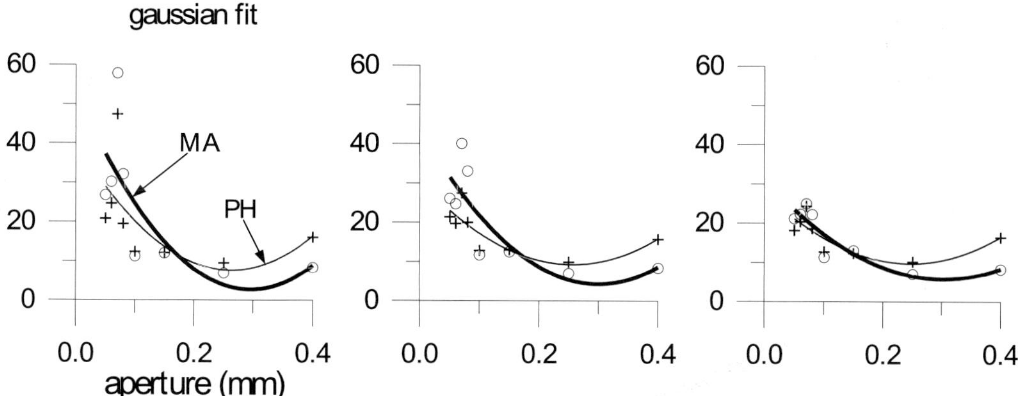

Fig. 6. Coefficient of variation versus fracture aperture for peak height and missing attenuation coefficient (+ = PH, ○ = MA).

these CV values increases in the sense that small fractures are more accurately determined using PH. PH will increase more rapidly for small fractures than MA does (Fig. 6), thus increasing the signal-to-noise ratio. Larger fractures (with apertures >0.15 mm), however, will be more accurately determined using MA. For low fracture apertures, the CV of the MA calculated using the fitting procedures is larger than the CV value determined from the crude data. An explanation for this feature could be that small fractures are usually not visible in the data. The fitting procedure then determines a high value of the peak width, which leads to an overestimation of the MA. It can also be observed that at low fracture apertures, the CV of PH for the single Gaussian fit is higher than for the other two methods. This indicates that fracture aperture determination for low fracture apertures can be improved by taking noise into account in the fitting procedure.

The main conclusions of this part can be summarized as follows:

- Peak height (PH) is the better parameter when fractures are small (apertures <0.15 mm). An additional advantage of PH is its independence of fracture direction.
- Missing attenuation (MA) is the better parameter when determining larger fractures (with apertures >0.15 mm), but it is dependent on the direction of the fracture plane. For non-vertical fractures, MA has to be multiplied with the cosine value of the inclination of the fracture plane.
- For large fractures (with apertures >0.15 mm), the coefficient of variation (CV) of MA and PH for the fitting procedures and the crude calculation are approximately the same.
- For small fractures (with apertures <0.15 mm), the fitting procedure that incorporates noise in the host rock performs better than the fitting procedure that does not account for noise. The CV of PH, for the fitting procedure incorporating noise, is equal to that for the crude data. However, it is important to note here that the images used in this study were images with a high signal-to-noise ratio due to the homogeneity of the material and efficient artefact removal. It can be expected that in less homogeneous materials the fitting procedure will improve results, compared to crude methods. Also, noise is described by a sine function, which is only an approximation based on the images.

Calculating variable aperture fractures

In this part, the derived calibration relationships are used to determine the fracture apertures in a naturally fractured sample. Microscope analysis showed that the mean fracture aperture at the surface is less than 0.2 mm. Figure 7 visualizes the position of the minimum ($y_{min}(x)$) in a slice of the naturally fractured sample, as determined by the crude method and by the fitting procedure (value of x_p in equation (1)) (smooth curve). Because fractures are smooth and linear structures, it is expected that a smooth curve should be obtained. When determining the minimum by a crude method errors are more likely, due to image noise and a finite pixel width. This can be important when

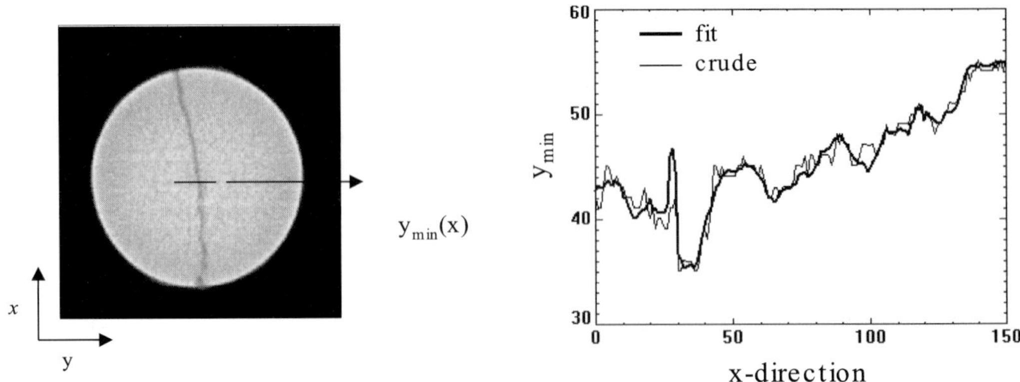

Fig. 7. Position of the minimum in a slice determined using the crude method and the fitting procedure.

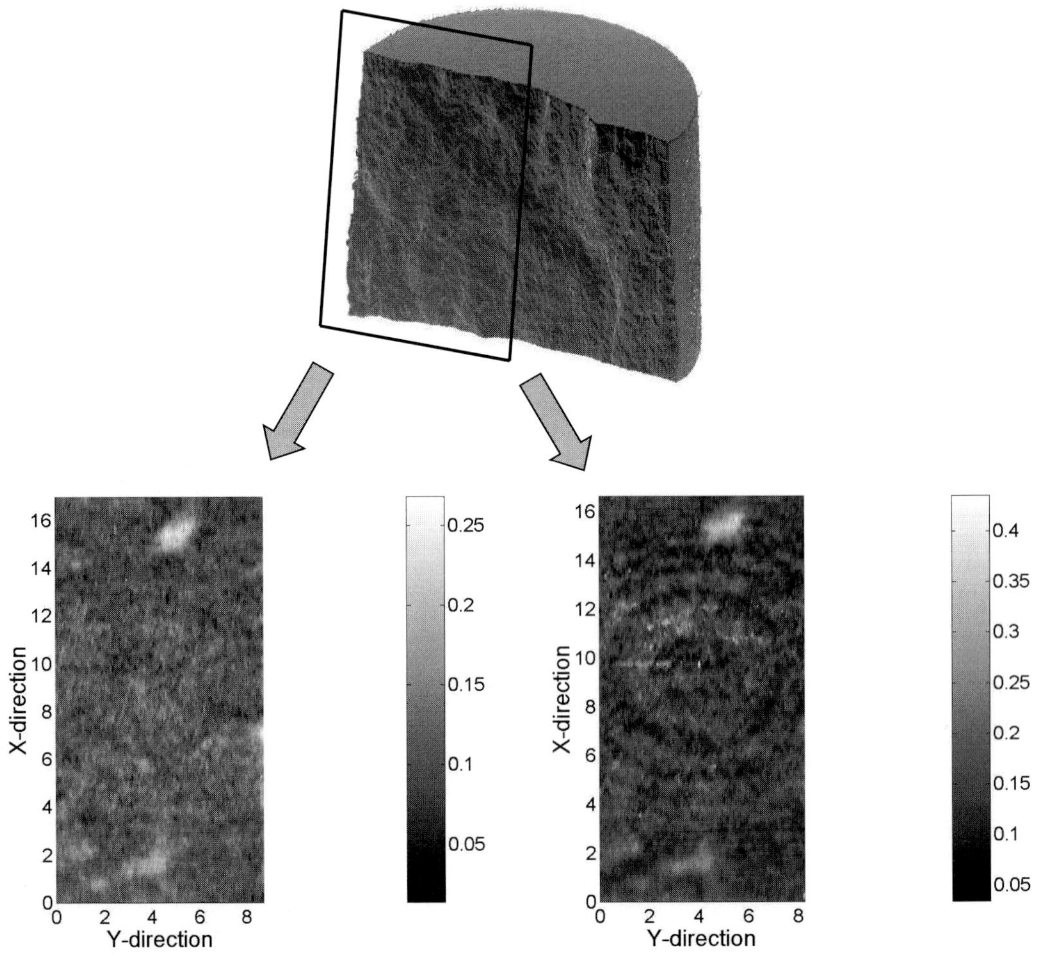

Fig. 8. Segment of the fractured sample as determined using microfocus X-ray CT (top). The fracture plane is visible. Fracture aperture as calculated from peak height (bottom left) and missing attenuation (bottom right).

working with fractures that are small or have a complex shape (in order to determine the direction of the fracture plane).

From the above observations, we concluded that fitting the data first is often useful, especially for small fracture apertures. The fracture apertures were determined in a naturally fractured sample using PH and MA, while first fitting the data with the combination of a Gaussian and a sine function. In Figure 8 the fractured sample is visualized with the fracture plane visible. The fracture apertures in the plane of the fracture are shown as calculated using PH (left) and MA (right), respectively. An additional artefact in the form of concentric circles is present in the image on the right. It appears that these circles are present as three-dimensional concentric spheres in the μCT images. The reason for this artefact is as yet unclear, but it is assumed that this is related to the cone beam configuration of the CT system or to a systematic variation of the point spread function. In the attenuation profiles, this artefact is visible as a systematic noise. The fitting procedure, including the sine function, removes part of the artefact. This artefact is currently being investigated in more detail using a μCT simulator (Busselen *et al.* 2001). In MA, the effect of the artefact is reinforced because both PH and peak width are involved in the calculation of MA. Because of this systematic error in the data, MA is probably a less reliable parameter.

Finally, the fracture apertures determined from the μCT images with aperture measurements on microscope images were compared with incident light taken at the surface of the sample (Fig. 9). These images were taken at high magnification (200×). The general trend resembles the microscope data (left). From the cumulative frequency curve, it is concluded that the apertures from the μCT data are slightly lower than those of the microscope data. Discrepancies can be explained by the fact that the comparison is not performed at the exact same location, because a μCT slice always has a certain finite thickness. Errors can also be introduced by the microscope technique. A visual thresholding technique was used in order to delineate the boundary of the fracture, which could lead to additional errors.

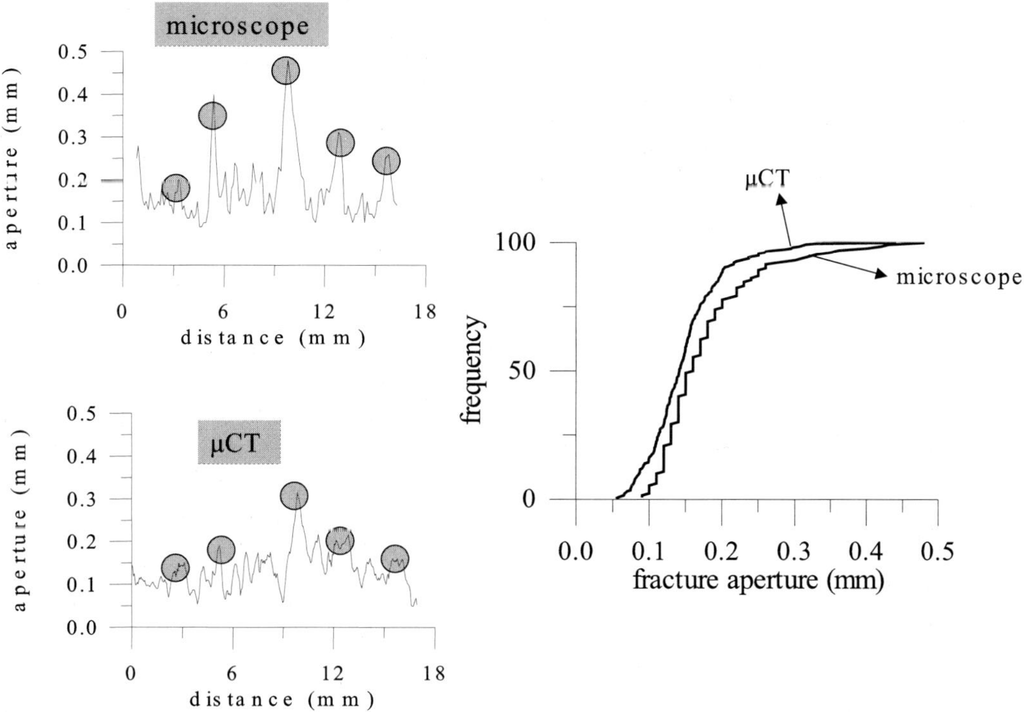

Fig. 9. Apertures determined by optical microscopy compared to those determined by μCT. Profile along the fracture, with indication of corresponding peaks (left) and cumulative frequency in function of fracture aperture (right).

Conclusions

Fracture apertures were determined using μCT in a sample of crinoidal limestone. First, the images were successfully corrected for artefacts (especially the severe beam hardening artefact). In order to improve the signal-to-noise ratio, a fitting procedure was applied to the fracture attenuation profiles, which consists of fitting either a Gaussian function through the fracture attenuation profile or a combination of a Gaussian function with a sine function describing the noise in the host rock. Peak height of the fracture attenuation profile and missing attenuation coefficient were found to perform differently as calibration parameters: peak height is the better parameter for small fractures (with apertures <0.15 mm), while missing attenuation coefficient performs better for larger fractures. For the images with high signal-to-noise ratios studied here, this ratio was not significantly improved by the fitting procedure. However, the fitting procedure enables a more accurate determination of the position of the minimum within the fracture. The usefulness of taking into account noise into the fitting procedure was demonstrated. In the naturally fractured sample, an additional μCT artefact in the form of concentric circles was discovered on the apertures determined using MA. This artefact was not visible for the PH data. Therefore, MA was considered a less reliable parameter for the material and fracture apertures that were studied. Comparison with microscopic data shows that the μCT results follow the same trend. Further research in this area will focus on developing techniques for quantifying complex fractures and fracture networks in three dimensions.

References

ABELIN, H., NERETNIEKS, I., TUNBRANT, S. & MORENO, L. 1985. Final report of the migration in a single fracture: Experimental results and evaluation. *Stripa Project Technical Report* **85–03**. SKB, Stockholm, Sweden.

BERTELS, S.P., DICARLO, D.A. & BLUNT, M.J. 2001. Measurement of aperture distribution, capillary pressure, relative permeability, and in situ saturation in a rock fracture using computed tomography scanning. *Water Resources Research*, **37**, 649–662.

BROWN, A.R., KRANZ, R.L. & BONNER, B.P. 1986. Correlation between the surfaces of natural rock joints. *Geophysical Research Letters*, **13**, 1430–1433.

BUSSELEN, B., BASTIAENS, W., DE MAN, B., VAN GEET, M., NUYTS, J., SWENNEN, R.,WEVERS, J. & CARMELIET, J. 2001. A microfocus computed tomography (μCT) simulator for quantitative exploitation of μCT in material research. *EUG XI Programme and Abstracts*. Cambridge Publications, Cambridge, 779.

DETWILER, R.L., PRINGLE, S.E. & GLASS, R.J. 1999. Measurement of fracture aperture fields using transmitted light: an evaluation of measurement errors and their influence on simulations of flow and transport through a single fracture. *Water Resources Research*, **35**, 2605–2617.

GENTIER, S., BILLAUX, D. & VLIET, L.V. 1989. Laboratory testing of the voids of a fracture. *Rock Mechanics*, **22**, 149–157.

HAMMERSBERG, P. & MANGARD, M. 1998. Correction of beam hardening artefacts in computerized tomography. *Journal of X-ray Science and Technology*, **8**, 75–93.

JOHNS, R.A., STEUDE, J.S., CASTANIER, L.M. & ROBERTS, P.V. 1993. Nondestructive measurements of fracture aperture in crystalline rock cores using X-ray computed tomography. *Journal of Geophysical Research*, **98**, 1889–1900.

KELLER, A. 1998. High resolution, non-destructive measurement and characterization of fracture apertures. *International Journal of Rock Mechanics and Mining Sciences*, **35**, 1037–1050.

KUMAR, A.T.A., MAJORS, P.D. & ROSSEN, W.R. 1995. Measurement of aperture and multiphase flow in fractures using NMR imaging. *In*: *Proceedings of the SPE Annual Technical Conference and Exhibition, Dallas, Texas*. Paper **SPE-30558**.

PEYTON, R.L., HAEFFNER, B.A., ANDERSON, S.H. & GANTZER, C.J. 1992. Applying X-ray CT to measure macropore diameters in undisturbed soil cores. *Geoderma*, **53**, 329–340.

PYRAK-NOLTE, L.J., MYER, L.R., COOK, N.G.W. & WITHERSPOON, P.A. 1987. Hydraulic and mechanical properties of natural fractures in low permeability rock. *In*: HERGET, G. & VONGPAISAL, S. (eds) *Proceedings of the 6th International Congress of Rock Mechanics*. Balkema, Rotterdam, 225–231.

RENSHAW, C.E., DADAKIS, J.S. & BROWN, S.R. 2000. Measuring fracture apertures: a comparison of methods. *Geophysical Research Letters*, **27**, 289–292.

TIMMERMAN, A., VANDERSTEEN, K., FUCHS, T., VANCLEYNENBREUGEL, J. & FEYEN, J. 1999. A flexible and effective pre-correction algorithm for non medical applications with clinical X-ray CT scanners. *In*: *Proceedings of the International Workshop on Modelling of Transport Processes in Soils at Various Scales in Time and Space, Leuven, Belgium, 24–26 November*, 121–131.

VAN GEET, M. 2001. *Optimisation of microfocus x-ray computer tomography for geological research with special emphasis on coal components (macerals) and fractures (cleats) characterisation*. PhD thesis, K.U. Leuven, Belgium.

VERHELST, F., VERVOORT, A., DE BOSSCHER, PH. & MARCHAL, G. 1995. X-ray computerized tomography: determination of heterogeneities in rock samples. *In*: FUJII, T. (ed.) *Proceedings of the 8th International Congress on Rock Mechanics*. Balkema, Rotterdam, 105–108.

Three-dimensional visualization of fractures in rock test samples, simulating deep level mining excavations, using X-ray computed tomography

E. SELLERS[1], A. VERVOORT[2] & J. VAN CLEYNENBREUGEL[3]

[1] *CSIR Division of Mining Technology, P.O. Box 91230, Auckland Park 2006, South Africa*
[2] *Department of Civil Engineering, K.U. Leuven, Kasteelpark Arenberg 40, B-3001 Heverlee, Belgium (e-mail: andre.vervoort@bwk.kuleuven.ac.be)*
[3] *Department of Electrical Engineering, K.U. Leuven, Kasteelpark Arenberg 10, B-3001 Heverlee, Belgium*

Abstract: A series of experiments on cubic blocks of quartzite were performed to create fractures with three-dimensional characteristics so that they could serve as verification examples for numerical models that are being developed to analyse the fracture processes around mining excavations. The experiments proved to be very successful for creating 3D fracture patterns that have characteristics similar to those observed underground. The shape and position of the fracture surface is determined by the mining geometry and by the interaction with pre-existing discontinuities. However, variations of the fracture planes within the sample could not be determined from the visual study of the block surfaces. The application of a state-of-the-art medical X-ray computed tomography scanner and the development of automatic surface reconstruction software provided a method of producing a full three-dimensional, digital view of fractures within laboratory test samples. Software was developed to provide an interactive graphical method for studying the scans in three orthogonal planes simultaneously. By contouring below a selected density threshold, three-dimensional images of the fracture surfaces were produced. X-ray computed tomography was found to provide a unique means of visualizing the fractures within rock test samples, which can greatly assist the study of rock fracture processes.

Mining and tunnelling at great depths (e.g. more than 1000 m) leads to considerable fracturing of the rock surrounding the excavation, as is the case for the deep level gold mines in South Africa. The presence of large blocks caused by mining-induced fracturing, or the intersection of these fractures with pre-existing geological discontinuities increases the potential for rockfalls, which are hazardous to workers and can interrupt production (Jager & Ryder 1999). Rockbursts may occur when the stresses induced around the mining excavation exceed the strength of the solid rock, or when they are sufficiently high to trigger slip on pre-existing discontinuities such as faults or dykes. The violent nature of the formation processes may expel rock into the stope during a rockburst (Ortlepp 1997; Durrheim *et al.* 1998).

The ability to predict the rock mass response to planned excavation shapes depends upon an understanding of three-dimensional fracture processes around deep level gold mines. One of the challenges with regards to this is the development of numerical models for representing rock failure phenomena so that the future stability of a mine excavation can be accurately predicted. Selection of the correct models and verification of the material parameters can best be done by comparison with specific experiments conducted under controlled conditions. X-ray computed tomography (CT) can be used to obtain a view into the samples and to provide a digital reconstruction of the fracture pattern for comparison with numerical models. Apart from the benefits as an input for the development and verification of numerical models, three-dimensional visualization of the fracture patterns around experimentally modelled excavations provides additional insight into fracture processes.

An outline of mining methodology combined with the layout of the laboratory experiments and preparation of the test samples is presented first, followed by a brief description of the CT scanning procedures and the data analysis software. Analysis of the results of scans for several samples are shown and are compared with the preliminary results from numerical modelling.

From: MEES, F., SWENNEN, R., VAN GEET, M. & JACOBS, P. (eds) 2003. *Applications of X-ray Computed Tomography in the Geosciences*. Geological Society, London, Special Publications, **215**, 69–80. 0305-8719/03/$15.

Longwall mining at great depth

The gold-bearing ore in South Africa is found in the Witwatersrand basin (e.g. Tainton 1994). The Carbon Leader reef is the deepest of the gold bearing strata and is contained in a succession of quartzitic strata that vary in strength and stiffness. Mining of this reef is carried out at depths of up to 3800 m. The shallowest gold-bearing reefs occur in the Ventersdorp Contact reef, a conglomerate rock along the contact between the underlying layered Witwatersrand quartzites and the overlying massive doleritic lavas of the Alberton sequence (e.g. Henning *et al.* 1994).

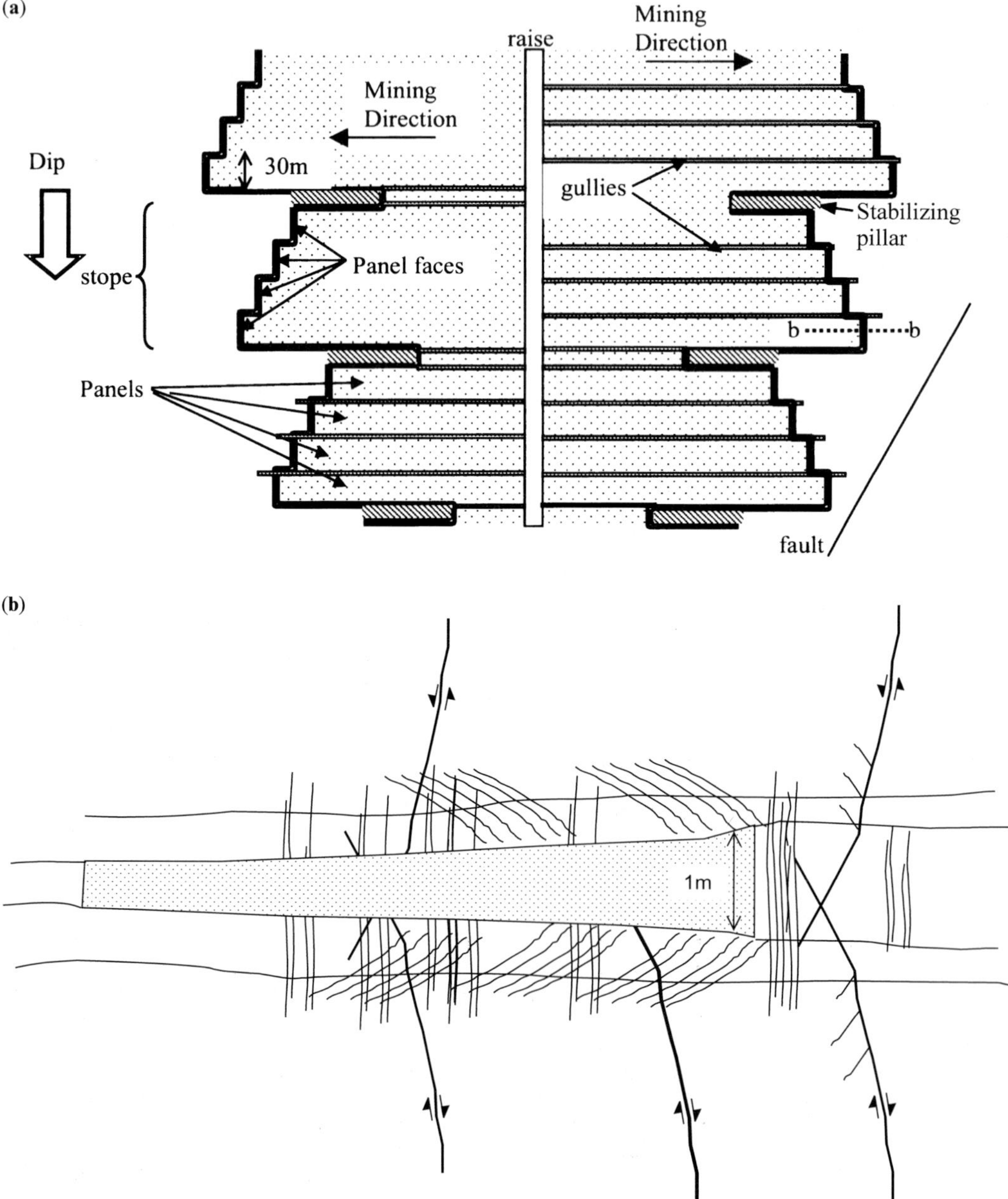

Fig. 1. (**a**) Schematic plan view of a South African longwall gold mining operation at depth (after Jager & Ryder, 1999) (mined-out region is shaded lightly, unmined stabilizing pillars are cross-hatched). (**b**) Schematic section of fracturing around the longwall face.

The ore is generally mined using scattered or longwall mining methods, although new mining sequences are being investigated for ultra deep mining (Johnson & Schweitzer 1996). The gold ore occurs in thin reef layers mostly less than 1 m in height. The large extent of the reefs, in the order of hundreds of square kilometres, implies that large regions of rock are mined. The mining process on a single longwall operation is carried out as shown in Figure 1a. A tunnel, known as a raise, is normally developed updip from an access tunnel. Mining extends along strike, outwards from the raise line in a staggered pattern. Each mining stope consists of several panels with an individual face length of 30 m in the dip direction. Neighbouring panels are advanced with a lag of about 10 m. The height of the mined out zone (stope width) incorporating the reef is usually about 1 m. The gully is an access way that runs alongside the panel on the downdip side and is used to transport the ore from the stope face. The gullies are usually 2 m in height and are extended slightly ahead of the leading face to improve the rock mass conditions. A number of different lead-lag configurations have been developed in order to minimize the damage to any panel due to rockfalls or seismic events (COMRO 1988; Jager & Ryder 1999).

Stabilizing pillars are left between the stopes to transfer load from the hanging wall (roof) to the footwall (floor) and bracket pillars are often left around faults or dykes to prevent slip on any potentially weak interfaces. The mined out areas are either backfilled or a slow convergence takes place until there is contact between the hanging wall and the footwall. This occurs at a distance of about 100 m from the face. Most fractures are induced either around the panel faces or around the pillars (e.g. the stabilizing pillars). Prior to complete closure, the weight of the overburden rock directly above the mined out area has to be carried by the surrounding rock mass, mainly the area ahead of the face and the stabilizing pillars. Due to the large extent of the excavation and the great depth, the stress concentration is very significant and normally exceeds the strength of the rock. This results in the formation of shear and extension fractures around the excavation, as shown in Figure 1b (after Jager & Ryder 1999; Adams *et al.* 1981; Lenhardt 1990). The seismic events associated with the formation of these fractures are known as rockbursts (Legge & Spottiswoode 1987; Lenhardt 1990; Ortlepp 2000; Stewart *et al.* 2001). These can expel large volumes of rock into the stope and pose a significant hazard to the miners. As can be inferred from this simplistic description of the deep tabular mining method and the associated stress redistribution processes, it is imperative to properly understand the three-dimensional nature of fracture formation. Most fractures induced around the face appear to be parallel to the panel face. However, in the neighbourhood of the gullies the fracture planes turn and interconnect. The way in which these fractures interact is not well understood, but it is probably significantly influenced by the lag distance between adjacent panel faces (Adams *et al.* 1981). Reducing the width of the stabilizing pillars leads to an increase in the stress concentration and further fracturing of the pillar or even foundation failure, that extends into the hanging wall or the footwall, or both (Lenhardt 1990). An improved insight into the distribution of damage should assist in the redesign of support and mine layouts to minimize the hazard of rockfalls and rockbursts.

Experimental programme

The aim of the experimental programme was to design tests that would improve the understanding of fracture processes around the mine layouts discussed in the previous section. It was anticipated that the tests would produce fractures with three-dimensional characteristics so that they could serve as verification examples for numerical models that are being developed. The experimental configuration consisted of cubic samples that were tested under poly-axial stress states. Slots cut into the samples simulated tabular mine openings and acted as stress raisers for the initiation of fractures.

Two rock types were selected. The first was Elsburg quartzite, a metamorphosed quartz wacke, which is associated with the footwall of the Ventersdorp contact reef. This rock would be subjected to the pillar loading from mining on that reef plane. Characterization of the rock fracture processes is, therefore, important for mine design and the determination of support strategies. Elsburg quartzite is a light green rock that contains fine- to medium-sized grains (0.24–1.52 mm) set in a light beige matrix, consisting mostly of mica. The quartz content is about 84% and the rock texture is weakly schistose and stylolitic. The unconfined compression strength ranges from 120 MPa to 180 MPa, the Young's modulus varies between 55 GPa and 65 GPa and the Poisson's ratio is 0.2. The second rock type selected was Marble Bar quartzite. This light grey rock forms part of the Timeball Hill series of the Witwatersrand supergroup but does not occur near gold reefs. The rock is a fine grained (0.072–0.56 mm) metamorphosed quartz-arenite with a quartz content of 95%. Small clots of muscovite are coarsely scattered throughout the rock. The rock has an unconfined compression strength of about 350 MPa, a Young's modulus of 85 GPa and a Poisson's ratio equal to 0.15. This material is very similar to the quartz-arenite found in the hanging wall of the Carbon Leader reef. The Marble Bar and Elsburg quartzite can be obtained from surface outcrops, where they are not affected by the micro-fracturing observed in rocks obtained from excavations at depth. The finer grained Marble Bar quartzite provides a good contrast with the coarser grained Elsburg quartzite to investigate the effect of grain size and microstructure. Due to the age of the Witwatersrand strata and the metamorphism to the lower greenschist facies, there is very little fluid present in the rock mass, apart from lenses associated with certain fault structures and the water used for drilling. Thus, the samples are tested in a dry state.

The blocks are cut to a nominal 80 mm side length in two directions and 81 mm in the third direction, creating a nearly cubic sample. A special vice is used to hold the sample. The sides are ground flat and parallel to within 0.1°. Two methods are used to prepare the slots in the samples. The first, and simplest, is to cut partly through the sample using a diamond-tipped blade. The second method is used when the slot is thinner than the available saw blades or when the slot is required to have a more complex shape that, in plan view, is similar to parts of the mining layout shown in Figure 1. The longer side is cut in half over the full section and each of the new surfaces is ground to the same tolerances. One side of the sample is placed in the vice of a milling machine so that the top surface is exactly horizontal. A 6 mm diameter diamond-tipped shaft that rotates at high frequency is used to grind out a shape in the rock surface. For example, to represent two panels in the mining layout shown in Figure 1 the shape ground into the surface consists of two adjacent rectangles (see Fig. 2). The grinding is continued until the slot has the required depth (e.g. 1 mm).

Blocks are loaded in a poly-axial testing cell to simulate the *in-situ* stresses. The sample is placed into the cell as shown in Figure 2. A bi-axial cell applies confining pressures to the sides of the block via steel platens and a MTS stiff testing machine is used to apply the load to the top surface of the block. The block is loaded equally in all three directions until the required confining pressure is reached (usually 20 MPa). The valves of the pistons acting on the sides of the samples are then closed and the axial load is increased to a selected load level, or until the sample has failed.

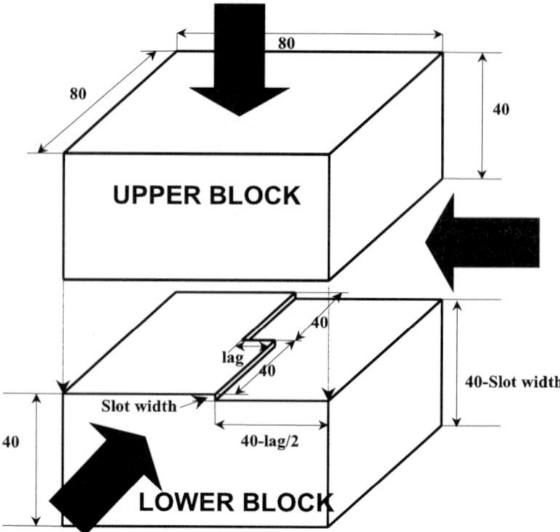

Fig. 2. Cubic sample with mine layout tested in a poly-axial cell (dimensions in mm).

In order to study the effect of the lag-distance on the induced fractures, a series of samples were prepared with lags of 0 mm, 5 mm, 10 mm, 20 mm and 30 mm using Elsburg quartzite blocks. The centre of the block was always placed halfway between the panels in order to keep the loaded area constant for all samples. Because the reefs usually dip at angles of 20–30° in the Witwatersrand basin, two tests were performed with the slot at an angle of 20° to compare with the results from the horizontal slots. In one case, the slot was parallel to the strike direction (equivalent to updip mining) and in the other case parallel to the dip direction. Thus, the samples contain a pre-existing discontinuity that is similar to the underground situation where the reef occurs in the layered Witwatersrand sequence (e.g. Adams *et al.* 1981). A Marble Bar quartzite sample was prepared without cutting through the sample. The sample contained a slot cut halfway through the samples (i.e. a lag of 0 mm).

Visual inspection shows that the samples with different lead-lag layouts produced fracture patterns, seen from the outside of the blocks, that were similar in terms of the overall position and direction of fractures. The fracture traces can be clearly seen on the surface of the sample, as shown in Figure 3a. On the sides of the sample, it can be seen that the fracture pattern is not symmetrical about the slot, due to the slot being ground into one side only. On the top surface of the lower block, shown in Figure 3b, there are no fracture traces cutting across the face. Therefore, the fractures must be initiated at the face position and follow the shape ground into the surface. In this case, another fracture appears to have

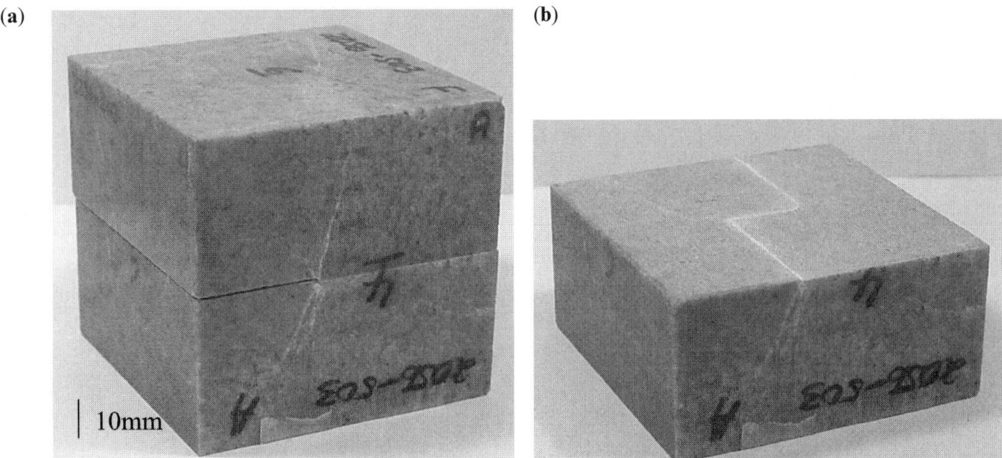

Fig. 3. Fracture traces on the surfaces of the Elsburg quartzite sample for a 20 mm lead-lag layout showing (**a**) the entire sample (upper and lower block) and (**b**) the ground surface (footwall).

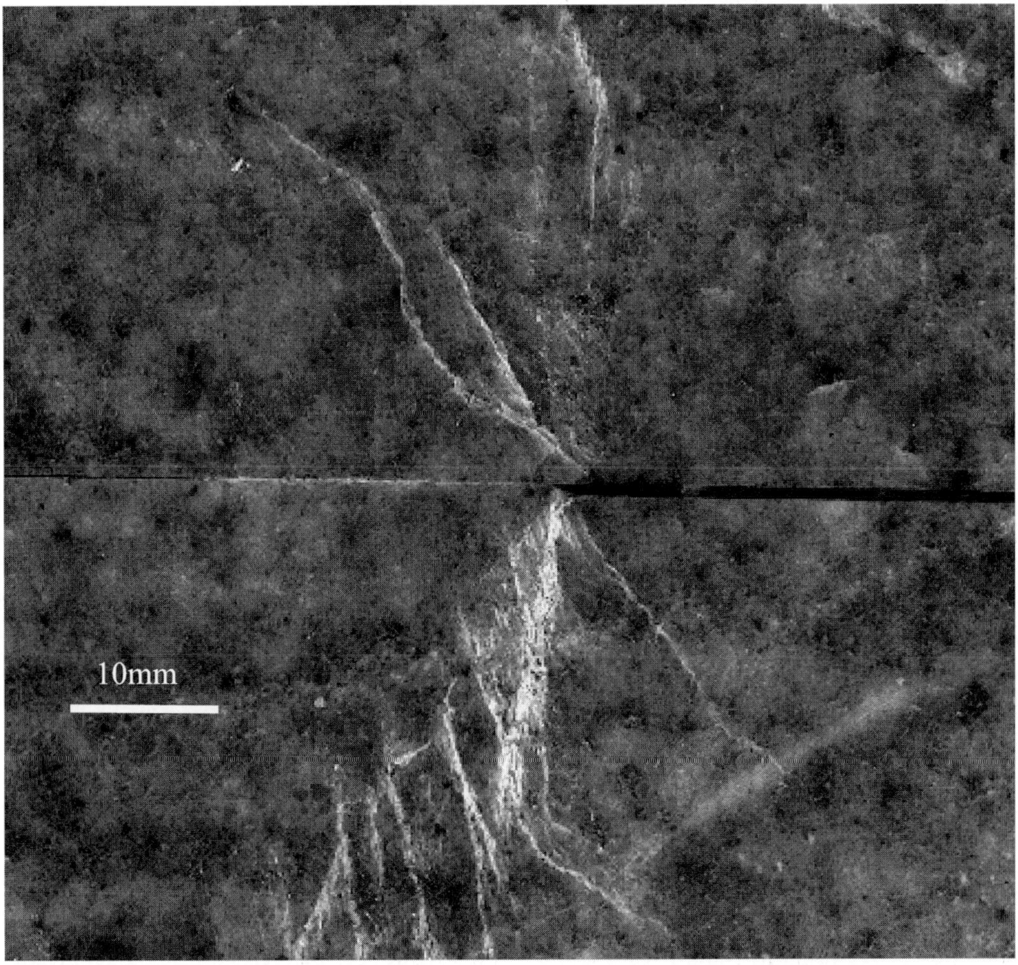

Fig. 4. Fractures observed within a tested Elsburg quartzite sample having a slot width of 0.25 mm. The section is taken along the line separating the two panels.

initiated at the corner of the lagging face and extends across the leading panel. This alignment of fracturing parallel to the excavation face is also observed in petroscope observation of fracturing underground (Adams *et al.* 1981) and can be inferred from seismic observations (Legge & Spottiswoode 1987). The actual shape of the face and the relative position of bedding planes or stratigraphic layers will determine the symmetry of fracturing above and below the stope (Tomlin *et al.* 1997). However, the information from the fracture traces on the outside of the block is insufficient to describe in detail the fracture paths within the sample.

To study three-dimensional paths of the fractures within the sample, sections can be cut out of the samples. This is not only a time-consuming process, but the amount of sections that can be made is limited, additional damage can be introduced by the cutting and the view of the fracture pattern remains two-dimensional. A section through one of the blocks, with a 20 mm lead-lag layout, is shown in Figure 4. The section is taken along the line separating the two panels. The fracture consists of a shear zone that appears to be very similar to large scale tectonic fault structures and also to the mining-induced fault zones associated with rockbursts (Ortlepp 1997). The fracture consists of a number of en echelon segments orientated approximately parallel to the largest compressive principal stress. Sheared zones that appear to contain a fault gouge connect the en echelon segments. The thickness of the sheared zone depends on the stress level and varies between samples. Samples with thinner slots sustain higher pressures and exhibit more shear fracturing. When the slots are thin enough to close, as was the case for the sample in Figure 4 the fractures occur into the solid material in front of the slot. Wider slots do not close and induce fractures that curve above and below the opening, as shown in Figure 3.

To overcome the practical problems related to cutting the samples in sections, but to still accurately describe the fracture paths in three dimensions, the tested and fractured samples were scanned using medical X-ray computed tomography.

X-ray tomography procedures

Brief description of the technique

X-ray computed tomography uses the attenuation of an X-ray beam to reconstruct a section through a specimen. The details of the theory of X-ray CT scanning and applications to materials that are relevant to this study are presented elsewhere (e.g. Verhelst *et al.* 1995; Keller 1998) and are not repeated here. CT analysis can provide a pattern of density variations in a slice. Taking a number of slices across the sample leads to a three-dimensional grid of volume elements (voxels), each having a value related to the density in that volume. This information can be obtained in digital form and reconstructed for visualization. A multi-slice medical CT scanner (Siemens VolumeZoom) was used for this study. The scan parameters were selected to have a radiation dose determined by 140 KeV and 180 mAs, a slice width of 0.5 mm, a collimation of 0.5 mm and a rotation time of 1 s. A standard medical reconstruction algorithm designed for the human inner ear was used. The reconstructed slice width was 0.5 mm with an increment of 0.2 mm. Two Siemens reconstruction kernel settings were used – the ultra sharp u90u and the medium smooth u30u.

Viewing of three perpendicular sections

Visualization software was written based on standard visualization techniques (Schroeder *et al.* 1997). Three orthogonal sections of the scanned grid can be viewed simultaneously to enable the researcher to follow fracture paths within the sample. The density is represented in various shades of grey within a window imposed by the users. Contour values are manipulated by setting the central value and range of the CT numbers. Dark values represent low density regions and hence correspond to fracture paths. The light values represent high density regions. The CT scans clearly show the fractures within the sample (Figs 5 and 6). The views in Figure 5 indicate that the fracture is not symmetrical about the slot and they compare well with the external fracture traces discussed earlier. As shown in Figure 5a, the ultra sharp u90u kernel provided a high contrast, high noise image that enabled the main fractures to be clearly defined. The u30u scan appears to be less sharp, but the larger variation in grey scales increases the number of finer features that can be resolved. The direction of scanning also influenced the quality of the output. The axis of rotation of the scanner should be parallel to the strike of the fracture plane for best results.

Using the visualization software to step through the sample at 0.2 mm intervals permits an excellent understanding to be made of the fracture paths. The plan views of the fractures for a lag length of 10 mm are shown in Figure 6 for three distances above the slot, confirming the way

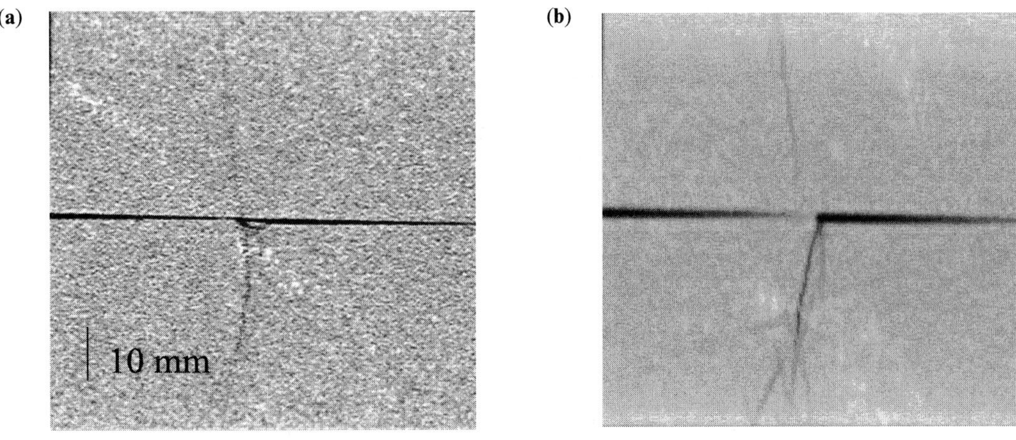

Fig. 5. Comparison of reconstruction algorithms for CT scans for the Elsburg quartzite rock sample with 10 mm lag, showing a fracture trace (dark): (**a**) ultra sharp, (**b**) medium smooth.

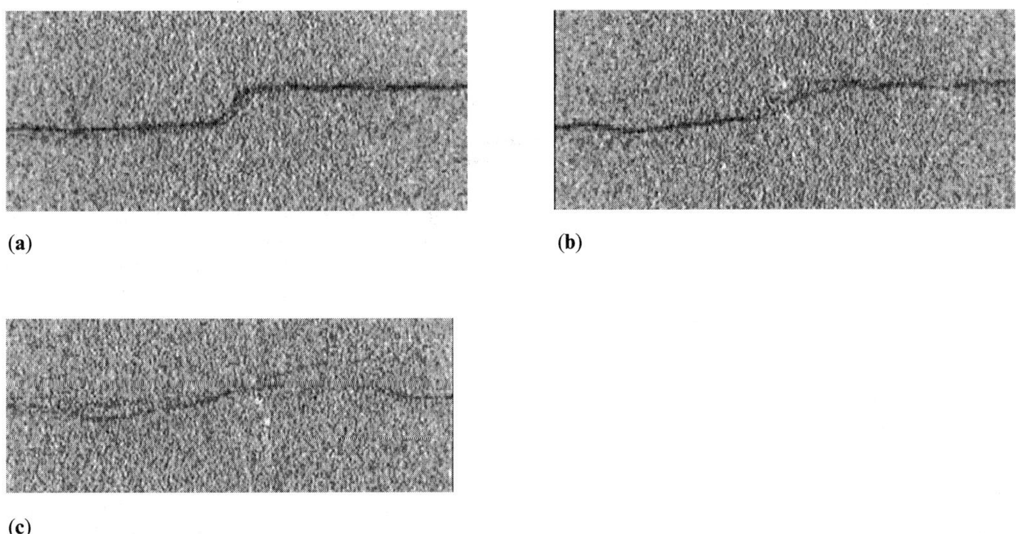

Fig. 6. CT scans of rock sample showing how the fracture traces (dark lines) follow the general shape of the layout at vertical distances of (**a**) 0.5 mm, (**b**) 10 mm, and (**c**) 30 mm above the slot with a lag of 10 mm.

that the fracture surface transforms inside the sample. The ability to exactly determine the position and shape of the fracture within the sample allows qualitative comparisons with the fracture surfaces predicted using three-dimensional numerical models.

Three-dimensional views

Three-dimensional views of the fracture planes can be reconstructed using image processing and visualization techniques (Schroeder *et al.* 1997). The VTK visualization package (Schroeder *et al.* 1997) is an open source software library that can be accessed using scripting languages to provide interactive visualization modules. The data file containing the image representing the voxel grid of a scanned sample was exported from the CT scanner and imported into the visualization software. This process requires a large amount of memory and disk storage space, as there are 1.25×10^8 voxels in the scan of a single 80 mm cubic sample. Approximately 250 Mb of storage

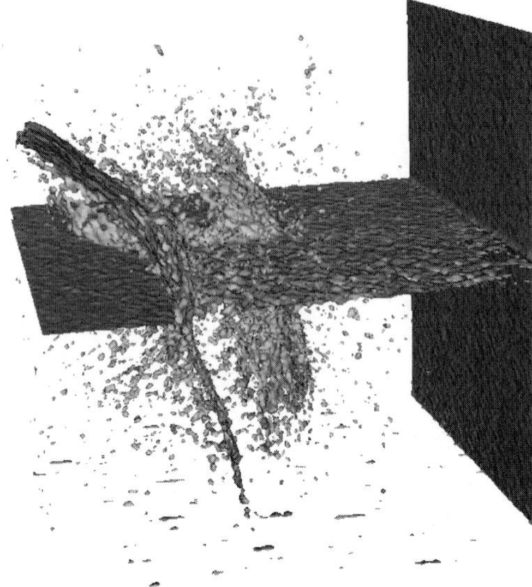

Fig. 7. Three-dimensional reconstruction of the fracture pattern in an Elsburg quartzite block with a lead-lag of 20 mm.

and 1.1 Gb of memory are required. The resolution of the scan is, therefore, selected based on the amount of data that can be manipulated with the available computer resources. A 'marching cubes' algorithm was used to obtain an isosurface contour of densities below a selected threshold value of CT number from the voxel grid.

The three-dimensional fracture pictures in Figures 7 to 10 demonstrate the insight that can be obtained when viewing a digital reconstruction of the entire fracture pattern. Figure 7 shows the fracture pattern observed in the sample with a lag of 20 mm. The figure shows that the fracture surface follows the edge of the slot in the centre of the sample. Further into the hanging wall and footwall, the surface becomes s-shaped. Similarly, for the sample with a 10 mm lag (Fig. 8), the sharp edge of the slot initiates the fracture, which gradually transforms into a curved surface. Figure 8 also illustrates how the fracture curves below the leading panel, especially in the region adjacent to the other panel. In the underground situation, this would represent a great hazard. Due to the relatively low dip of the fracture at this point, the process of fracture formation and any subsequent slip displacement in this region would transmit seismic waves directly upwards at the edge of the panel and the gully. As the gully is a main access way and a region with a high probability that workers are present, there is considerable risk of injury or loss of life.

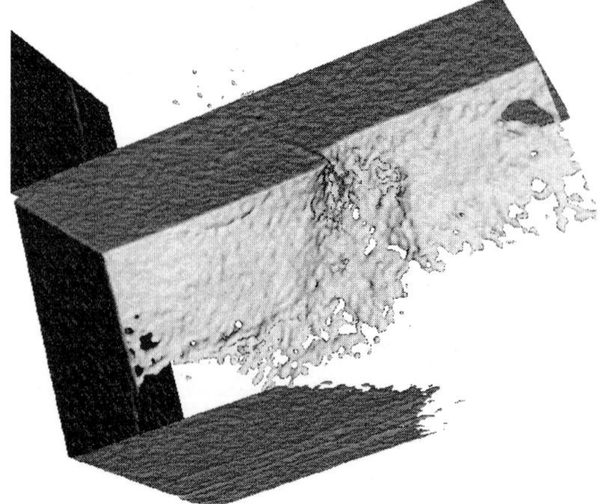

Fig. 8. Three-dimensional reconstruction of the fracture pattern in an Elsburg quartzite block with a lead-lag of 10 mm.

Figure 9 shows the two-dimensional nature of the fractures when there is no lag between the panels. The fractures are reconstructed especially well in the Marble Bar quartzite. The slot does not close in this very stiff rock and the fracture surfaces curve above and below the excavation. Vertical tensile fractures form behind the curved fracture surface. These are specific to the experimental set-up. The hanging wall and

Fig 9. Three-dimensional reconstruction of the fracture pattern in a Marble Bar quartzite block with no lag.

footwall form cantilevers that bend towards the open slot, inducing tensile stresses at the top and bottom surfaces of the block and driving the fractures towards the centre of the block. The automatic algorithm identified only one of the two observed tensile fractures. This may be due to the position of the second fracture close to the surface, where the effect of beam hardening is strongest. Another possible explanation is that the fracture has not opened sufficiently and does not cause sufficient contrast in density for it to be resolved by the automatic algorithm.

Beam hardening, presenting the sample as being denser at the edges, poses a problem in the automatic fracture reconstruction procedure. Contouring fractures, as CT numbers below a single threshold value, is only possible for a region in the centre of the sample. Increasing the threshold value selects large portions of the regions in the centre and does not isolate the fractures near the edge. The isosurface contouring algorithm presents these as a number of small volumes distributed in space, as can be seen in Figure 7. As the threshold density is increased, the number and volume of these spurious artefacts increases and they tend to fill the sample volume. The CT number artefact decreases rapidly from the edges and so fractures within the central 60% of the sample can be automatically identified.

An alternative method of determining the fracture traces was developed. This method involves a manual delineation of the fractures. The section viewing software was modified to allow the user to step through the voxel grid at regular intervals and to trace out the fractures that were observed. Sections 1 mm apart were used for the initial study of the method. On each section, the fracture is digitized by manually selecting points that lie on the dark line representing the fracture trace. The points for all the sections are then saved to file. The interactive visualization software was modified to reconstruct the set of points as a single surface. The method is time-consuming and fractures that are sub-parallel to the viewed section may appear as thick lines. Thus, if the fracture is not almost perpendicular to a viewing plane, the digitization of the fracture is not unique in that section. Fractures must be traced in whichever of the three orthogonal sections is closest to being perpendicular to the fracture plane. Fracture traces may then be added together to obtain a complete picture of the fracture pattern. The interactive viewer was modified to be able to plot both the automatic reconstruction and the planes obtained from manual delineation. An example is shown in Figure 10. The automatic reconstruction is the same as used in Figure 9 but the missing fracture plane has been added using the manual technique. The detail of the fracture surface is greater in the automatic isosurface, as the number of sections that are used in the manual delineation process is limited. Sections were taken 1 mm apart, whereas the original scans were reconstructed with a grid size of 0.2 mm. The triangulation procedure used to interpolate between sections obscures some of the finer variations in the surface. Combining both methods provides additional possibilities for representing the entire fracture plane, or for adding features such as other heterogeneities that can be observed in the scan, but are not included in the isosurface contour.

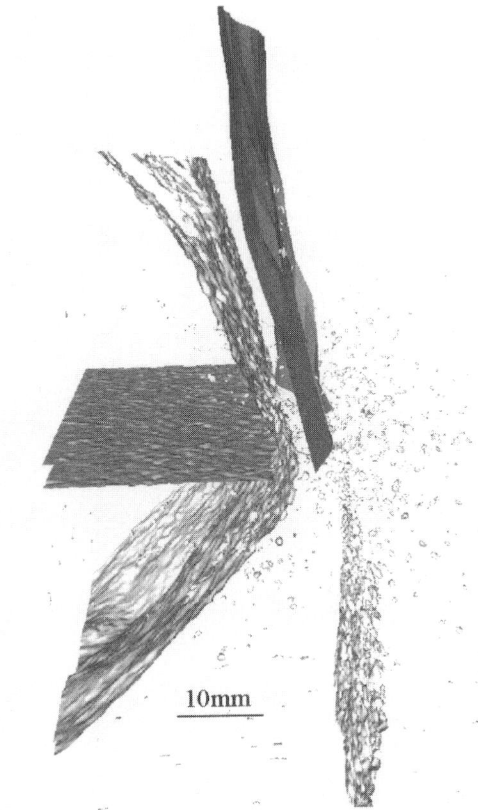

Fig 10. Combination of automatic and manual delineation methods for reconstructing three-dimensional fracture patterns (Marble Bar quartzite block with no lag; see Fig. 9).

Comparisons with numerical modelling

Large-scale mine modelling is carried out regularly for mines, using elastic boundary element

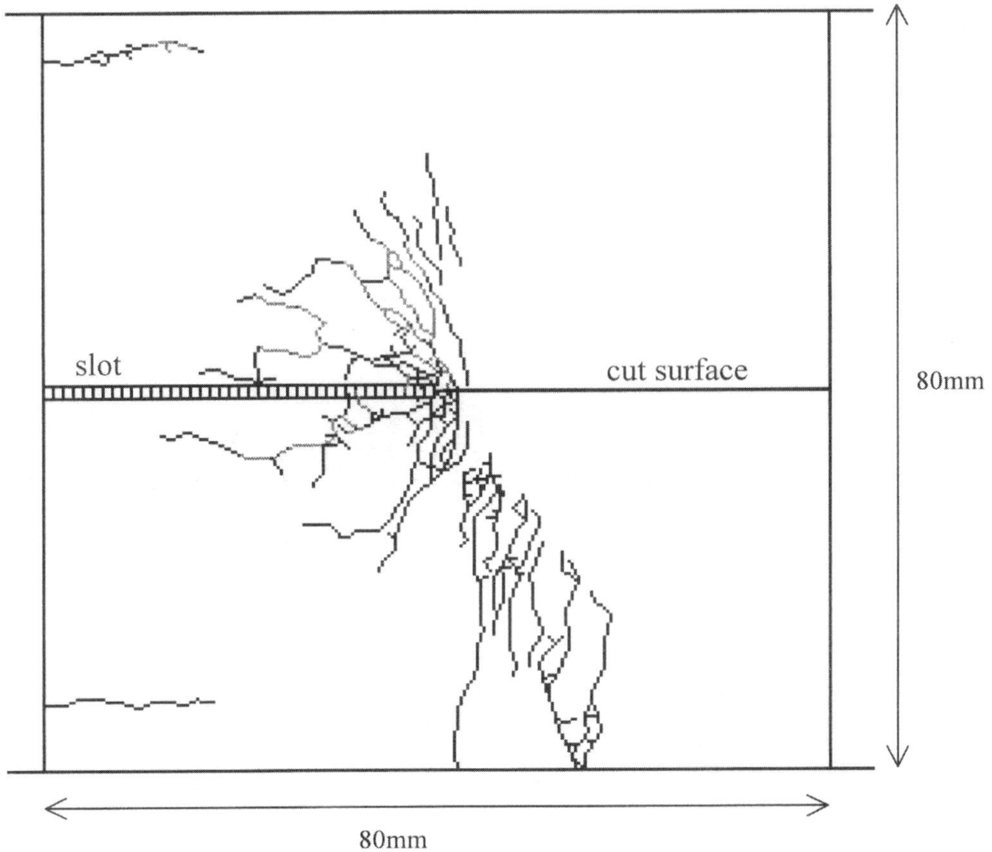

Fig 11. Example of numerical prediction of the fracture pattern in the experimental samples using the two-dimensional boundary element program DIGS.

methods (Napier & Stephansen 1987). These models can identify areas of high stress concentration, but they cannot describe the rock failure processes and subsequent stress re-distributions. Considerable effort has been made to develop numerical models and constitutive equations that can represent the non-linear fracture processes. These have currently only been applied to laboratory test situations (e.g. Sellers & Napier 1997) and some two-dimensional models of the fracture processes around deep level mining excavations (Sellers *et al.* 1998; Van de Steen 2001; Van de Steen *et al.* in press; Klerck 2000).

As an example for comparison with the observed fracture patterns, the boundary element program DIGS (Napier 1990) has been applied to solve a two-dimensional problem. A square sample with a side length of 80 mm is constructed. The slot is represented by a line of boundary elements through the centre of the block, as shown in Figure 11. These elements have zero normal or shear stresses acting on the surfaces and closure is limited to the specified width of the slot. The cut surface forming the contact between the upper and lower blocks is represented with boundary elements having a friction angle of 30°. A tessellation approach (e.g. Napier 1990; Sellers & Napier 1997) is applied to represent fracture growth. A Delaunay triangulation of the sample was created with an average triangle side length of 1 mm. The side of each triangle is assumed to be a potential fracture path and is assigned a failure criterion based on the Mohr-Coulomb model. At each loading increment, all potential fracture sites are searched. The site with the highest excess shear stress is activated and can then transmit its influence to the other elements. The inclusion of an active element causes stress redistribution and other sites may activate. Once all sites have been activated for a given load, the load is incremented and the process is repeated. The final fracture pattern is shown in Figure 11. The pattern is representative of the fracture induced in the Elsburg quartzite shown in Figure 4.

The modelling of fracturing in three dimensions is still at an early stage (Napier 1998; Potyondy & Cundall 1998; Sellers 2001). Preliminary studies (e.g. Sellers 2001) have shown that the three-dimensional variations in the fracture surface may be modelled using the 3DIGS boundary element program, the Particle Flaw Code and the ELFEN finite/discrete element technique.

Conclusions

A set of experiments was performed to create fractures with three-dimensional characteristics so that they could serve as verification examples for numerical models that are being developed to analyse fracture processes around mine excavations. The aim of the experimental program was to design tests that would be relevant for learning more about fracturing around the mine layouts to be studied. The experimental configuration consisted of cubic samples that were tested under poly-axial stress states. Slots cut into the samples simulate the tabular mine openings of deep gold mines in South Africa. By creating configurations that have some similarity with mining layouts, the study of the experimental results can also provide an additional understanding of mining induced rock fracture processes.

The experiments proved to be very successful for creating 3D fracture patterns. However, when only a visual inspection can be made significant variations of the fracture planes within the sample can not be determined based on observations of the block surfaces. The fracture traces on the sample provide an indication of the paths of the fracture planes, but they cannot provide information about the shape and extent of fracturing within the sample. The conventional method for obtaining three-dimensional views is to section the sample at regular intervals. However, this method is destructive, time-consuming and still does not give a complete picture of the fracturing. Medical computed X-ray tomography was successfully applied to reconstruct a digital representation of the three-dimensional fracture. The scanned image was imported into specially developed software that provides an interactive graphical method for studying the scan in three orthogonal planes simultaneously. Further software developments took place to contour a particular density threshold and so to produce a three-dimensional image of the fracture surface. Some artefacts in the reconstruction of the image lead to the limitation that only fractures within the central (60%) region of the block can be automatically processed. A manual digitizing method was developed to add other fractures to the automatically generated planes.

The reconstructed image can be visually compared with fracture patterns predicted by numerical models or acoustic emission event locations. Analysis of the images has suggested that the lag between the faces results in considerable distortions in the fracture plane between two panels. This information is valuable for focusing underground studies that can be used to confirm the effect of mining geometry on fracture formation. Three-dimensional numerical models for rock fracture processes are still under development, but examples are shown that can reproduce the observed fracture patterns. Whilst this study has concentrated on South African gold mining conditions, it shows that X-ray CT provides a unique means of visualizing fracture patterns within rock samples tested in a context of mining simulations. The methodology presented provides new avenues for the study of rock fracture processes in this field.

Funding for this project was provided by the Safety in Mines Research Advisory Committee, as part of project GAP 601, and is gratefully acknowledged. The authors thank G. Marchal and his team for the assistance during the scanning of the rock samples at the Gasthuisberg University Hospital in Leuven, Belgium. G. Schreurs is thanked for helpful comments on the manuscript.

References

ADAMS, G.R., JAGER, A.J. & ROERING, C. 1981. Investigations of rock fracture around deep-level gold mine stopes. *In*: *Proceedings of the 2nd, US Symposium on Rock Mechanics*. Massachusetts Institute of Technology, 213–218.

COMRO 1988. *Industry Guide to Methods of Amelioration of Rockfalls and Rockbursts*. Research Organisation of the Chamber of Mines of South Africa, Johannesburg.

DURRHEIM, R., HAILE, A., ROBERTS, M.C.K., SCHWEITZER, J.K., SPOTTISWOODE, S.M. & KLOKOW, J.W. 1998. Violent failure of a remnant in a deep South African gold mine. *Tectonophysics*, **289**, 105–116.

HENNING, L.T., ELS, B.G. & MEYER, J.J. 1994. The Ventersdorp contact placer at Western Deep Levels Gold Mine – an ancient terraced fluvial system. *South African Journal of Geology*, **97**, 308–318.

JAGER, A.J. & RYDER, J. 1999. *A Handbook on Rock Engineering Practice for Tabular Hard Rock Mines*. The Safety in Mines Research Advisory Committee (SIMRAC), Johannesburg.

JOHNSON, R.A. & SCHWEITZER, J.K. 1996. Mining at depth: evaluation of alternatives. *In*: *Proceedings of the 2nd North American Rock Mechanics Symposium*. Balkema, Rotterdam, 359–366.

KELLER, A. 1998. High Resolution, non-destructive measurement and characterization of fracture apertures. *International Journal of Rock Mechanics and Mining Sciences*, **35**, 1037–1050.

KLERCK, P.A. 2000. *The finite element modelling of discrete fracture in quasi-brittle materials*. PhD Thesis, University of Wales, Swansea.

LENHARDT, W.A. 1990. Seismic events associated with large mining induced fractures. *In*: ROSSMANITH, H.P. (ed.) *Proceedings of the 2nd Conference on Mechanics of Jointed and Faulted Rock, MJFR-2*. Balkema, Rotterdam, 727–731.

LEGGE, N.B. & SPOTTISWOODE, S.M. 1987. Fracturing and microseismicity ahead of a deep gold mine stope in the pre-remnant and remnant stages of mining. *In*: *Proceedings of the 6th ISRM Congress*. Balkema, Rotterdam, 1071–1077.

NAPIER, J.A.L. 1990. Modelling of fracturing near deep level mine excavations using a displacement discontinuity approach. *In*: ROSSMANITH, H.P. (ed.) *Proceedings of the 2nd Conference on Mechanics of Jointed and Faulted Rock, MJFR-2*. Balkema, Rotterdam, 709–715.

NAPIER, J.A.L 1998. Three dimensional modelling of seismicity in deep level mines. *In*: ROSSMANITH, H.P. (ed.) *Proceedings of the 3rd Conference on Mechanics of Jointed and Faulted Rock, MJFR-3*. Balkema, Rotterdam, 285–290.

NAPIER, J.A.L. & STEPHANSEN, S.J. 1987. Analysis of deep level mine design problems using the MINSIM-D boundary element program. *In*: *APCOM '87, Proceedings of the 20th International Symposium*. SAIMM, Johannesburg, 3–19.

ORTLEPP, W.D. 1997. *Rock Fracture and Rockbursts*. South African Institute of Mining and Metallurgy, Johannesburg.

ORTLEPP, W.D. 2000. Observation of mining-induced faults in an intact rock mass at depth. *International Journal of Rock Mechanics and Mining Sciences*, **37**, 423–436.

POTYONDY, D.O. & CUNDALL, P.A. 1998. Modeling notch-formation mechanisms in the URL Mine-by test tunnel using bonded assemblies of circular particles. *In*: *Proceedings of the 3rd North American Rock Mechanics Symposium, NARMS '98*. International Journal of Rock Mechanics and Mining Sciences, 35(4–5), Paper No. **067**.

SCHROEDER, W., MARTIN, K. & LORENSEN, B. 1997. *The Visualization Toolkit. An Object-Oriented Approach to 3D Graphics*. Prentice Hall, Upper Saddle River, New Jersey.

SELLERS, E. & NAPIER, J. 1997. A comparative investigation of micro-flaw models for the simulation of brittle fracture in rock. *Computational Mechanics*, **20**, 164–669.

SELLERS, E., BERLENBACH, J. & SCHWEITZER, J. 1998. Fracturing around deep level stopes: comparison of numerical simulations with underground observations. *In*: ROSSMANITH, H.P. (ed.) *Proceedings of the 3rd Conference on Mechanics of Jointed and Faulted Rock, MJFR-3*. Balkema, Rotterdam, 425–430.

SELLERS, E.J. 2001. Modelling of continuum to discontinuum transitions for deep level mining. *In*: BICANIC, N. (ed.) *Proceedings of the 4th International Conference on the Analysis of the Deformation of Discontinuous Media*. University of Glasgow, 63–72.

STEWART, R.A., REIMOLD, W.U., CHARLESWORTH, E.G. & ORTLEPP, W.D. 2001. The nature of a deformation zone and fault rock related to a recent rockburst at Western Deep Levels Gold Mine, Witwatersrand Basin, South Africa. *Tectonophysics*, **337**, 173–190.

TAINTON, S. 1994. A review of the Witwatersrand basin and trends in exploration. *In*: *Proceedings of 15th CMMI Congress*. SAIMM, Johannesburg 19–45.

TOMLIN, W.B., SELLERS, E.J. & OZBAY, U. 1997. Modelling of the influence of interfaces on the fracture pattern during sequential mining. *In*: *Proceedings of the 1st International Conference on Damage and Failure of Interfaces*. Balkema, Rotterdam, 451–458.

VAN DE STEEN, B. 2001. *Effect of heterogeneities and defects on the fracture pattern in brittle rock*. PhD thesis, Katholieke Universiteit Leuven, Belgium.

VAN DE STEEN, B., VERVOORT, A., NAPIER, J.A.L. & DURRHEIM, R.J. in press. Implementation of a flaw model to the fracturing around a vertical shaft. *Rock Mechanics and Rock Engineering*.

VERHELST, F., VERVOORT, A., DE BOSSCHER, PH. & MARCHAL, G. 1995. X-ray computerized tomography: Determination of heterogeneities in rock samples. *In*: FUJII, T. (ed.) *Proceedings of the 8th ISRM Congress*. Balkema, Rotterdam, 105–108.

Geostatistics and the representative elementary volume of gamma ray tomography attenuation in rock cores

J. R. VOGEL[1] & G. O. BROWN[2]

[1] *US Geological Survey, Water Resources Division, 100 Centennial Mall North, Federal Building Room 406, Lincoln, NE 68508 USA*
Biosystems and Agricultural Engineering, Oklahoma State University, Stillwater, OK 74078, USA (e-mail: jrvogel@usgs.gov)

Abstract: Semivariograms of samples of Culebra Dolomite have been determined at two different resolutions for gamma ray computed tomography images. By fitting models to semivariograms, small-scale and large-scale correlation lengths are determined for four samples. Different semivariogram parameters were found for adjacent cores at both resolutions. Relative elementary volume (REV) concepts are related to the stationarity of the sample. A scale disparity factor is defined and is used to determine sample size required for ergodic stationarity with a specified correlation length. This allows for comparison of geostatistical measures and representative elementary volumes. The modifiable areal unit problem is also addressed and used to determine resolution effects on correlation lengths. By changing resolution, a range of correlation lengths can be determined for the same sample. Comparison of voxel volume to the best-fit model correlation length of a single sample at different resolutions reveals a linear scaling effect. Using this relationship, the range of the point value semivariogram is determined. This is the range approached as the voxel size goes to zero. Finally, these results are compared to the regularization theory of point variables for borehole cores and are found to be a better fit for predicting the volume-averaged range.

Computed tomography (CT) is being increasingly used in porous media research as a non-destructive tool for measurement of interior properties of geological cores. Petrovic *et al.* (1982), Crestana *et al.* (1985), Tollner & Verma (1989), Phogat *et al.* (1991), Brown *et al.* (1993), Hsieh *et al.* (1998*b*) and others have met with increasing success in producing images of the spatial distribution of solids, fluids and voids within porous media columns. These spatial characteristics are directly related to many porous media phenomena. As a result, CT images have been used to analyse permeability, pore space distribution, connectivity and relative elementary volume (Coskun & Wardlaw 1993; Grevers & de Jong 1994; Ioannidis *et al.* 1999; Brown *et al.* 2000).

CT images characterize porous media content at a scale smaller than is possible with most sampling methods. These data provide a unique opportunity to use geostatistics, specifically the semivariogram, to determine the spatial structure and anisotropy of porous media at these smaller scales. Schafmeister-Spierling & Pekdeger (1989) have shown that dispersion can be modelled more accurately when small-scale spatial variability is considered. Flühler *et al.* (2001) also show that mixing processes at the millimetre scale exerts a pronounced effect on flow and transport behaviour of water and solutes. For porous media in which macropore dispersion is a significant pathway of contaminant transport, such as the Culebra Dolomite, this small-scale geostatistical characterization of the rock may yield improved aquifer characterization.

The large amount of spatial data provided by CT images also allows for investigation of the effect of resolution and scale on semivariogram parameters. Gelhar (1986) states that 'spatial correlation scales should not be viewed in any absolute sense, but rather will depend on the problem at hand'. Openshaw & Taylor (1979) also show that aggregating fixed observations on small areas in various ways can produce a large range of correlation coefficients.

CT images of gamma ray attenuation are used here to characterize the semivariogram properties and anisotropy of Culebra Dolomite bulk density structure at core scales. With this information, the semivariogram parameters of adjacent cores and cores at two different scales are compared. Using relative elementary volume (REV) concepts, the disparity scale required for stationarity of spatial data is quantified. The

From: MEES, F., SWENNEN, R., VAN GEET, M. & JACOBS, P. (eds) 2003. *Applications of X-ray Computed Tomography in the Geosciences*. Geological Society, London, Special Publications, **215**, 81–93. 0305-8719/03/$15.

modifiable areal unit problem is addressed and the scaling factor associated with increasing scanning resolution is determined. Finally, the results are compared to the theory involving the regularization of point variables.

Porous media descriptions

Semivariograms

The semivariogram measures the average degree of dissimilarity between an unsampled value and a nearby data value (Deutsch & Journel 1998). The traditional semivariogram is defined as half of the average squared difference between two data points approximately separated by a vector h:

$$\gamma(h) = \frac{1}{2N(h)} \sum_{i=1}^{N(h)} (x_i - y_i)^2 \tag{1}$$

where $\gamma(h)$ is half of the average squared difference; $N(h)$ is the number of pairs; x_i is the data value at the start, or tail; and y_i is the data value at the end, or head. Definitions of basic semivariogram characteristics can be found in Isaaks & Srivastava (1989).

A semivariogram can be fit by analytical relationships to ensure that its model is positive definite. Researchers in stochastic hydrology often use the exponential model,

$$\gamma(h) = 1 - e^{(-(3h/a))} \tag{2}$$

where a is the range and h is the lag distance (Woodbury & Sudicky 1991). This model reaches its sill asymptotically, with the practical range a defined as that distance at which the variogram value is 95% of the sill. Another common semivariogram model is the spherical model:

$$\gamma(h) = \begin{cases} 1.5\dfrac{h}{a} - 0.5\left(\dfrac{h}{a}\right)^3 & \text{if } h < a, \\ 1 & \text{otherwise} \end{cases} \tag{3}$$

The spherical model has a linear behaviour near the origin, flattens out at larger distances, and reaches the sill at a (Isaaks & Srivastava 1989).

Multiple, or nested, sills in the semivariogram may be associated with physical phenomena occurring at different scales. Detection of ranges nested within larger ranges requires sample spacing shorter than the minimum detected range. In general, nested sills are only considered when they can be associated with physical phenomena (Solie *et al.* 1999). Nested sills can be modelled by using a linear combination of positive definite semivariogram models with positive coefficients. This property results in the family of positive definite models:

$$\gamma(h) = \sum_{i=1}^{n} |w_i| \gamma_i(h) \tag{4}$$

which are positive definite as long as the n individual models are all positive definite. A weighting function, w_i, is defined for each individual model subject to $\sum_{i=1}^{n} w_i = 1$ (Isaaks & Srivastava 1989).

Interpretation of 3D semivariograms of geological data will yield information on subsurface anisotropy. Geometric anisotropy exists when the directional semivariograms show the same shape and sill, but different range values. An anisotropy factor may be used to characterize the geometric anisotropy and is defined as the ratio of the ranges in the major and minor direction. Another type of anisotropy is zonal anisotropy, where both the sill and range change with direction. A lower horizontal sill represents anisotropy due to geological strata. The distance between the horizontal (lower) and vertical (upper) sills represents the variability between different geological strata and may be related to laterally continuous sedimentary features. When the vertical semivariogram reaches a lower sill than the horizontal semivariogram, zonal anisotropy is present and is due to areal trends. Geometric and both types of zonal anisotropy typically exist simultaneously in the field and complicate analysis (Goggin *et al.* 1988; Kupfersburger & Deutsch 1999).

Recent studies have also shown that the size of the sampled area (or volume) causes a scaling effect on semivariogram correlation lengths. This occurs when the sampled area might is too small to capture a long-range structure (Kupfersburger & Deutsch 1999). Gelhar (1986) and Schafmeister & Pekdeger (1993) show that correlation length increases with increasing scale. Thus, it is not an absolute property of the subsurface, but rather related to the size of the investigated area. Larger scales generally lead to an increased anisotropy ratio. Comparison of semivariograms should, therefore, be done with caution if similarly sized sampling and investigated areas are not used.

Geostatistics and the REV

Conditions required for geostatistical interpretation can be related to the REV. Brown *et al.* (2000) used REV concepts to evaluate CT images. The REV is defined as the volume range, $V_{min} < V_0 <_{max}$,

$$\left.\frac{\partial Z(x, V_i)}{\partial V}\right|_{V_i=V_0} = 0, \quad (5)$$

where $Z(x, V_i)$ is the value of a hydrologic property measured on a volume V_i with a centroid at x, with a given domain $\Re$. Bear & Bachmat (1990, pp. 16–28) employ the notion that a volume size, V_i, that falls within the REV region can be treated as an ergodic stationary random function in $\Re$. They further define a stationary random function as ergodic if any statistical characteristic (spatial moment) of the function, taken over a sufficiently large volume in a single realisation, is an unbiased and consistent estimate of the same characteristic over the entire set of possible realisations in the field. An estimate of a population parameter is unbiased if its expected value is equal to the value of the parameter. An estimate of a population parameter is said to be consistent if it approaches, probabilistically, the value of the parameter as the sample size increases. Bear & Bachmat (1990, pp. 16–28) also describe a practical test for the REV as being the volume range where the first two moments of a property are an unbiased estimate of the property. This corresponds to a second-order stationary function in $\Re$. By definition, a function is second-order stationary in $\Re$ if

$$E[Z(x, V_i)] = \mathbf{constant} \quad (6)$$

and

$$\mathrm{Cov}[Z(x+h, V_i), Z(x, V_i)] = f(h) \rightarrow 0 \quad \mathbf{as} \quad h \rightarrow 0 \quad (7)$$

where $E[Z(x, V)_i)]$ is the expected value of Z and $\mathrm{Cov}[Z(x+h), Z(x)]$ is the covariance of two points and is a function of the separation distance, h, only, for all $x \in \Re$. Then,

$$\mathrm{Var}(Z(x)) = f(0) = \mathbf{constant}. \quad (8)$$

Cressie (1991) states that when assuming ergodicity in the geostatistical setting, one can make the weaker ergodic assumptions of Eqs 6 and 7. This recommendation is limited by the assumption that the sample $Z(x,V_i)$ consists of a finite number of observations from a continuous process $Z(\cdot)$. The REV in porous media, therefore, corresponds to these weaker ergodic assumptions.

In a strict sense, Eq. 6 may be met by a linear function in space. However, as discussed by Brown *et al.* (2000), the limited scanning domains possible in CT preclude random, stationary volume averaging procedures. Thus, in the practical sense, any V_0 within the REV range will correspond to the volume required for ergodicity. This makes it possible to make statistical inferences from the semivariogram.

Scale disparity

Gelhar (1986) states that 'to be meaningful and estimable, the correlation scale must be small compared with the scale of the problem. Essentially, a disparity in scale is required'. To quantify this scale disparity, Λ, we define the standard element volume (SEV) as the volume represented by the product of the correlation length in the x, y and z directions. By considering a REV_{min} with the same proportional dimensions as the SEV such that

$$\mathrm{REV}_{min} = (a_x\Lambda)(a_y\Lambda)(a_z\Lambda) \quad (9)$$

where a_x, a_y, and a_z are the correlation length in the x, y and z direction, respectively, the scale disparity factor, Λ, is defined as

$$\Lambda = \sqrt[3]{\frac{\mathrm{REV}_{min}}{\mathrm{SEV}}}. \quad (10)$$

Λ is the ratio of the geometric mean length of the REV_{min} to the characteristic length of the porous media's features and would define how many times larger a sample must be than the correlation length for statistical inferences to be made from the sample.

Clausnitzer & Hopmans (1999) show that CT scanning of uniformly packed beads produces a REV_{min} with side length about two times the bead diameter, d. That implies a REV volume of $8d^3$. A semivariogram of that packing would produce a correlation length equal to d and thus a Λ of 2. This is probably the minimum Λ that can be expected in porous media.

REV analysis is based on volume averaging concepts, which are traditionally scalar in nature and based upon a cubic volume. Conversely, semivariograms are fundamentally based on vector analysis. In highly anisotropic porous media, therefore, a REV dimensionally proportional to the SEV would give lower V_{min} values than V_{min} calculated using traditional volume averaging concepts.

Regularization

Point data is rarely available in geostatistics. Most often the data is on a volume $V_i(x)$ centred on point x. This is the case for gamma ray attenuation data in CT images. The volume averaged value $Z_i(x)$ is said to be the regularization of the point variable Z over the volume $V_i(x)$. Journel & Huijbregts (1978) show that the regularization of a second-order stationary point random function is also second-order stationary. It is problematic, however, to discuss a property's point value in CT applications because some techniques reach the scale where continuum principles of fluid mechanics no longer apply. Thus, the semivariogram parameters of point values should be considered an abstraction and not a true property. The regularized semivariogram is derived similarly to the point semivariogram, but the sill of the regularization is less than for the point. This result is somewhat intuitive as the variance (sill value) of volume-averaged values will be less than that of point values from the same population. All semivariograms derived in this study were regularized with the volume V being the volume of the voxel. Journel & Huijbregts (1978) further describe the regularized range for borehole core samples (the range derived from semivariogram analysis of volume-averaged data) as the sill from the point semivariogram model a plus the sample length l. The range reported for the best-fit models used in this study is the regularized range. However, for consistency and clarity of terminology, when specifically referring to this value we will refer to it as the volume-averaged range.

Materials and methods

Culebra Dolomite

Core samples were taken from the Culebra Dolomite Member of the Rustler Formation from drill holes near the Waste Isolation Pilot Plant (WIPP) site in SE New Mexico. The WIPP is a US Department of Energy research and development facility designed to demonstrate safe disposal of transuranic radioactive waste resulting from the United State's defence programs. As the most transmissive unit above the repository, the Culebra Dolomite is considered the most likely transport path by which radionuclides could reach the environment (Lappin *et al.* 1989).

The Culebra is a finely crystalline, vuggy dolomite ($CaMg(CO_3)_2$), which is often argillaceous and fractured. Kelley & Saulnier (1990) report that the mean horizontal and vertical hydraulic conductivity for the Culebra Dolomite at this site is 6.2×10^{-9} m/s and 5.1×10^{-9} m/s, respectively. However, these estimates exhibit a six order-of-magnitude variation in the vicinity of the sampling site. For the horizontal cores of the sedimentary Culebra, bedding generally runs parallel to the axis of the core. Density frequency analysis by Hsieh *et al.* (1998*b*) divides Culebra Dolomite into four components: solid dolomite, gypsum infills, silty dolomite and mixed regimes (dolomite, gypsum and/or voids) and voids.

Sample description

Three horizontal cores were imaged and analysed. C2611a had a diameter of 147 mm and length of 450 mm. C2611a had been used in numerous actinide transport experiments and was scanned saturated for operational reasons. The other two samples (C1AV and C2AV) were scanned dry and had diameters of 37 mm and lengths of 52 mm. These two samples were subsamples of a larger core that visually appeared to have different compositions.

Computed tomography

Each core was scanned in a three-dimensional grid using the custom pencil-beam gamma ray CT scanner of Brown *et al.* (1993). The x, y and z axes on the horizontal cores are defined as shown in Figure 1. The bedding planes of the Culebra Dolomite are parallel to the x and z axes of the cores. Cores C1AV and C2AV were imaged at 71 uniformly spaced slices, while C2611a was imaged at 150 positions, not all of which were used here. The z direction voxel spacing was double the spacing in the x and y direction for these cores. The images from C1AV and C2AV have also been used for analyses in Hsieh *et al.* (1998*b*) and Brown *et al.* (2000).

CT scanning parameters are listed in Table 1. For the analysis here, core C2611a has been separated into two different sections to illustrate differences in slices 30 to 90 and 90 to 150 of the same core. Semivariograms are completed on data in rectangular prisms in the centre of the

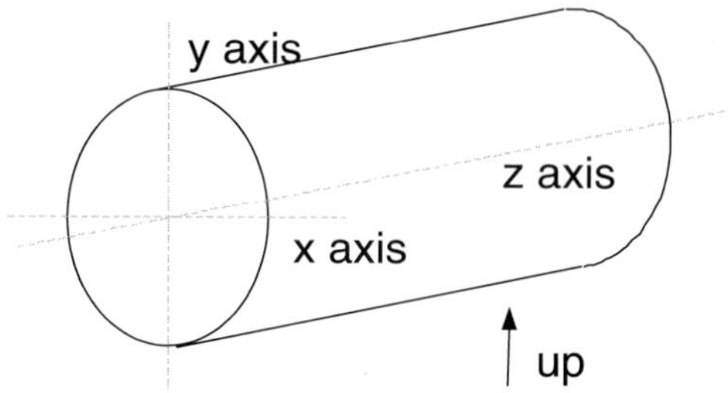

Fig. 1. Orientation of the cores.

Table 1. *Attenuation statistics of samples*

Sample	Dimensions, $X \times Y \times Z$ (voxels)	Voxel length[(1)] (mm)	Attenuation mean (mm^{-1})	Attenuation std. dev. (mm^{-1})	Bulk density mean (g/ml)	Bulk density std. dev. (g/ml)
C2611a (30–90)	$61 \times 61 \times 81$	1.5	0.0174	0.0013	2.57	0.19
C2611a (90–150)	$61 \times 61 \times 81$	1.5	0.0171	0.0013	2.53	0.19
C1av	$71 \times 71 \times 46$	0.37	0.0183	0.0030	2.32	0.38
C2av	$71 \times 71 \times 46$	0.37	0.0192	0.0024	2.44	0.30

(1) z direction voxel spacing is double the x and y direction voxel spacing listed.

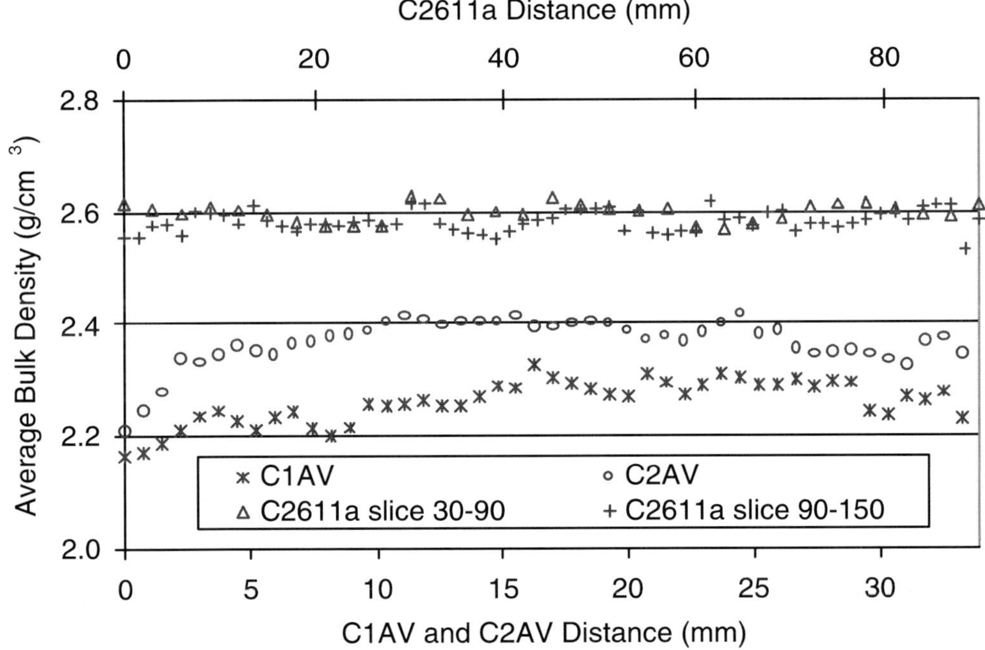

Fig. 2. Individual slice average bulk density along each core.

cylinders for ease of computation. The bulk density in Table 1 is calculated by:

$$\rho_b(x, y, z) = C\mu(x, y, z) \quad (11)$$

where C is the calibration factor and $\mu(x, y, z)$ is the voxel attenuation value (Luo & Wells 1992). The calibration coefficient, which varies for different collimation, was $127\,g\,mL^{-1}\,mm^{-1}$ for cores C1AV and C2AV and $147.8\,g\,mL^{-1}\,mm^{-1}$ for core C2611a.

Cores C1av and C2av were scanned at 0.10 mm^3 voxel volume, while C2611a was scanned at $6.8\,mm^3$ voxel volume. The two samples from core C2611a exhibit larger bulk density means due to their being scanned wet, but smaller standard deviations than cores C1av and C2av. Core C1av has the smallest mean dry bulk density and highest standard deviation. As a demonstration of the variability in the cores, the mean bulk density along the z axis of each sample is shown in Figure 2.

Table 2 uses four structural components and the statistical segregation threshold of Hsieh *et al.* (1998*a*) to differentiate components of the bulk density frequency distribution in each of the cores. The two samples from core C2611a indicate predominantly silty and solid dolomite. Core C1AV is similar to samples C2611a, except there is a detectable percentage of solid gypsum.

Table 2. *Component content by percent volume in samples*

Sample	Voids	Silty dolomite	Solid gypsum	Solid dolomite
C2611a (30–90)	<1	38	<1	62
C2611a (90–150)	<1	46	<1	54
C1av	<1	30	4	66
C2av	<1	9	9	82

Core C2AV is different from the other cores, being mostly solid dolomite with a smaller percentage of silty dolomite areas. Core C2AV also has the largest percentage of solid gypsum areas.

Semivariogram analysis

Deutsch & Journel (1998) provide a package of FORTRAN 77 geostatistical programs, which were used here. The GAM code was recompiled using Visual Fortran 6.6 to increase the maximum number of allowable data points. Each semivariogram was modelled by spherical, exponential and combination models of the two using the curve-fitting program TableCurve 2D 4.0.

Results and discussion

Geostatistics

A reconstruction of an x-y slice from each of the cores and standardized semivariograms for each direction are shown in Figures 3 and 4 and parameters of each are listed in Table 3. The

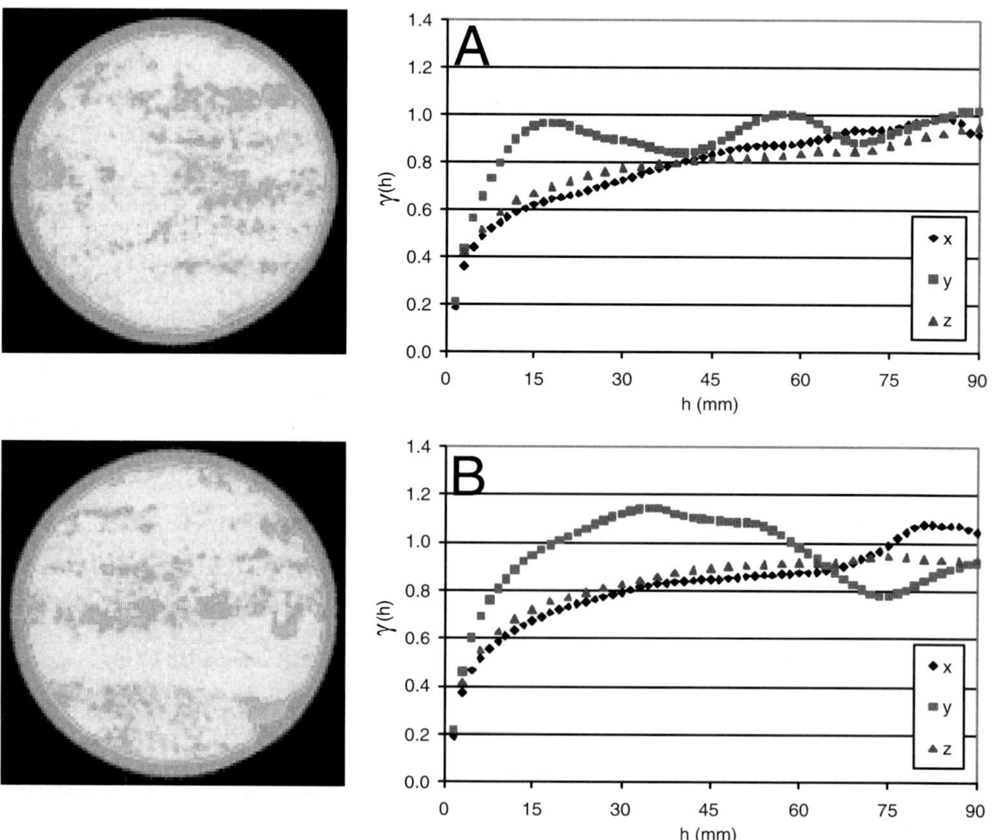

Fig. 3. Computer tomography images and semivariograms of core C2611a, (**a**) slice 30–90 and (**b**) slice 90–150.

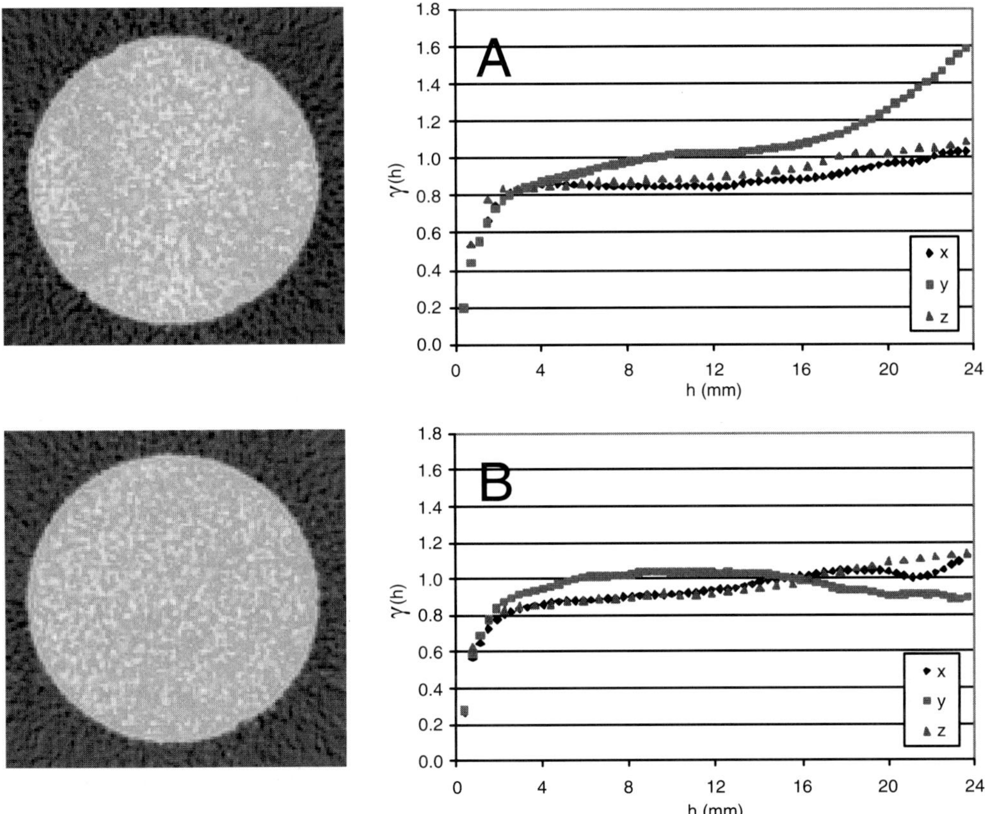

Fig. 4. Computer tomography images and semivariograms of (**a**) core C1AV and (**b**) core C2AV.

directional semivariograms listed are simply the best-fit model for each data. Due to the complex bedding of Culebra Dolomite, combination models may be well justified. However, additional analysis would be required to determine if these fittings are appropriate for a given application. Results are presented as three separate directional semivariograms, but they may be translated to three-dimensional models using the methods described by Isaaks & Srivastava (1989).

For the two larger resolution samples (C2611a) the small-scale correlation lengths ranged from 3.3 to 17.3 mm and the large-scale correlation lengths ranged from 24.0 to 52.0 mm. For the two smaller resolution samples (C1AV and C2AV) the small-scale correlation lengths ranged from 1.6 to 3.1 mm and the large-scale correlation lengths ranged from 13.7 to 28.7 mm. Two semivariograms were best-fit with exponential models. The rest of the semivariograms were best-fit with a combination model using exponential and/or spherical models so as to remain positive definite. The use of a combination model indicates the presence of nested sills. Best fit models were chosen by maximizing the coefficient of determination, r^2, between predicted and measured semivariogram values. The best-fit models all show a r^2 value of 0.98 or greater from $h = 0$ to the lag indicated. The best-fit model was not determined for the entire semivariogram since at large lags the semivariogram often begins to drift up or down. This is a result of a smaller number of comparisons near the edges of the sample and is not necessarily considered representative of the porous media. The best-fit model was also not determined for the entire semivariogram if a hole effect was exhibited. None of the semivariograms for the cores shown in Figure 3 and 4 and described in Table 3 have been modelled with a nugget. This indicates that at these resolutions, the spatial variability in the cores is fully represented in the semivariogram. Large-scale correlation lengths in the x and z direction of core C2AV have values that approached the sample length. This may be indicative of a large-scale trend in the mean

Table 3. *Standardized semivariogram parameters*[(1)]

Sample	Direction	Semivariogram model[(2)]	Range (mm)	Sill, w_i	r^2, mm[(3)] (lag mm)
C2611a (30–90)	x (nested)	S	6.6	0.49	0.992
	x	S	52.0	0.47	(90)
	y	E	17.3	1.02	0.997
	y (hole)	V	60	0.84	(18)
	z (nested)	E	4.8	0.41	0.997
	z	E	52.0	0.44	(72)
C2611a (90–150)	x (nested)	S	3.5	0.37	0.999
	x	E	35.0	0.52	(63)
	y (nested)	S	24.0	0.40	0.995
	y	E	7.4	0.72	(35)
	z (nested)	S	3.3	0.43	0.997
	z	E	37.4	0.51	(66)
C1av	x	E	3.1	0.86	0.981 (16.3)
	y (nested)	S	1.9	0.58	0.997
	y	E	13.7	0.48	(14.1)
	z (nested)	S	1.6	0.12	0.983
	z	S	20.1	0.81	(15.5)
C2av	x (nested)	S	2.3	0.83	0.991
	x	E	25.9[(4)]	0.22	(13.0)
	y (nested)	S	9.1	0.17	0.992
	y	E	2.3	0.85	(15.9)
	z (nested)	S	28.7[(4)]	0.16	0.993
	z	E	1.6	0.83	(14.8)

(1) None of the cores were modelled with a nugget.
(2) Semivariogram models: S = spherical; E = exponential; V = visually estimated
(3) The coefficient of determination and the largest lag value used for calculation of the coefficient of determination is reported.
(4) Range approaching sample width.

Table 4. *Anisotropy of samples*

Sample	Direction	An. Ratio (small scale)	An. Ratio (large scale)	Small scale anisotropy	Large scale anisotropy
C2611a (30–90)	x–y	0.38	0.87[(1)]	Zonal due to strata	na[(2)]
	z–y	0.28	0.87[(1)]	Zonal due to strata	na
	x–z	1.38	1.00	Zonal[(3)]	Zonal[(3)]
C2611a (90–150)	x–y	0.47	1.46	Zonal due to strata	Zonal due to strata
	z–y	0.45	1.56	Zonal due to strata	Zonal due to strata
	x–z	1.06	0.94	Isotropic	Isotropic
C1av	x–y	1.63	na	Zonal areal trends	na
	z–y	0.84	1.47	Zonal due to strata	Zonal due to strata
	x–z	1.94	na	Zonal[(3)]	na
C2av	x–y	1.00	2.84	Isotropic	Geometric
	z–y	0.65	3.15	Geometric	Geometric
	x–z	1.53	0.90[(4)]	Geometric	Isotropic[(4)]

(1) The range of the hole effect is used to calculate the large scale x–y and z–y anisotropy ratio for this sample.
(2) na = not available
(3) Further characterization of zonal anisotropy is not included in this characterization because neither direction is vertical.
(4) Range approaching sample width.

(Gelhar 1986). The C2AV curve in Figure 2 may indicate a slight trend in the mean along the z direction. Core C2611a (30–90) exhibits a hole effect in the y direction. This cannot be modelled by nested sills, but is estimated from the semivariogram and listed in Table 3.

The three-dimensional anisotropy ratios and anisotropy in the large- and small-scale are characterized in Table 4. The anisotropy characterizations are based on the definitions of Kupfersberger & Deutsch (1999) for well log analysis. Anisotropy ratios describe the geometric anisotropy of a sample (i.e. if sill values are equal). Therefore, the ratios listed in Table 4 for cores where geometric anisotropy does not exist are not true anisotropy ratios, but simply a ratio of the range values in the two directions of interest. Because nested ranges have been modelled on these cores, small-scale and large-scale anisotropy ratios have been calculated corresponding to the structure within the core.

Three anisotropy characterizations are provided for each core to demonstrate if the samples are isotropic or geometrically anisotropic in the x–z plane (two horizontal directions), as expected. Core C2611a (90–150) and C2AV exhibit these anisotropies, while C2611a (30–90) and C1AV exhibit zonal anisotropy in the x–z plane. The zonal anisotropy is not further defined in this instance because neither direction is vertical. Core C2611a (30–90) was zonal because of different sill values between the x- and z-directions. The x direction of core C1AV was modelled by a basic model and did not exhibit a large-scale structure. Nearly all of the x–y and z–y (horizontal–vertical) anisotropy characterizations are zonal due to strata, with the exception of C1AV in the x–y direction, which is zonal due to areal trends, and core C2AV, which exhibits isotropy and geometric anisotropy.

Geostatistical comparisons of adjacent cores

Two adjacent sections of core C2611a have been analysed for semivariogram parameters with a resolution of 1.5 mm per voxel in the x and y directions and 3.0 mm in the z direction. For these cores, range values differ in all three primary directions. A nested sill is indicated in the y direction of core C2611a, slice 90–150, while a hole effect is shown in the slice 30–90 core. The hole phenomenon may be representative of strata or horizontal bedding planes in the core and this difference is likely a result of a change in stratification between the two core subsections. These results indicate that different semivariogram parameters can be easily identified for adjacent cores at this scanning resolution.

Cores C1AV and C2AV were extracted from visually different sections of a larger core and were located within 70 mm of each other. The main structural difference between these two samples is the absence of a nested sill in the x-direction on core C1AV. A higher percentage of silty dolomite in C1AV (Table 2) could be partially responsible for this result. If there is more silty dolomite, there conversely will be smaller continuous regions of solid dolomite. This would account for the absence of a larger-scale range in this core and may also explain the anisotropy differences exhibited by these two cores. As with the larger resolution, these results indicate that at this resolution samples obtained next to each other can exhibit clearly different semivariogram parameters.

Geostatistical comparisons at different resolutions

In general, when comparing the semivariogram parameters of smaller resolution cores (C1AV and C2AV) to the larger resolution cores (C2611a), the two resolutions show similar range values in the vertical direction when comparing the nested ranges in the large resolution to the ranges indicated in the smaller resolution. In the horizontal direction, the ranges are larger when comparing equal scales of the large resolution to the small resolution. However, the smaller resolution cores have larger ranges than the nested ranges in the larger resolution. This could be a result of changing horizontal geological structure of the Culebra Dolomite, comparison of semivariograms from differing sample sizes and resolutions, and/or axes of anisotropy that may not be exactly horizontal. The larger range values associated with the larger scales are not detected in the smaller resolution cores because of the smaller sample size used in the smaller resolution analysis. Nested range values on the smaller scale cores are not detectable in the larger resolution because the values approach the resolution limit in the CT images. Comparison of the small-scale and large-scale anisotropy ratios on these cores shows the ratio increasing at increasing scales. These results reinforce the results of Gelhar (1986) and Schafmeister & Pekdeger (1993): that increasing scales show increasing ranges and anisotropy ratios.

Scale disparity quantification

Table 3 lists the small-scale correlation lengths for cores C1AV and C2AV, ranging from 1.6 mm to 3.1 mm. SEV volumes were thus 9.4 mm^3 and 8.5 mm^3 for these two cores, respectively. Brown *et al.* (2000) report a REV$_{min}$ of 1000 mm^3 and

2200 mm^3 for cores C1AV and C2AV, respectively. Therefore, the scale disparity factor, Λ is 4.7 and 6.4 for cores C1AV and C2AV, respectively. Since the REV_{min} corresponds to the volume required for ergodicity, the sample length should be at least five or six times larger than the correlation lengths to detect the spatial structure of the Culebra Dolomite at this small scale. The sample volume used for semivariogram analysis is also larger than the minimum REV for both samples.

The SEV can be viewed as the characteristic size of the density features (pore, solid dolomite, or silty regions) within the core. Conversely, the REV_{min} is the minimum volume required to obtain a statistically significant average of all these dominant features. The K measured for the cores thus imply that Λ^3 density features are within V_{min}, or 104 and 262 features for C1AV and C2AV, respectively.

Effect of changing resolutions on the volume-averaged range

Comparison of CT images at different resolutions also lends itself to an analysis of the modifiable areal unit problem. Cressie (1991) discusses the situation known as the modifiable areal unit problem where averaging, or aggregating, of regionalized variables leads to larger sample correlation lengths. Openshaw & Taylor (1979) show that by averaging fixed observations on small areas in various ways, a large range of correlation coefficients can be produced.

Gamma ray CT is similar to the modifiable areal unit problem, as the image attenuation value is simply the average attenuation through the area encompassed by the voxel. Therefore, semivariograms were calculated for the same CT images of core C2AV with voxels modified by increasing voxel lengths by 2×, 3×, 4×, and 5×, and averaging the attenuation values within these new volumes. This increased the voxel volume by a factor of 8 27, 64 and 125, respectively. The results of these exercises are shown in Figure 5 and Table 5.

As the voxel size increases, the correlation length increases in all directions. All of these semivariograms become more linear as voxel size increases. Larger resolution semivariograms would likely be best-fit with a nugget. Regularization theory states that the regularized range for borehole cores would be the range a of the point semivariogram plus l, the length of the sample volume (Journel & Huijbregts 1978). However, as normally applied this is for a borehole core that increases in only one dimension along its axis. In this instance all three dimensions increase. For this reason, the scaling effect of increasing resolution is presented as a linear relationship between voxel volume and the volume-averaged range in Figure 6. Also shown is the best linear fit to each relationship. These linear fits all showed an R^2 of at least 0.9 and were found to be much better than linear fits to the $a + l$ range model used in regularization theory applied to borehole cores. The improvement of the fits in the y (vertical) direction for the voxel volume relationship is especially pronounced. The intercept of the trend line in Figure 6 represents the range of the point value

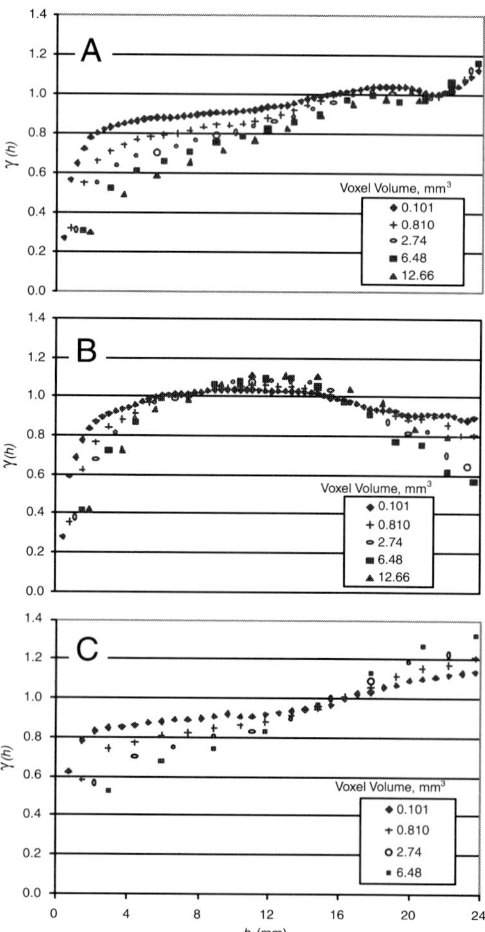

Fig. 5. Semivariograms of computer tomography images for core C2AV with different voxel volumes in (**a**) the x direction, (**b**) the y direction and (**c**) the z direction.

Table 5. *Parameters of standardized semivariograms for core C2AV using different aggregations for the modifiable area unit problem*[(1)]

Aggregation and voxel volume	Direction	Semivariogram model[(2)]	Range (mm)	Sill w_1	r^2, mm[(4)] (lag mm)
1×	*x* (nested)	S	2.3	0.83	0.991
(0.10 mm³)	*x*	E	25.9[(3)]	0.22	(13.0)
	y (nested)	S	9.1	0.17	0.992
	y	E	2.3	0.85	(15.9)
	z (nested)	S	28.7[(3)]	0.16	0.993
	z	E	1.6	0.83	(14.8)
2×	*x* (nested)	S	2.3	0.46	0.994
(0.81 mm³)	*x*	E	21.2[(3)]	0.58	(13.3)
	y (nested)	S	2.3	0.64	0.996
	y	S	9.1	0.40	(14.1)
	z (nested)	S	2.4	0.69	0.992
	z	S	25.9	0.32	(14.8)
3×	*x* (nested)	S	3.1	0.47	0.964
(2.74 mm³)	*x*	E	25.9[(3)]	0.54	(15.5)
	y (nested)	S	3.1	0.60	0.998
	y	E	9.7	0.47	(14.4)
	z (nested)	S	3.1	0.53	0.997
	z	S	25.9[(3)]	0.41	(11.1)
4×	*x* (nested)	S	3.9	0.46	0.999
(6.48 mm³)	*x*	S	22.6[(3)]	0.51	(14.8)
	y (nested)	S	4.6	0.29	0.997
	y	E	10.5	0.83	(13.3)
	z (nested)	S	3.7	0.53	0.978
	z	S	25.9[(3)]	0.41	(14.8)
5×	*x* (nested)	S	4.3	0.36	0.999
(12.7 mm³)	*x*	S	20.2[(3)]	0.56	(13.0)
	y (nested)	S	5.6	0.48	0.998
	y	S	12.6	0.63	(14.8)
	z[(5)]	–	–	–	–

(1) The initial sample spacing was 0.37 mm in the *x* and *z* directions and 0.74 mm in the z direction.
(2) Semivariogram models: S = spherical; E = exponential; V = visually estimated
(3) Range approaching sample width.
(4) The coefficient of determination and the largest lag value used for calculation of the coefficient of determination is reported.
(5) Not enough points in the semivariogram to fit a meaningful model at this lag distance.

semivariogram, while the slope indicates the scaling effect caused by increasing voxel volume.

Summary and conclusions

Geostatistics has been used to determine the spatial structure and anisotropy of Culebra Dolomite at small scales. This analysis was possible because of characterization of the rock from CT images utilizing gamma ray attenuation. Horizontal range values for the four samples ranges from 1.6 to 17.3 mm for small-scale structure and from 13.7 to 52.0 mm for large-scale structure. The smaller-resolution analysis did not detect the largest-scale structure since the sample size was smaller than the range associated with this structure. Likewise, the larger-resolution analysis did not detect the smallest-scale structure because the correlation lengths approached the resolution limit.

Results suggest that cores sampled adjacent to each other can show different semivariogram parameters at these resolutions. In general, correlation lengths were similar at different resolutions in the vertical direction. These vertical direction results could be a result of fairly uniform stratification in the Culebra Dolomite. However, since four different samples are being compared, this is not conclusive enough evidence to discount a scaling effect in the vertical direction. Correlation lengths in the horizontal direction, however, changed with resolution and sample size. Possible reasons for these changes include comparing semivariogram parameters

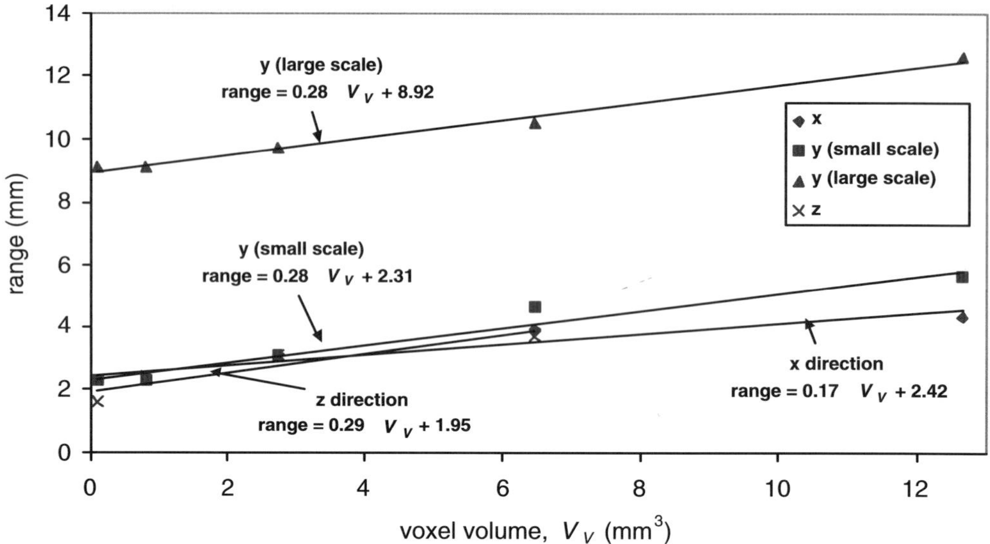

Fig. 6. Voxel volume versus range for core C2AV.

for different size samples with different resolutions, changing horizontal geological structure of the Culebra Dolomite, and/or axes of anisotropy that may not be exactly horizontal.

Relative elementary volume (REV) concepts have been shown to be analogous to the ergodic property of second-order stationarity. Using the REV and standard element volume (SEV), a scale disparity factor has been quantified to determine sample size necessary to determine a meaningful and estimable semivariogram with specified correlation lengths.

The modifiable areal unit problem has also been addressed with this data to look at the effect of changing resolutions on the volume-averaged range. A linear relationship between voxel volume and volume-averaged range was determined. The regularization between voxel volume and volume-averaged range is better for this data than for the voxel length model used to regularize the range of borehole cores because the voxels were aggregated in three dimensions. The y-intercept of the voxel volume/volume-averaged range relationship is the range of the point value semivariogram and is the range unaffected by the scaling effect of increasing resolution. Overall, these results show that sample size and resolution selection for future core analysis should be a function of the scale of interest for the intended data use.

The procedures presented here are relatively complex, but could be automated into most CT instruments. They will provide reasonable and reproducible numerical indexes for the complex structures shown in scans of typical geological materials. Thus, just as we currently use density and porosity as surrogates for other properties, these methods can be used to interpret and/or predict changes in transport and other processes.

References

BEAR, J. & BACHMAT, Y. 1990. *Introduction to Modelling of Transport Phenomena in Porous Media*. Kluwer, Norwell, Massachusetts.

BROWN, G.O., HSIEH, H.T. & LUCERO, D.A. 2000. Evaluation of laboratory dolomite core size using representative elementary volume concepts. *Water Resources Research*, **36**, 1199–1207.

BROWN, G.O., STONE, M.L. & GAZIN, J.E. 1993. Accuracy of gamma ray computerized tomography in porous media. *Water Resources Research*, **29**, 479–486.

CLAUSNITZER, V. & HOPMANS, J.W. 1999. Determination of phase-volume fractions from tomographic measurements in two-phase systems. *Water Resources Research*, **22**, 577–584.

COSKUN, S.B., & WARDLAW, N.C. 1993. Estimation of permeability from image analysis of reservoir sandstones. *Journal of Petroleum Science and Engineering*, **10**, 1–16.

CRESSIE, N.A.C. 1991. *Statistics for Spatial Data*. John Wiley and Sons, New York.

CRESTANA, S., MASCARENHAS, S. & POZZI-MUCELLI, R.S. 1985. Static and dynamic three-dimensional studies of water in soil using computed tomography scanning. Soil Science, **140**, 326–332.

DEUTSCH, C.V. & JOURNEL, A.G. 1998. *GSLIB: Geostatistical Software Library and User's Guide, 2nd edition*. Oxford University Press, New York.

FLÜHLER, H., URSINO, N., BUNDT, M., ZIMMERMAN, U. & STAMM, U. 2001. The preferential flow syndrome – A buzzword or a scientific problem. *In*: *Proceedings of the 2nd International Symposium on Preferential Flow, Water Movement and Chemical Transport in the Environment*. American Society of Agricultural Engineers, St Joseph, Michigan, 21–24.

GELHAR, L.W. 1986. Stochastic subsurface hydrology from theory to applications. *Water Resources Research*, **22**, 1358–1458.

GOGGIN, D.J., CHANDLER, M.A., KOCUREK, G. & LAKE, L.W. 1988. Patterns of permeability in eolian deposits: Page Sandstone (Jurassic), Northeastern Arizona. *Society of Petroleum Engineering, Formation Evaluation*, 297–306.

GREVERS, M.C.J. & DE JONG, E. 1994. Evaluation of soil-pore continuity using geostatistical analysis on macroporosity in serial sections obtained by computed tomography scanning. *In*: ANDERSON, S.H. & HOPMANS, J.W. (eds) *Tomography of Soil-Water-Root Processes*. Soil Science Society of America Special Publication, **36**, 73–86.

HSIEH, H.T., BROWN, G.O. & STONE, M.L. 1998*a*. Quantification of porous media using computerized tomography and a statistical segregation threshold. *Transactions of the American Society of Agricultural Engineers*, **41**, 1697–1706.

HSIEH, H.T., BROWN, G.O., STONE, M.L. & LUCERO, D.A. 1998*b*. Measurement of porous media component content and heterogeneity using gamma ray tomography. *Water Resources Research*, **34**, 365–372.

IOANNIDIS, M.A., CHATZIS, I. & KWIECIEN, M.J. 1999. Computer enhanced core analysis for petrophysical properties. *Journal of Canadian Petroleum Technology*, **38**, 18–24.

ISAAKS, E.H. & SRIVASTAVA, R.M. 1989. *An Introduction to Applied Geostatistics*. Oxford University Press, New York.

JOURNEL, A.G. & HUIJBREGTS, CH.J. 1978. *Mining Geostatistics*. Academic Press, London.

KELLEY, V.A. & SAULNIER, G.J. 1990. *Core Analyses for Selected Samples from the Culebra Dolomite at the Waste Isolation Pilot Plant Site*. Contractor Report SAND90-7011, Sandia National Laboratories, Albuquerque, New Mexico.

KUPFERSBERGER, H. & DEUTSCH, C.V. 1999. Methodology for integrating analog geologic data in 3-D variogram modeling. *American Association of Petroleum Geologists Bulletin*, **83**, 1262–1278.

LAPPIN, A.R., HUNTER, R.L., GARBER, D.P. & DAVIES, P.B. (eds) 1989. *Systems Analysis, Long-Term Radionuclide Transport, and Dose Assessments, Waste Isolation Pilot Plant (WIPP), Southeastern New Mexico*. SAND89–0462, Sandia National Laboratories, Albuquerque, New Mexico.

LUO, X. & WELLS, L.G. 1992. Evaluation of gamma ray attenuation for measuring soil bulk density, 1, Laboratory investigation. *Transactions of the American Society of Agricultural Engineers*, **35**, 17–26.

OPENSHAW, S. & TAYLOR, P.J. 1979. A million or so correlation coefficients: three experiments on the modifiable areal unit problem. *In*: WRIGLEY, N. (ed.) *Statistical Applications in the Spatial Sciences*. Pion, London, 127–144.

PETROVIC, A.M., SIEBERT, J.E. & RIEKE, P.E. 1982. Soil bulk density analysis in three dimensions by computed tomography scanning. *Journal of the Soil Science Society of America*, **46**, 445–450.

PHOGAT, V.K., AYLMORE, L.A.G. & SCHULLER, R.D. 1991. Simultaneous measurement of the spatial distribution of soil water content and dry bulk density. *Soil Science Society of America Journal*, **55**, 908–915.

SCHAFMEISTER-SPIERLING, M.TH. & PEKDEGER, A. 1989. Influence of spatial variability of aquifer properties on groundwater flow and dispersion. *In*: KOBUS, H.E. & KINZELBACH, W. (eds) *Contaminant Transport in Groundwater*. Balkema, Rotterdam, 215–220.

SCHAFMEISTER, M.TH. & PEKDEGER, A. 1993. Spatial structure of hydraulic conductivity in various porous media – Problems and experiences. *In*: SOARES, A. (ed.) *Geostatistics Troia 1992*. Kluwer Academic Publishers, Dordrecht, 733–744.

SOLIE, J.B., RAUN, W.R. & STONE, M.L. 1999. Submeter spatial variability of selected soil and bermudagrass production variables. Soil *Science Society of America Journal*, **63**, 1724–1733.

TOLLNER, E.W. & VERMA, B.P. 1989. X-ray CT of quantifying water content at points within a soil body. *Transactions of the American Society of Agricultural Engineers*, **32**, 901–905.

WOODBURY, A.D. & SUDICKY, E.A. 1991. The geostatistical characteristics of the Borden aquifer. *Water Resources Research*, **27**, 533–546.

Porosity and fluid flow characterization of granite by capillary wetting using X-ray computed tomography

Y. GÉRAUD[1], F. SURMA[1] & F. MAZEROLLE[2]

[1] *Université Louis Pasteur, Ecole et Observatoire des Sciences de la Terre, Centre de Géochimie de la Surface, UMR 7517, 1 rue Blessig, F-67084 Strasbourg cedex, France (e-mail: ygeraud@illite.u-strasbg.fr)*

[2] *CNRS, Laboratoire de Mécanique et d'Acoustique, Chemin Joseph Aiguier, F-13008 Marseille cedex 20, France*

Abstract: The porosity and transfer properties of a very low porosity material (granite) are measured. A new procedure is defined using a capillary test and X-ray computed tomography (CT) scanning. Injected volumes are very low, i.e. a few cm^3 for a sample volume of 1 dm^3, using a fluid/rock ratio lower than 0.1%. This technique allows monitoring of the anisotropy of fluid flow during the test. Flow along the injection direction is higher than along the perpendicular direction. Saturation depends on the specific saturation of each mineral zone. Multiscale analysis allows defining the flow conditions as being controlled at both the mineral and the sample scale. Results indicate the specific role for various constituting parts of the material. High speed flow occurs in the crack network of K-feldspar, while the storage function is localized in the reaction zone forms by quartz and muscovite.

Strain in geological materials is induced by various processes at different depths relative to the earth surface, e.g. weathering under surface conditions, fault activity, diagenetic processes in sedimentary basins or processes at nuclear wastes repositories (Tsang *et al.* 2000). These processes involve a large number of factors, both external (stress, temperature) and internal (porosity, grain size, mineralogical composition). The aim of this paper is to identify mineral and porosity networks for modelling of coupled processes. For a realistic representation, a 3D network must be defined. Classical tools provide only 2D information by direct imaging (optical microscopy, scanning electron microscopy) or by indirect imaging (electric, thermal or acoustic conductivity measurements). Indirect techniques generally give information about only one or two parameters of the porous network; models, or at least some assumptions, about the rock structure are needed (Chatzis & Dullien 1977). Only a few techniques can give direct 3D images within specific limits, e.g. confocal microscopy, which can be applied for a few 100 microns of depth (Demarty *et al.* 1996) and NMR imaging, which is sensitive to the fluid content without providing information about the mineral network (Guillot *et al.* 1991). One of the more efficient techniques for pore and mineral distribution studies is X-ray computed tomography (CT) (Wellington & Vinegar 1987; Withjack 1988; Anderson *et al.* 1992). Several types of CT scanners have been developed, with resolutions from 1 mm to a few μm for metre- to millimetre-sized samples, allowing rock structure images to be obtained at different scales (Ketcham & Carlson 2001). An experimental procedure to couple CT scanning with capillary tests was developed by Géraud *et al.* (1993). Interpretation of CT images using radiological densities can define mineral and pore networks and the fluid flux associated with each part of the sample. This forms the first step in an analytical approach, leading towards a physico-chemical fluid-rock interaction model.

Techniques

The experimental procedure developed for these investigations uses: (i) X-ray CT to image sample porosity saturation and to measure fluid volume; and (ii) a capillary test to determine transport parameters. Theoretical interpretation of the radiological density, in terms of mineralogical composition and porosity, aims to define a pore network and the variations in saturation of the pore volume (Géraud *et al.* 1999).

A medical CT CGR ND8000 scanner with a tungsten anticathod tube was used, operated at

From: MEES, F., SWENNEN, R., VAN GEET, M. & JACOBS, P. (eds) 2003. *Applications of X-ray Computed Tomography in the Geosciences*. Geological Society, London, Special Publications, **215**, 95–105. 0305-8719/03/$15.

120 kV and 18 mA. Images were reconstructed with a resolution of 256 × 256 voxels, with a voxel size of 0.7 × 0.7 × 2 mm. The value of each voxel represents the radiological density. For a discussion of technical aspects of X-ray CT refer to Wellington & Vinegar (1987) and Ketcham & Carlson (2001).

Samples

Analyses were performed on granite from the Gueret massif, near Dun le Palestel, NW of the Massif Central, France. All samples of this sheared batholith come from a low strain zone (Rey *et al.* 1994; Géraud *et al.* 1995). Two types of heterogeneity are present. The first is a mineralogical heterogeneity with four main mineral phases characteristic of granite and the second is a structural heterogeneity with a strain-controlled phase distribution. Bands with quartz and muscovite can be present. The experiment explores the effect of both types of heterogeneity on the flow conditions. The samples were cored with the core axis along the strain direction in the C-plane.

Mineralogical interpretation of CT density

The relationship between mineral density and CT density has previously been established, especially for heterogeneous materials, following an empirical and numerical approach (Raynaud *et al.* 1989; Géraud *et al.* 1993; Géraud *et al.* 1999). Theoretical radiological densities are inferred from the stochiometric composition of each mineral using X-ray attenuation data for the elements.

The radiological density of a material is defined as (Hounsfield 1973):

$$\mathrm{CT} = \alpha\,\frac{\mu(E) - \mu_{\mathrm{H_2O}}(E)}{\mu_{\mathrm{H_2O}}(E)} + K \tag{1}$$

where CT is the radiological density, $\mu(E)$ is the linear attenuation of the material for the energy E of the beam (in cm^{-1}), $\mu_{\mathrm{H_2O}}(E)$ the linear attenuation of the water for the same energy E

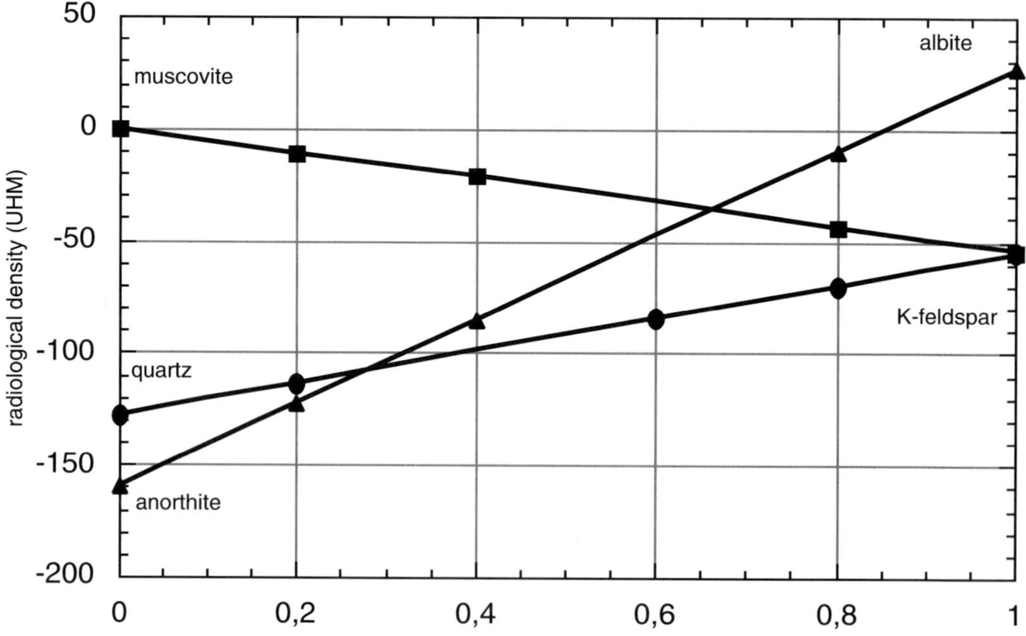

Fig. 1. Radiological density of minerals occurring in granite, inferred using Eq. 2. Pure mineral phases are used: quartz and plagioclase (oligoclase) have similar (low) densities, K-feldspar has an intermediate density and muscovite has the highest density. These minerals form end members of mixing lines used to analyse the measured radiological densities.

(in cm^{-1}) and α and K are constants depending on the visualization scale. For granite visualization, α is 300 and K is -700 and the measured radiological density is expressed in Hounsfield Modified Units (HMU) (Raynaud *et al.* 1989). Reference density values are -1000 HMU for air and -700 HMU for water.

The linear attenuation (μ) is defined as:

$$\mu = \rho\mu_m, \quad (2)$$

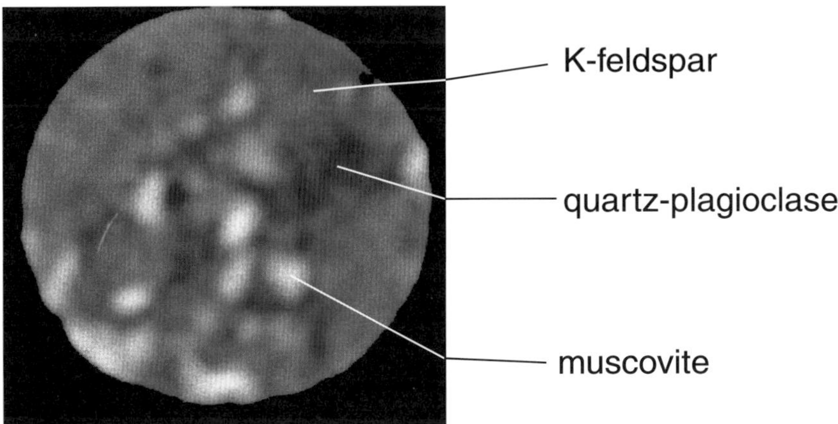

Fig. 2. CT image for section 1 under dry conditions, showing the mineral network. Three sets of minerals are defined, using the inferred values shown in Fig. 1: quartz and plagioclase in dark grey tones, muscovite in light grey tones and K-feldspar with intermediate grey values. Two phenocrystals of K-feldspar occur at the edge of the section and surround a zone with quartz and muscovite.

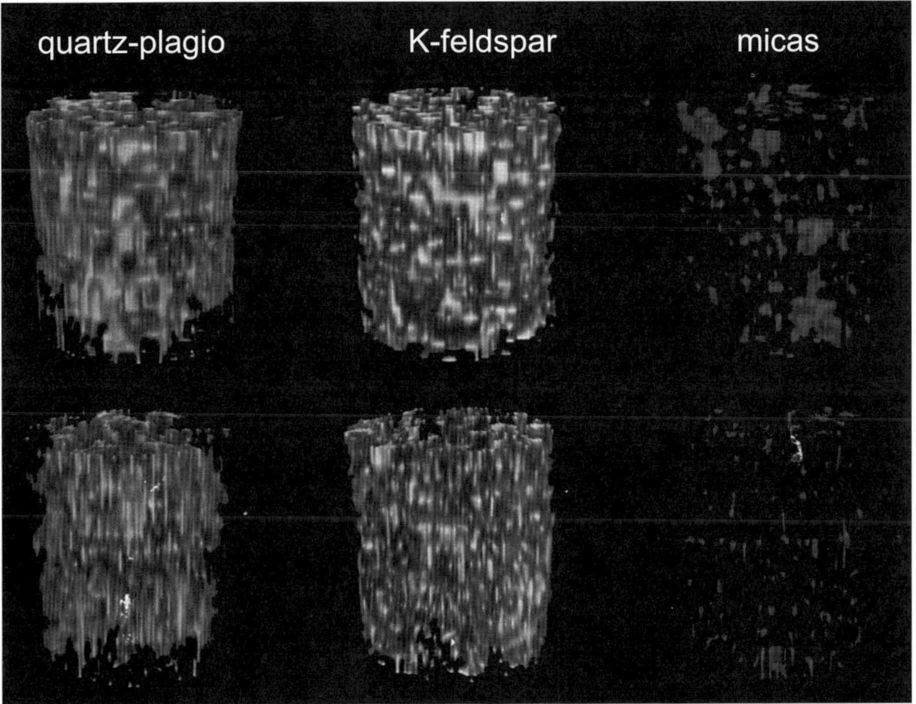

Fig. 3. 3D images of the mineral content. The three sets of minerals are shown from two directions. Section 1 is at the bottom of the reconstructed volumes. The two K-feldspar phenocrystals are recognized in the lower part of the core. A reaction zone composed of quartz and muscovite occurs between these two crystals.

where ρ is the gravimetric density (characteristic of mineralogical composition) and μ_m the mass attenuation, for a molecular species, μ_m defined as:

$$\mu_m = \sum_i (f_i \mu_{mi}) \quad (3)$$

where f_i is the mass fraction of each element and μ_{mi} is its mass attenuation. Radiological densities of the main mineral phases in granite material (quartz, K-feldspar, plagioclase, mica) are inferred using equation (1) and attenuation coefficient tables (Joffre & Page 1968). These values are compared to measured values on macroscopic crystals. Both types of data nearly coincide (Géraud *et al.* 1992), confirming the accuracy of the approach that was used.

Three density zones can be defined (Fig. 1). In the first zone, with low CT densities, quartz and plagioclase (anorthite) are dominant. These two minerals have similar radiological densities. The second zone mainly corresponds to K-feldspars, whereas the third zone, with the highest density values, is characteristic for micas. Based on these findings, a map of the mineral distribution was obtained for each image (Fig. 2). The set of 40 available images was then used to create a 3D map of the sample (Fig. 3).

Porosity determination

Two methods may by used to define porosity. First, porosity can be inferred from gravimetric density values, but this requires mineral identification and quantification. This limitation can be partly removed if double energy analysis is performed (Van Geet 2001). Second, by a comparison between acquired images from the same sample under dry conditions and in different states of wetting defines the porosity and its location. The radiological density by wetting is defined by:

$$CT = 300 \frac{[(\mu_m(1-n)\rho_s) + (S \cdot n \cdot \mu_l \cdot \rho_l)](E) - \mu_{H_2O}(E)}{\mu_{H_2O}(E)} - 700 \quad (4)$$

where n is the porosity and μ_l and ρ_l are the mass attenuation and gravimetric density of the injected liquid. In these tests, a fluid with a higher mass attenuation than water was used (Telebrix®), resulting in a lower detection limit (Fig. 4). Images of the same part of the sample obtained under different conditions of wetting are compared. Under variable saturated conditions, differences in the radiological density between the different states provide saturated porosity values. Subtraction of images illustrates the location of the fluid and thus porosity.

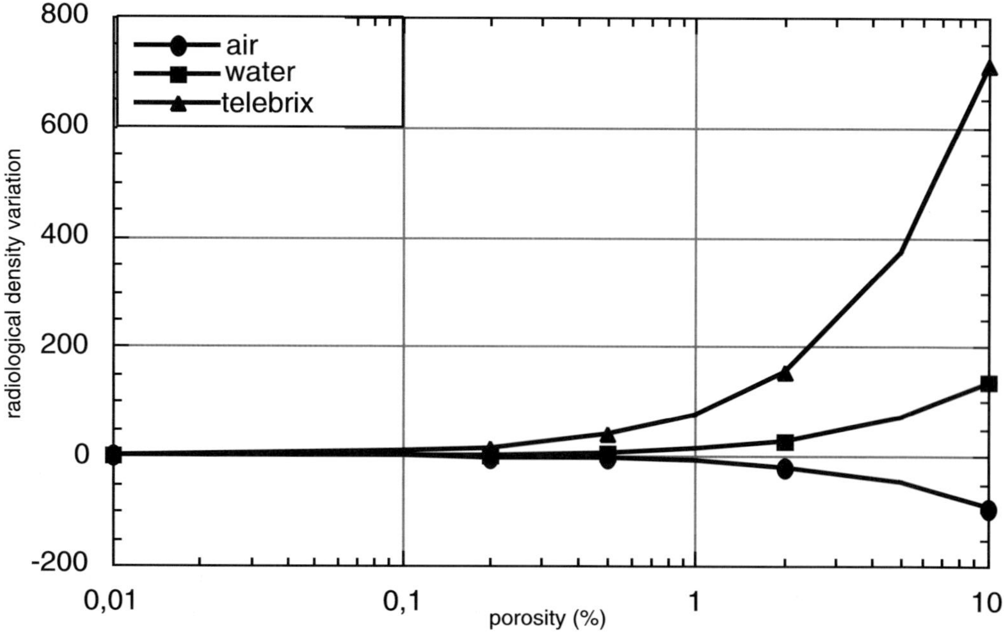

Fig. 4. Radiological density variation induced by different fluids saturating the porosity. If a detection threshold of 1 MHU is assumed, the lowest detected porosity variation is 0.01% using telebrix and 0.1% using water.

Because the porosity may be partly saturated, a saturation coefficient, S is defined:

$$S = \alpha/\Phi$$

were α is the volumetric fluid content and Φ is the total connected porosity.

Capillary test

Hydraulic characteristics of a sample are defined by a capillary test. Under these conditions, fluid flow is associated with a variation of fluid content inducing variations in radiological density.

As stated by Laplace, 'wetting fluid displacement in a fine capillary tube is controlled by the difference in pressure measured on both side of a meniscus'. This capillary pressure is:

$$Pc = 2\gamma \cos\theta / R \quad (5)$$

where R is the capillary tube radius, γ is the surficial tension and θ is the contact angle between fluid/gas/solid. Jurin's law shows that the weight variation, w and the meniscus location, L against the bottom of the core sample evolve lineary according to $t^{1/2}$

$$\frac{W}{S} = \sqrt{\frac{\Pi R^5 \gamma \cos\theta}{2\eta}\, t} = A\sqrt{t} \quad (6)$$

$$L = \sqrt{\frac{R\gamma \cos\theta}{2\eta}\, t} = B\sqrt{t} \quad (7)$$

where Π is a constant, R is the meniscus radius, γ is the surficial tension, θ is the contact angle, η is fluid viscosity and t is time. The inferred equivalent radius, (R), is representative of transport properties and takes into account effective radius, diameter variations, tortuosity and roughness.

Two coefficients are defined, A and B for the weight and meniscus location variations, respectively, proportional to $R^{5/2}$ and $R^{1/2}$. Although the morphology of rock porosity is generally more complex than that of a capillary tube, experimental measurements show that a simple relationship exists and that one B parameter and two A parameters can be defined. The first A parameter, with a high value, is assumed to follow the Laplace relationship during capillary saturation. The second A parameter, with a lower value, is associated with diffusion into the largest voids. The break point is defined as capillary porosity; a relationship between A and B incorporating this parameter can be proposed (Hamecker 1993):

$$B = (A * 100)/(N_{cap} * \rho_e) \quad (8)$$

where N_{cap} is the capillary porosity and ρ_e is the gravimetric density of the fluid. As a result, the first part of a curve describing capillary absorption is typically associated with capillary processes, whereas its second part is associated with diffusion processes.

Experimental procedure

Cylindrical samples 35 mm in diameter and 80 mm long were coated on the sides with fast setting resin to limit the drying effect. The cores were set between two end-plates. At the bottom, a wetting chamber allowed the percolation fluid to wet the core. Fluid moved through the core by capillary suction. The top plate had an opening for free air to flow out. During the wetting phase, a set of 40 CT images of 2 mm thick sections was acquired to describe fluid repartition in the sample. Five sets of 40 slices were obtained. The experimental procedure consists of five steps (Géraud *et al.* 1998): (i) the sample is dried until the weight is constant; (ii) sample and injection cell are placed in the acquisition area; (iii) an initial set of images is acquired under dry conditions to obtain a mineral distribution map; (iv) the wetting chamber is filled and the fluid is in contact with the sample; (v) fluid movement is monitored by acquiring CT images at different locations and times. Scans were made after 0 18, 80, 116, 142 and 172 minutes. Between acquisition of these sets, the lowest section (section 1) was imaged, allowing a better description of the saturation.

Results

The mineral content and pore volume of each voxel determine its measured radiological density. The first step is the creation of a mineral distribution map, then porosity is determined and finally the transport properties are deduced.

Mineralogy

CT thresholds inferred from petrographical characteristics allow identification of mineralogical assemblages for the tested sample, as described above. Radiological densities lower than −120 UHM are characteristic for quartz and plagioclase. Pixels with values higher than −90 UHM contain muscovite as the main mineral phase; values between these two thresholds correspond with K-feldspar occurrences.

The image of section 1 in Figure 2 includes two homogeneous zones with intermediate grey values, which consist of K-feldspar. A central heterogeneous zone with light grey pixels (high density) and dark grey pixels (low density) localizes the area of strain in the sample, with muscovite and quartz that formed by recrystallization along *C* and *S* planes (Hippert 1994; FitzGerald & Stunitz 1993; Géraud *et al.* 1995). In the 3D maps of Figure 3 the shapes of the mineral phases can be recognized.

Porosity

Total connected porosity could be inferred if a constant radiological density value would be reached at the end of the experiment. At the sample scale, the radiological density did not reach a constant value and only 0.5% of the sample volume was saturated because of the short duration of the test. Also, under capillary conditions the pore volume is only partly saturated and only the largest voids are saturated during short duration experimentation. Only the density of sections 1 and 2 reaches a constant value at the end of the test (Fig. 5). The other sections, from 3 to 13, show an increase in radiological density with time, but a constant density was not reached. Above section 14, the radiological density does not change, indicating that the fluid did not reach this section even after 172 minutes.

At the section scale, a constant density is only reached for section 1 at the end of the test, with 0.7% of connected porosity saturated at that time (Fig. 6). For this section, only the capillary phase is obtained. Its porosity distribution is rather heterogeneous; it seems to be more homogeneous in the K-feldspars than in the quartz zone, even though the porosity is higher in that zone (Fig. 7). The K-feldspar regions reach a constant density value after 25 minutes, with a characteristic capillary porosity value of 0.6% (Fig. 6). The quartz zone is not completely saturated and has an injected fluid volume of 1.2% at the end of the test. The connected porosity value is probably the highest in the core.

Flow parameters

Analysis of the data was conducted at different scales, from sample scale over section scale to mineral scale. Using the capillary test, flow parameters were determined from measurements of the fringe location and from the variation in weight during the imbibition phase. The fringe location is measured at the sample scale and the weight variation is inferred from the CT density variations at the sample to mineral scale.

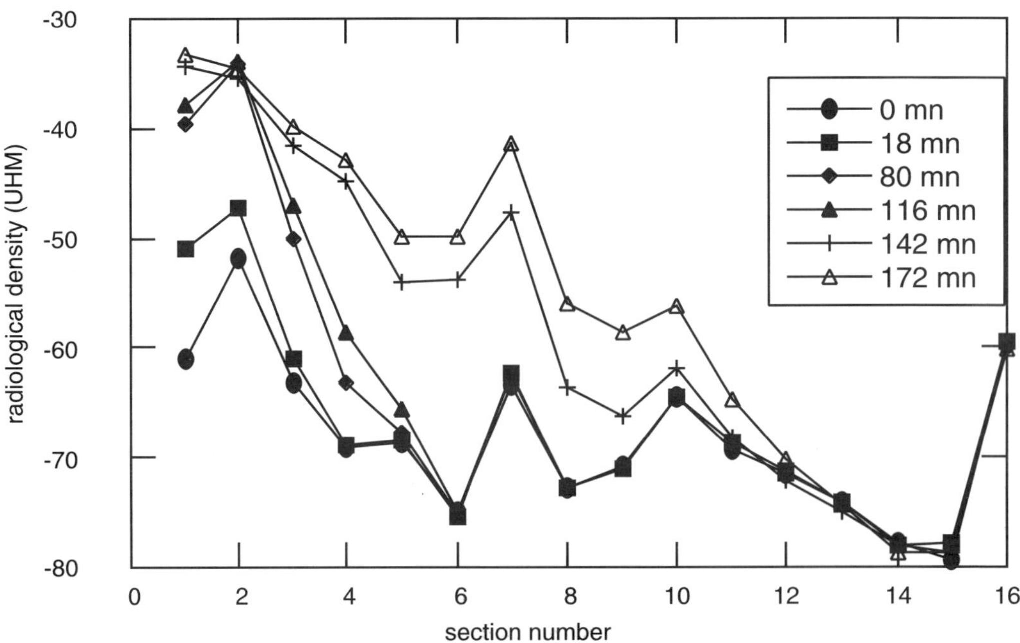

Fig. 5. Radiological density profiles along the core sample. Profiles are built using the average radiological density of each section. The profile acquired at 0 minutes represents the dry state.

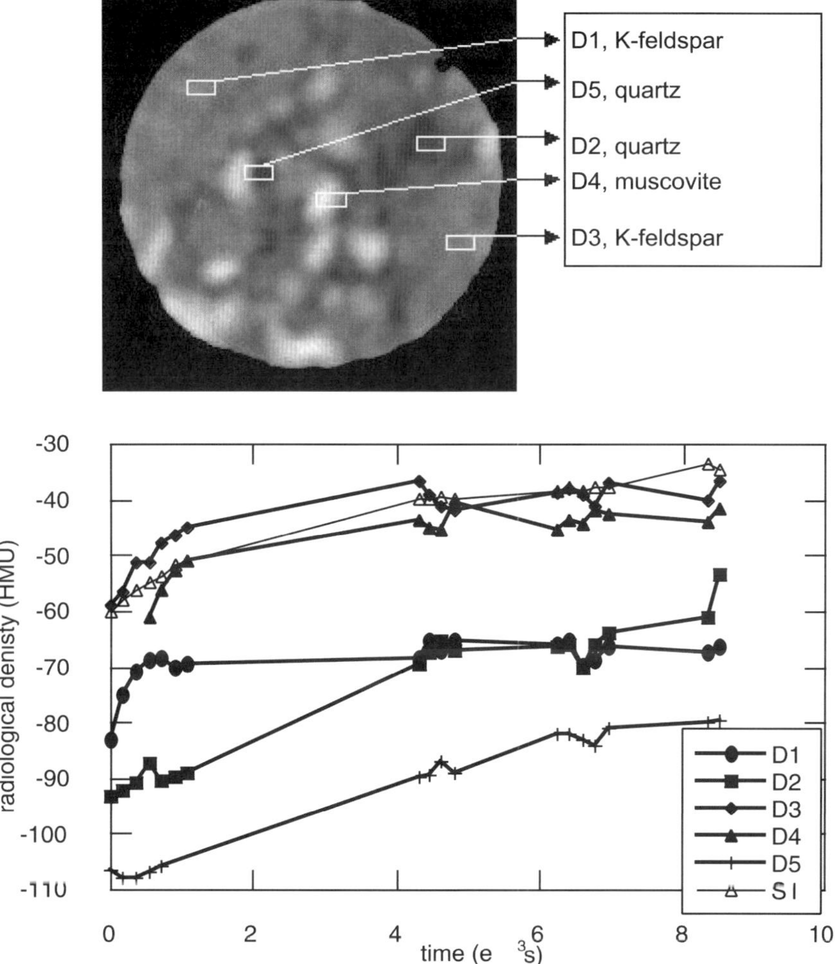

Fig. 6. Radiological density measured during the test for different volumes: S1 is section 1 D1 to D5 are different areas for which densities were determined within this section.

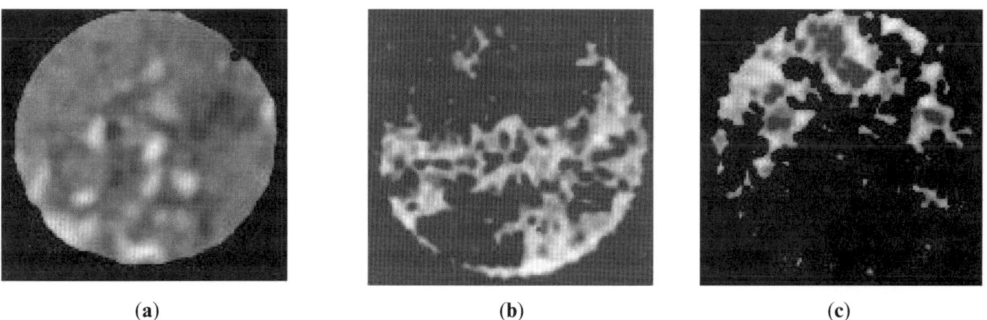

Fig. 7. (**a**) Mineral network in section 1 with two K-feldspar phenocrystals, and quartz and muscovite in the central part. (**b**) Subtraction between the CT image for the dry state and the CT image obtained after 1000 seconds, showing fluid distribution. (**c**) Subtraction between the CT images obtained after 1000 seconds and at the end of the test.

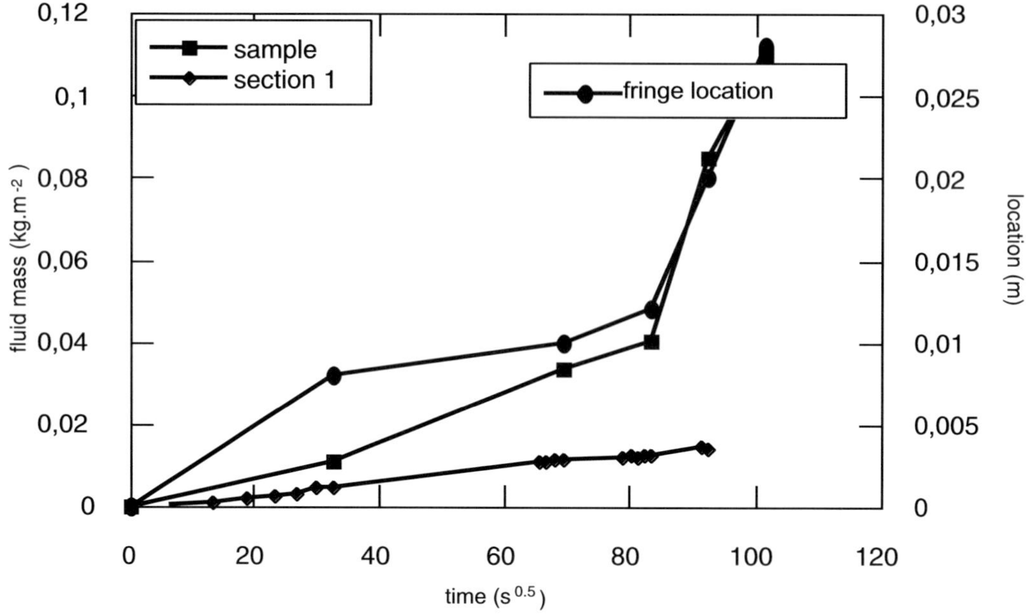

Fig. 8. Capillary curves for the sample and for section 1 and plot of the fringe location versus square root of time. Weight variation at the sample and the section scales, inferred from radiological density measurements. Parameters for weight variation and fringe location are listed in Table 1.

Table 1. *Capillary parameters inferred from radiological measurements for the different studied volumes*

Volume	*DH* (UHM)	*A*	*B* (m s^{-2})	N_{cap} (%)	N_{sat} (%)	B_{calc} (m s^{-2})	R_{eq} (m)
Sample	−84	9.7 e^{-4}	1.9 e^{-4}	0.51	0.42	2.3 e^{-4}	1.56 e^{-9}
Sample		3.8 e^{-3}	1.9 e^{-4}		0.42	9.04 e^{-4}	2.4 e^{-8}
Sample		5 e^{-4}			0.42	1.1 e^{-4}	4.1 e^{-10}
Section 1	−59	1.54 e^{-4}			0.5	3.08 e^{-5}	2.79 e^{-11}
D1	−83	2.3 e^{-4}		0.53	0.6	3.91 e^{-5}	4.5 e^{-11}
D1	−83	2.11 e^{-5}					
D2	−93	2.1 e^{-4}			1.08	1.9 e^{-5}	1.11 e^{-11}
D3	−59	2 e^{-4}		0.5	0.7	2.9 e^{-5}	2.47 e^{-11}
D3		2.1 e^{-4}		0.5	0.7	3 e^{-6}	2.6 e^{-13}
D4	−61	1.1 e^{-4}			0.4	2.75 e^{-5}	2.22 e^{-11}
D5	−106	1.9 e^{-4}			0.81	2.3 e^{-5}	1.6 e^{-11}

DH radiological density; A, *A* parameter, inferred from radiological density variations between two saturation states; B, *B* parameter, derived from the radiological profiles in Figure 5; N_{cap} capillary saturated porosity, inferred from *A* and *B* measurements using Eq. 8; N_{sat} saturated porosity at the end of the test; B_{calc} *B* parameter calculated from Eq. 8, using N_{sat} as N_{cap}; R_{eq} equivalent radius inferred from the calculated *B* parameter values.

An equivalent radius is inferred from the slope ‘B’ of the curve plotting the fringe location versus the square root of time (Fig. 8, Table 1). As mentioned before, this radius is a modelling parameter that takes into account several porosity parameters including size, tortuosity, volume and roughness.

At the sample scale. The studied volume is about 100 cm^3. The plot of the radiological density versus the section number (Fig. 5) allows monitoring of the progression of the fluid through the sample. After 18 minutes, density was increased in sections 1 to 4 while after 142 minutes, the fluid reached section 11 (Fig. 5). The slope of the wetting front versus the square root of time is 0.22×10^{-3} m s^{-2} and the inferred radius is 1.19×10^{-9} m (Fig. 8). Inaccuracy associated with the low variations around the slope is partly related to section thickness. On this curve, two parts can be distinguished. The slope value is low up to 80 $s^{0.5}$. This break point occurs when the fluid reaches a distance of 12 mm from the imbibition surface. This point characterizes variation of the flow condition in the feldspars, which controls the fringe location at the sample scale.

Radiological density variations are measured for each section at different saturation stages. When these density variations are converted into water weight variations, the capillary fluxes can be deduced (Fig. 8, Table 1). Both fluid mass curves show two slopes with a break point occurring at the same location for the B parameter. The A parameter of the second part is one order of magnitude greater than the A parameter of the first part. The difference in flux is two orders of magnitude. After 280 minutes, 1.07×10^{-7} m^3 of fluid is injected in the sample with a flow direction parallel to the core axis.

Imaging of the fluid distribution displays a heterogeneous porosity controlled by mineralogical composition and structural characteristics. Characterization of different parts of the sample is obtained by mineralogical and structural analysis of homogeneous areas.

At the section scale. Flow conditions could be only defined from radiological density measurements. The investigated volume is 1.9 cm^3. The curve shows only one slope and total saturation of the capillary porosity is not reached. At the section scale, the A parameter is lower than the one measured at the sample scale. The main flow direction of this scale analysis is perpendicular to the core axis, this is a radial direction for fluid propagation. In this plane of fluid propagation, the flux is lower than at the sample scale (Fig. 8).

At the mineral scale. Areas of a few mm^3 are selected in zones with specific mineralogical compositions to define the flow properties of each mineral phase or assemblage. The mineralogical composition of the analysed zones is deduced from the initial radiological density. Five zones were analysed: D5 and D2 are mainly composed of quartz, D4 represents quartz and muscovite, and D3 and D1 consist of K-feldspars. Two types of curve were obtained, namely curves with one slope and curves with two slopes. The first type characterizes the quartz zones and the second type is recognized for K-feldspars zones. In the quartz zones, the capillary porosity is not saturated after 280 minutes. The presence of muscovite seems to decrease the A parameter, in comparison with areas of pure quartz (Fig. 9). In the K-feldspars zones, a capillary porosity of 0.5% could be obtained. The total saturated porosity at the end of the test is 0.6% which is

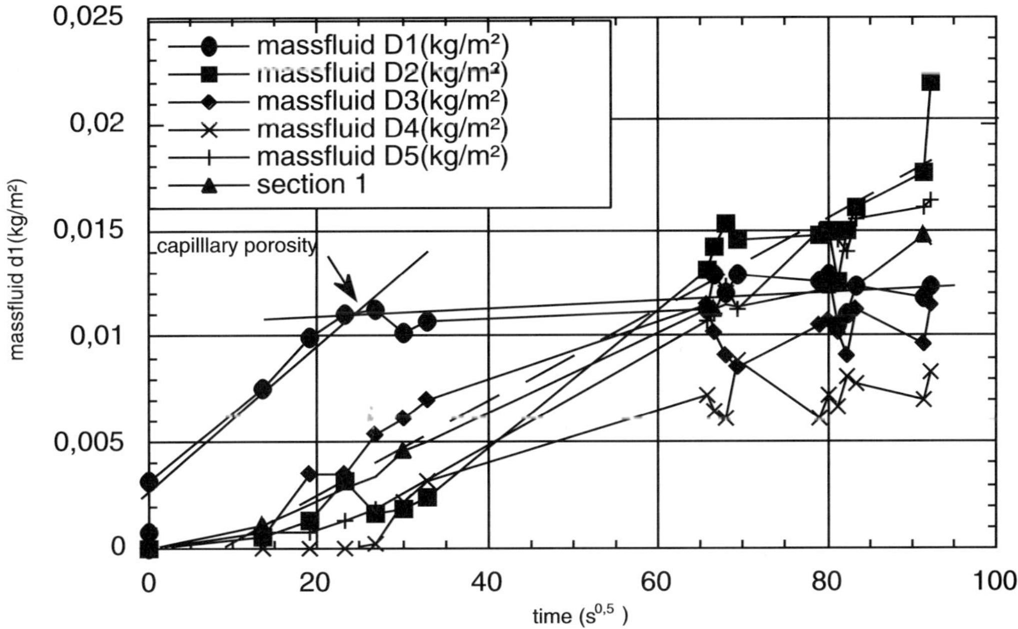

Fig. 9. Capillary curves for section 1 and mineral zones D1 to D5, showing weight variations inferred from radiological density measurements versus square root of time. Values of A parameters are listed in the Table 1.

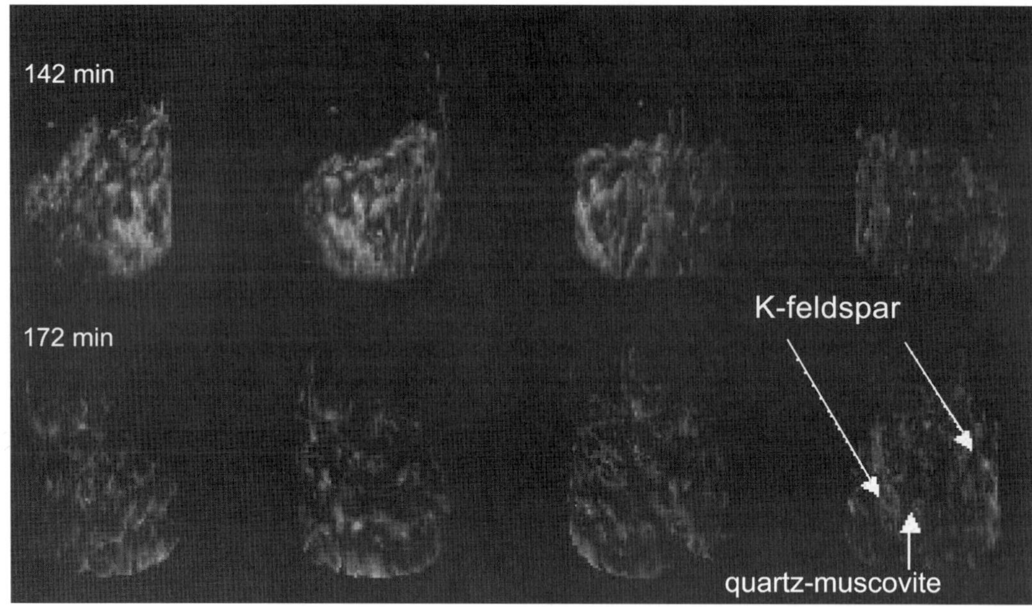

Fig. 10. Fluid location after 142 and 172 minutes, illustrated by 3D images. The injection surface is at the bottom. Fluid movement in the two K-feldspar phenocrystals at the edge of the core is faster than in the central part.

slightly higher than the capillary value. This value is close to the connected porosity value measured by mercury injection on K-feldspar phenocrystals. Imaging at different saturation stages (Fig. 10) demonstrates that fluid migration in K-feldspars is faster than in the quartz zone. Quantification of the flow properties is obtained from the CT density measurements. Each mineral assemblage has a particular capillary curve. The main difference between them is the speed of saturation of the capillary porosity. The observed speed in the K-feldspar is higher than in the quartz assemblage, even though its capillary volume is lower. This result seems to indicate a lower void diameter and/or a more tortuous network in the quartz assemblage. This is in agreement with the characteristics of the pore network developed at boundaries of quartz and muscovite grains. In K-feldspars, the pore network is composed of cracks with a large aperture and a low pore volume.

These data are summarized in Table 1. Flow properties and the pore volume depend on the investigated volume. Flow is governed by the faster network in the material, while at the sample scale it seems to be controlled by the crack network in the K-feldspars along the core axis and the strain direction (Fig. 10). In this direction, the equivalent radius is higher. Perpendicular to this direction, in the section plane, the radius reaches an intermediate value for each mineral zone. The reaction zone in the strain band controls the pore volume. These bands correspond to the location of recrystallized quartz and muscovite.

Conclusions

During the reported experiment, the porosity volume and transfer properties of a very low porosity material were measured. A new process using a capillary wetting and CT imaging was defined.

Detection of a very small fluid volume injected in granite is possible with a high attenuation fluid. Using the density sensitivity of the CT scanner, porosities lower than 0.1% could be detected, even with a voxel size of 1 mm^3 (Fig. 4). CT density measurements during wetting tests allows defining transport properties and connected pore volumes. This technique enables monitoring of the anisotropic flow that occurs during the test. Flow along the injection direction is greater than along the perpendicular direction. In the perpendicular direction, saturation depends on the specific saturation of each mineral zone. The capillary parameter of the lowest studied section had a mean value between that of the K-feldspar zone (high value) and quartz zones (low value). The lowest value for the connected porosity was measured in the K-feldspars, whereas the highest connected volume was measured in the quartz zone.

Multi-scale analysis allows defining the controlling flow conditions measured at the mineral scale and at the sample scale. Results demonstrate the specific role played by different parts of the material. In the case of granite, high-speed flow occurs in the crack network of the K-feldspars and storage is localized in the reaction zone. During the test, the saturation state depends on the specific porosity network associated with each mineral assemblage. K-feldspars are more rapidly saturated than recrystallization zones consisting of quartz and muscovite.

This is publication number EOST/2002-404-CGS and Forpro 2002-20.

References

ANDERSON, S.H., PEYTON, R.L., WIGGER, J.W. & GANTZER, C.J. 1992. Influence of aggregate size on solution transport as measured using computed tomography. *Geoderma*, **53**, 387–398.

CHATZIS, I. & DULLIEN, F.A. 1977. Modelling pore structure by 2D and 3D networks with application to sandstones. *Journal of Canadian Petroleum Technology*, **16**, 97–108.

DEMARTY, C.H., GRILLON, F. & JEULIN, D. 1996. Study of the contact permeability between rough surfaces from confocal microscopy. *Microscopy, Microanalyse, Microstructure*, 505–511.

FITZGERALD, R.D. & STÜNITZ, H. 1993. Deformation of granitoids at low metamorphic grade. I: reactions and grain size reduction. *Tectonophysics*, **221**, 269–297.

GÉRAUD, Y., MAZEROLLE, F. & RAYNAUD, S. 1992. Comparison between connected and overall porosity of thermally stressed granites. *Journal of Structural Geology*, **14**, 981–990.

GÉRAUD, Y., MAZEROLLE, F. & RAYNAUD, S. 1993. Essai de quantification de la porosité d'un granite altéré. Utilisation du scanner médical (tomodensitométrie X). *Bulletin de la Société Géologique de France*, **164**, 243–253.

GÉRAUD, Y., CARON, J.M. & FAURE, P. 1995. Porosity network of ductile shear zone. *Journal of Structural Geology*, **17**, 1757–1769.

GÉRAUD, Y., MAZEROLLE, F., CARDON, H., VIDAL, G. & CARON, J.M. 1998. Anisotropie de connectivité du réseau poreux d'une bande de cisaillement ductile. *Bulletin de la Société Géologique de France*, **169**, 645–654.

GÉRAUD, Y., TOURNIER, B. & MAZEROLLE, F. 1999. Detection of porosity and mineralogical variations in geological materials: radiological density measured by CT scan. *In*: *Proceedings of the International Symposium on Imaging Applications in Geology, Géovision 99, Liège, 6–7 May*, 109–112.

GUILLOT, G., KASSAB, G., HUGLIN, J.P. & RIGOLD, P. 1991. Monitoring of tracer dispersion in porous media by NMR Imaging. *Journal of Physics D*, **24**, 763–773.

HAMECKER, C. 1993. *Importance des Transferts d'Eau dans la Dégradation des Pierres en Œuvre*. PhD thesis, University Louis Pasteur, Strasbourg, France.

HIPPERT, J.F. 1994. Microstructures and C-axis fabrics indicative of quartz dissolution in sheared quartzites and phyllonites. *Tectonophysics*, **229**, 141–163.

HOUNSFIELD, G.N. 1973. Computerized transverse axial scanning (tomography). Part 1: Description of system. *British Journal of Radiology*, **46**, 1016–1022

JOFFRE, H. & PAGE, L. 1968. Coefficients d'Atténuation Massique et d'Absorbtion Massique en Energie pour les Photons de 10 keV à 10 MeV. CEA Rapport **3655**.

KETCHAM, R.A. & CARLSON, W.D. 2001. Acquisition, optimization and interpretation of X-ray computed tomographic imagery: applications to the geosciences. *Computers and Geosciences*, **27**, 381–400.

RAYNAUD, S., FABRE, D., MAZEROLLE, F., GÉRAUD Y. & LATIÈRE, H.J. 1989. Analysis of the internal structure of rocks and characterization of mechanical deformation by non destructive method: X-ray tomodensitometry. *Tectonophysics*, **159**, 149–159.

REY, P., FONTAIN, D.M. & CLEMENT, W.P. 1994. P waves velocity across a monoaxial ductile shear zone gradient: consequence for upper crustal reflectivity. *Journal of Geophysical Research*, **99**, 4533–4548.

TSANG, C.F., STEPHANSSON, O. & HUDSON, J.A. 2000. A discussion of thermo-hydro-mechanical (THM) processes associated with nuclear waste repositories. *International Journal of Rock Mechanics and Mining Sciences*, **37**, 397–402.

VAN GEET, M. 2001. *Optimisation of microfocus x-ray computer tomography for geological research with special emphasis on coal components (macerals) and fractures (cleats) characterisation*. PhD thesis, Catholic University of Leuven, Belgium

WELLINGTON, W.R. & VINEGAR, H.J. 1987. X-ray computorized tomography. *Journal of Petroleum Technology*, 885–898.

WITHJACK, E.M. 1988. Computed tomography for rock-property determination and fluid-flow visualization. *Society of Petroleum Engineers Formation Evaluation*, **3**, 696–704.

Direct imaging of fluid flow in fault-related rocks by X-ray CT

T. HIRONO[1], M. TAKAHASHI[2] & S. NAKASHIMA[1]

[1] *Interactive Research Center of Science, Tokyo Institute of Technology, Tokyo 152–8551, Japan (e-mail: t-hirono@spa.att.ne.jp)*

[2] *Rock Physics Team, Research Center for Deep Geological Environments, AIST, Tsukuba 305–8567, Japan*

Abstract: Faults profoundly affect patterns and rates of fluid flow and solute transport in the geological environment. They may act as conduits, barriers or combined conduit-barrier systems. In order to elucidate the relationship between fluid flow properties and deformation mechanisms of fault-related rocks, we applied X-ray CT during laboratory permeameter measurements for direct imaging of fluid flow during permeability testing. A KI solution, which has high X-ray attenuation values, was used as a contrast medium for the advection imaging. Three-dimensional fluid flow distributions were measured for the studied fault-related rocks. Fault zones characterized by independent particulate flow as deformation mechanism act as conduits for fluid flow, whereas cataclastic fault zones act as barriers.

Fluid flow in fractured rocks is currently being studied with regard to storage of high-level radioactive waste. The concept of waste disposal in Japan is based on a multibarrier system which combines an isolating geological environment with an engineered barrier system. Special consideration is given to the long-term stability of the geological environment, taking into account that Japan is located in a tectonically active zone. The characterization of such an active geological environment is important in terms of its barrier function with regard to groundwater flow rates, rock permeability, geochemical characteristics of groundwater, thermal and mechanical properties of rock formations and properties relevant for solute transport.

Mass transfer properties of deformed rocks are strongly influenced by heterogeneously distributed structures such as faults, cracks and dykes. With regard to the quantitative assessment of rock permeability, the impact of fault-related permeability anisotropy on fluid flow has been documented by Randolph & Johnson (1989) and Mozley & Goodwin (1995), amongst others. Faults profoundly affect patterns and rates of fluid flow and solute transport in geological formations.

Many laboratory-based permeability measurements have been performed on reconstituted fault material derived from crystalline rocks (Morrow *et al.* 1981; Morrow & Byerlee 1988; Chu & Wang 1988). Evans *et al.* (1997) presented experimentally determined permeabilities of natural fault-related rocks developed in granitic gneisses. However, these earlier experiments treat averaged bulk permeability values derived from pressure or volume differences between inflow and outflow by various methods (e.g. constant head, flow pump or transient pulse methods). The heterogeneity of the samples was not taken into consideration in these studies.

This paper focuses on visualizing fluid flow during permeability testing and aims to elucidate the relationship between fluid flow properties and deformation mechanisms of fault-related rocks.

X-ray CT scanner

In order to achieve the mentioned goals of this study, it was necessary to develop an apparatus with the capability not only of measuring the permeability but also of visualizing fluid advection. In this study an X-ray computed tomography (CT) medical scanner was used as a tool to non-invasively image three-dimensional flow patterns during permeability testing. The attenuation of a two-dimensional fan-beam of X-rays that crosses a sample is measured by an array of detectors. X-ray projection data from various directions are obtained during a 360° rotation of the X-ray source. A two-dimensional image representing the distribution of linear X-ray attenuation, determined by density and atomic number, is reconstructed using a backprojection algorithm. A three-dimensional data set of the sample is obtained by stacking contiguous two-dimensional images.

From: MEES, F., SWENNEN, R., VAN GEET, M. & JACOBS, P. (eds) 2003. *Applications of X-ray Computed Tomography in the Geosciences*. Geological Society, London, Special Publications, **215**, 107–115. 0305-8719/03/$15.

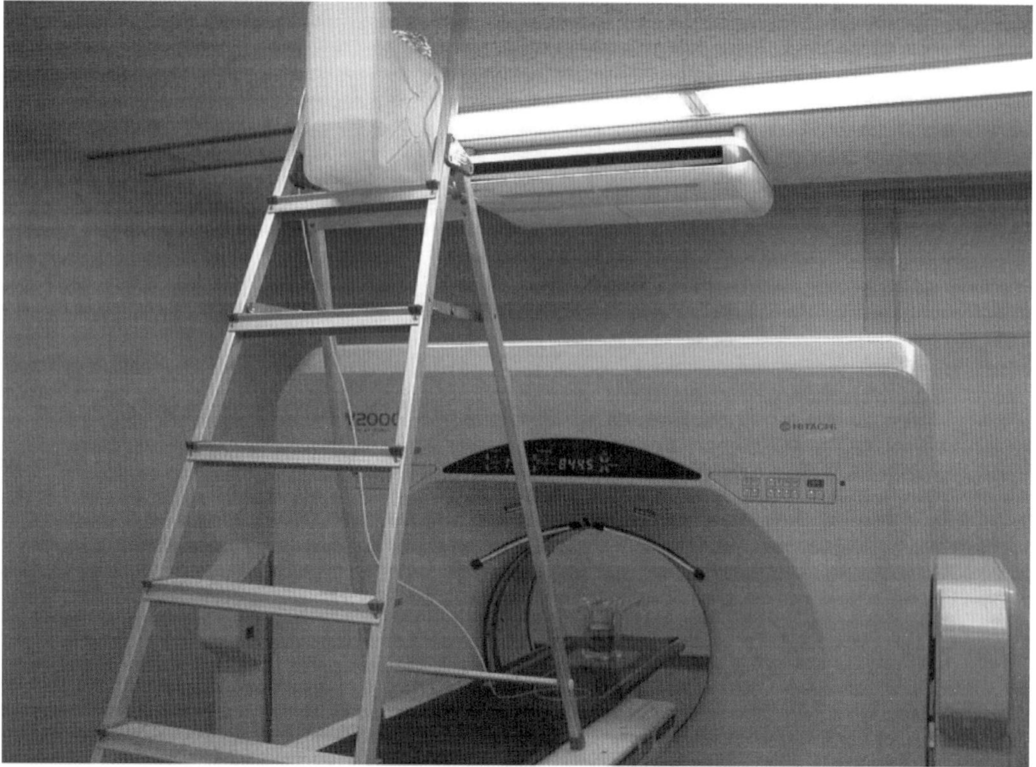

Fig. 1. Experimental arrangement of the medical X-ray CT scanner W2000.

An X-ray CT scanner (W2000, Hitachi Medical Co., Tokyo, Japan) of the Geological Survey of Japan was used for the present study (Fig. 1). An X-ray tube with a Mo-W alloy target was used, operated at 120 kV accelerating potential and 175 mA current. The attenuated X-ray beam after traversing the sample is measured by 768 detectors. The in-plane resolution (voxel size) is 0.31 mm. The CT data are reported as CT number (N_{ct}), defined as:

$$N_{\mathrm{ct}} = (\mu - \mu_{\mathrm{W}})/\mu_{\mathrm{W}} \times 1000$$

where μ is the linear X-ray absorption coefficient of the sample and μ_{W} is the linear absorption coefficient of pure water used as a standard reference. The CT number of water is defined as 0 and the CT number of air as -1000. The CT number is a function of density and chemical composition of the material in any voxel.

The output images are digitized as TIFF-formatted 16-bit image files (512×512 pixels). They are converted to 8-bit images with image processing software (UltimagePro). The 256 grey scale levels are linearly applied to a range of 32768 to 36768 in original images, which corresponds to CT numbers from 0 to 4000. One grey-scale level, therefore, has a 15.625 CT number range.

Experimental apparatus

The experiments used a permeameter cell made from acrylic plastics. The system consists of rigid-wall permeameters under atmospheric pressure. It is not possible to produce a greater confining pressure on the cell, because its strength is not sufficiently great. A schematic diagram and photo image of the permeameter cell is shown in Figure 2. It consists of acrylic vessels, six pillars, Teflon tubes and joints. The CT numbers of the constituting parts are fairly constant and range from 82 to 112 (Figs 2c and 2d). Because it has a low density (1.0128 to 1.0621 g/cm^3), acrylic plastic has a low CT number.

The container is set at the centre of the CT scanner and connected to a tank with a KI solution, positioned 135 cm above the container (Fig. 1). The samples are enclosed in the acrylic vessel and sealed by silicon-based sealing compounds to prevent lateral flow along the vessel

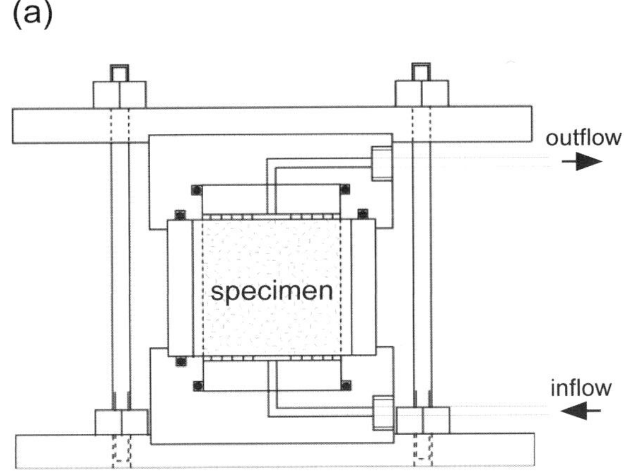

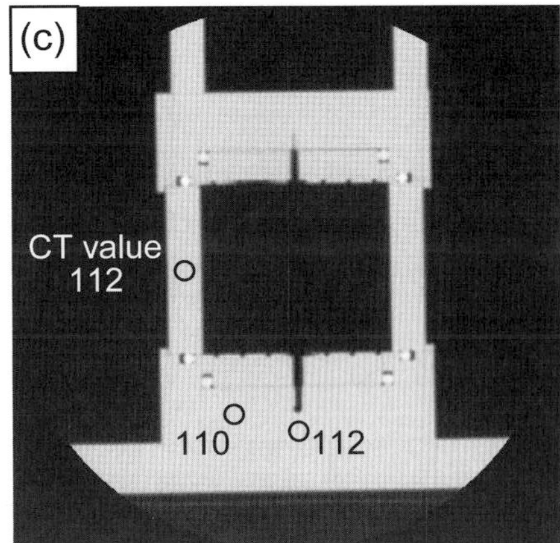

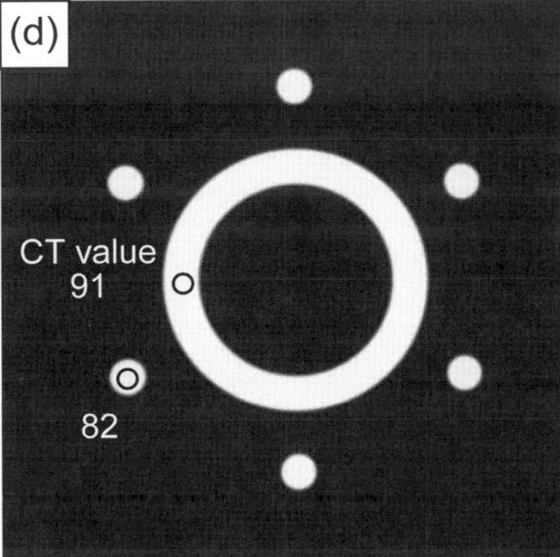

Fig. 2. (**a**) Schematic diagram of the permeameter cell. (**b**) Photo image of the cell. (**c**) Longitudinal CT image of the cell. (**d**) Transversal CT image of the cell.

wall. CT images are captured every 30 minutes during permeability testing.

Calibration between CT number and solute concentration

In this study the KI solution is used as a contrast medium for the advection imaging, because it has a high X-ray attenuation. Calibrations of CT numbers of KI concentration under different conditions were performed. First, the appropriate X-ray tube current and voltage settings had to be determined to obtain optimal results.

Beam hardening artefacts appeared in the CT images. This is recognized in Figure 3, which shows CT numbers for outer, middle and inner positions within a KI solution in a plastic bottle. The CT numbers for the outer position are higher than those for the middle and inner positions. This artefact results from the stronger attenuation of X-rays with a long wavelength, whose absorption in the outer part of a specimen results in apparently higher CT numbers for that part. This artefact can be reduced by the use of a T.B.C. compensation program, produced by the Hitachi Medical Co.

The degree of X-ray attenuation increases linearly with the molar concentration of electrolytes. The concentrations of KI solutions are plotted against the corresponding CT numbers under three different conditions –120 kV and 175 mA (black symbols), 120 kV and 100 mA (grey symbols), 100 kV and 100 mA (outlined symbols) (Fig. 3). For each condition, the CT number increases nearly linearly with the molar

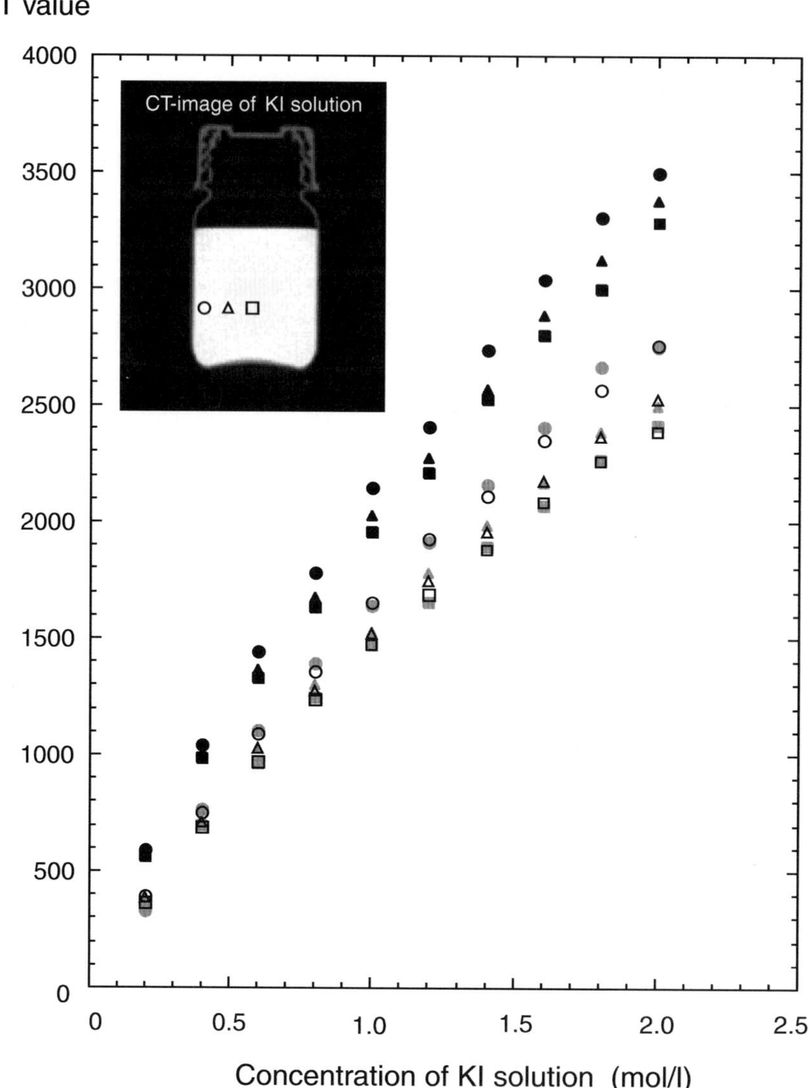

Fig. 3. Calibration of CT numbers under different conditions: black symbols – 120 kV, 175 mA; grey symbols – 120 kV, 100 mA; outline symbols – 100 kV, 100 mA. Circle, triangle and square symbols indicate the outer, middle and centre positions, respectively.

concentration of KI. A good linearity is found for a KI concentration between 0.2 and 0.8 mol/L.

Based on the results of these calibration tests, the best scanning condition were determined to be as follows: T.B.C. compensation program for beam hardening reduction; 120 kV and 100 mA as X-ray tube setting; 0.5 mol/L KI solution; 1.0 mm slice thickness; $0.313 \times 0.313 \times 1.0\,mm^3$ voxel size; 4.0 s scan time; 160 mm imaging diameter; 512×512 pixel matrix size.

Sample descriptions

Samples used in this study were collected from the Emi Group, occurring in outcrops on the Boso Peninsula, 100 km SE of Tokyo. The strata are composed of fine to coarse acidic tuffaceous clastic sedimentary rocks intercalated with tuff layers (Hirono & Ogawa 1998). The age of the sediments ranges from late Oligocene to mid-Miocene (Sawamura & Nakajima 1980; Suzuki *et al.* 1996). Two vein types are found in the deformed tuffaceous sandstone in this group. They stand out along weathered surfaces and occasionally display a few centimetres of displacement. One of the vein types commonly shows approximately 1–10 cm of displacement. For the second type, this is only 0.5–1 cm. Each vein can be considered as a fault zone and resembles a deformation band (Aydin & Johnson 1983; Antonellini & Aydin 1994).

(a)

(b)

(c)

Fig. 4. Photo images of the examined fault-related rocks. (**a**) IPF-F; (**b**) CF-P; (**c**) CF-V.

The microfabric of the first fault type is characterized by non-deformed grains, dimensional preferred orientations of sand grains parallel to veins and collapse of cemented material within the pores. These features relate the microscopic deformation mechanism to independent particulate flow following Knipe's definition (Knipe 1986). The fault zones show a gradual transition to the undisturbed rock. One core sample including this fault zone (IPF-P) was collected for our experiments (Fig. 4a). This core section was 5.0 cm in diameter and 2.5 cm in height. Two small oblique fault-related seams were included in this core sample.

The microfabric of the second fault type is characterized by cataclasis of sand grains and collapse of cemented materials within their pores. The boundaries between the fault zone and the host rock are very sharp. The host rocks are relatively undeformed. These features imply that the microscopic deformation mechanism is cataclastic flow, following Knipe's definition (Knipe 1986). Two core samples were selected with a variety of orientations relative to the fault plane. One sample included some cataclasis-related seams parallel to the axis of the core (CF-P). This sample was 5.0 cm in diameter and 3.5 cm high (Fig. 4b). The other sample, 5.0 cm in diameter and 1.5 cm in height, included some cataclasis-related seams perpendicular to the axis of the core (CF-V) (Fig. 4c).

Permeability measurements

Permeability can be described as a property of a porous rock, related to fluid flow through its pore space. Henri Darcy's empirical law (1856) for purely viscous fluid flow in a porous medium can be written for one-dimensional flow as:

$$Q = kAH/\Delta h \quad (1)$$

where Q is the volume flow rate (cm^3/s), A is the specimen's cross-sectional area perpendicular to the flow (cm^2), H is the hydraulic head (cm) and Δh is the specimen height (cm). The permeability, k (cm/s), is defined by this equation.

The permeability of core samples IPF-P, CF-P and CF-V were measured by the steady-state method, which entails measurement of volumetric flow-through rates under a constant fluid pressure gradient. The volumes of outflow are weighed directly using an electronic balance. The results show a linear relationship between the outflow volume and time. The permeabilities of IPF-P, CF-P and CF-V can be calculated using the outflow flow rate based on Darcy's

Table 1. *Densities, dynamic viscosities and viscosity coefficients of KI solutions and water*

	KI solution (0.5 mol)	KI solution (1.0 mol)	Water
Mass of 50 mL (g)	52.75	55.75	
Density (g/cm^3)	1.055	1.115	0.9982
Flow time (s)	387.3	353.3	
Constant of viscometer	0.002	0.002	
Dynamic viscosity (mm^2/s)	0.883	0.806	1.004
Viscosity coefficient (cP)	0.931	0.898	1.002

Table 2. *Intrinsic permeabilities and compensated permeabilities for the used KI solution*

	IPF-P	CF-P	CF-V
Measured permeability (cm/s)	5.10×10^{-7}	1.40×10^{-7}	7.40×10^{-8}
Intrinsic permeability (m^2)	5.22×10^{-16}	1.43×10^{-16}	7.57×10^{-17}
Compensated permeability (cm/s)	5.80×10^{-7}	1.59×10^{-7}	8.41×10^{-8}

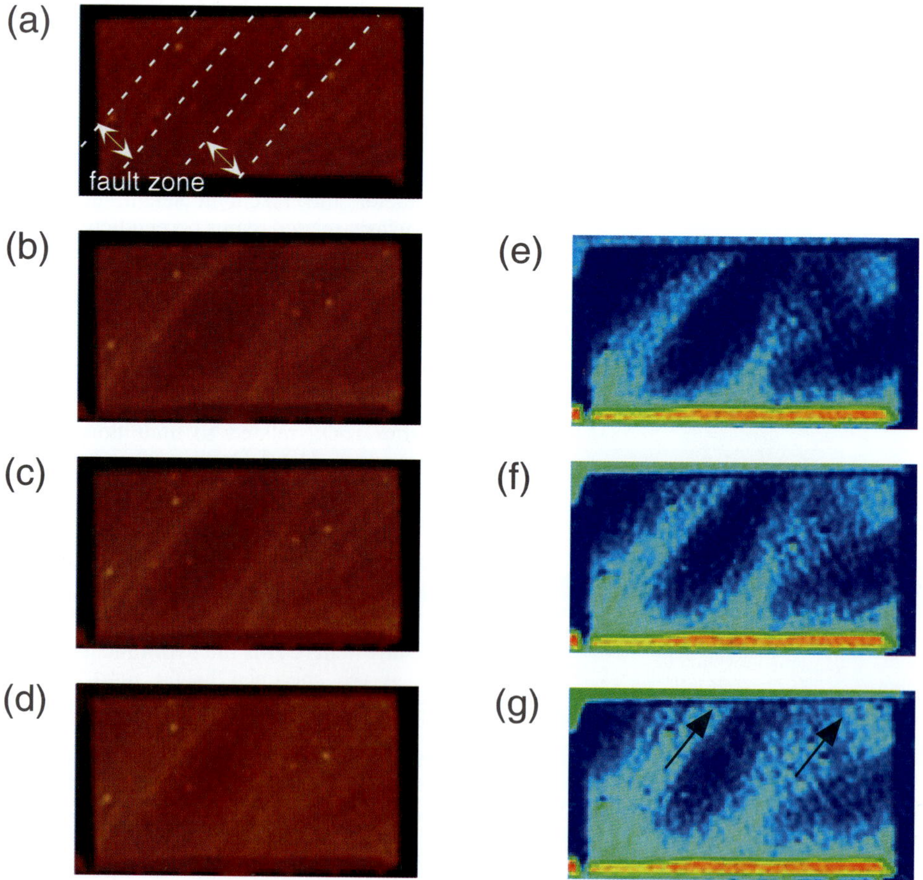

Fig. 5. (**a**) Converted 8-bit CT image of sample IPF-P at the initial state. (**b**) CT image after 150 minutes. (**c**) CT image after 210 minutes. (**d**) CT image after 270 minutes. (**e**) Differential image of CT scans recorded after 0 and 150 minutes. (**f**) Differential image of CT scans recorded after 0 and 210 minutes. (**g**) Differential image of CT scans recorded after 0 and between 270 minutes. Arrows indicate flow direction.

law. They are 5.10×10^{-7} cm/s, 1.40×10^{-7} cm/s and 7.40×10^{-8} cm/s, respectively.

The permeability, k is also related to physical properties of a fluid, such as viscosity and density. Taking this into account, the intrinsic permeability can be defined as follows:

$$k = K \times d \times g/\mu \qquad (2)$$

where k is the permeability (m/s), K is the intrinsic permeability (m^2), d is the fluid density (g/cm^3) and μ is the fluid viscosity (g/m.s). The density of the KI solution and pure water was measured using 50 mL volumetric flasks and an electronic balance. The mass of each 50 mL volume at a fluid temperature of 20°C and the calculated densities are shown in Table 1. The viscosity coefficients of the solutions are measured with a viscometer and they are calculated by the following equation:

$$d\mu = C \times t$$
$$\mu = d\mu \times d \qquad (3)$$

where $d\mu$ is the dynamic viscosity (mm^2/s), C is a constant of the viscometer (dimensionless), t is the fluid flow time through the viscometer (s), μ is the viscosity coefficient (cP) and d is the fluid density (g/cm^3). The measured flow time at a fluid temperature of 20°C and other calculated results are reported in Table 1. Higher concentrations of KI produce significantly lower viscosity coefficients. Using the values of these measured density and viscosity coefficients, intrinsic permeabilities were calculated (Table 2). For a 0.5 mol/L KI solution, the permeability can be compensated as reported in the bottom row of the table.

In-situ visualization of fluid flow

X-ray CT images of fluid flow within sample IPF-P, comprising two fault zones without cataclasis, are shown in Figure 5a. KI-permeated voxels are brighter in the CT images in Figure 5b, c and d. A three-dimensional flow image taken after 150 minutes is shown in Figure 6. With time, the KI solution gradually flows upward, in particular along the fault zone. The trend can be clearly recognized in the differential images (Fig. 5e, f and g), which are constructed by subtraction of pixel values for the initial and later images. This image conversion was carried out by: (1) exporting the TIFF-formatted 8-bit CT images as TEXT files using UltimagePro, which have matrix data of grey values for all pixels; (2) subtracting the matrix data of the later image from the initial image

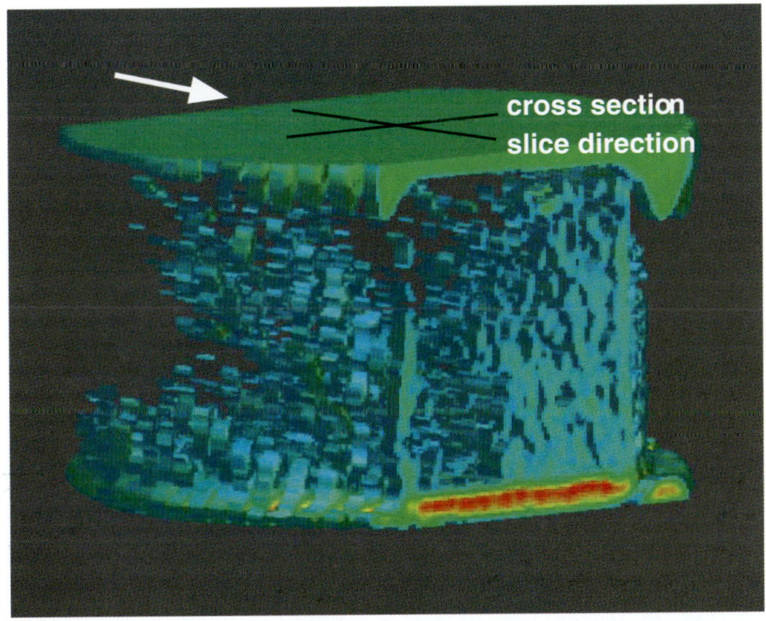

Fig. 6. Three-dimensional flow image of sample IPF-P, 150 minutes after initiation of the permeability test. Arrow indicates view direction of all X-ray CT images in Figure 5.

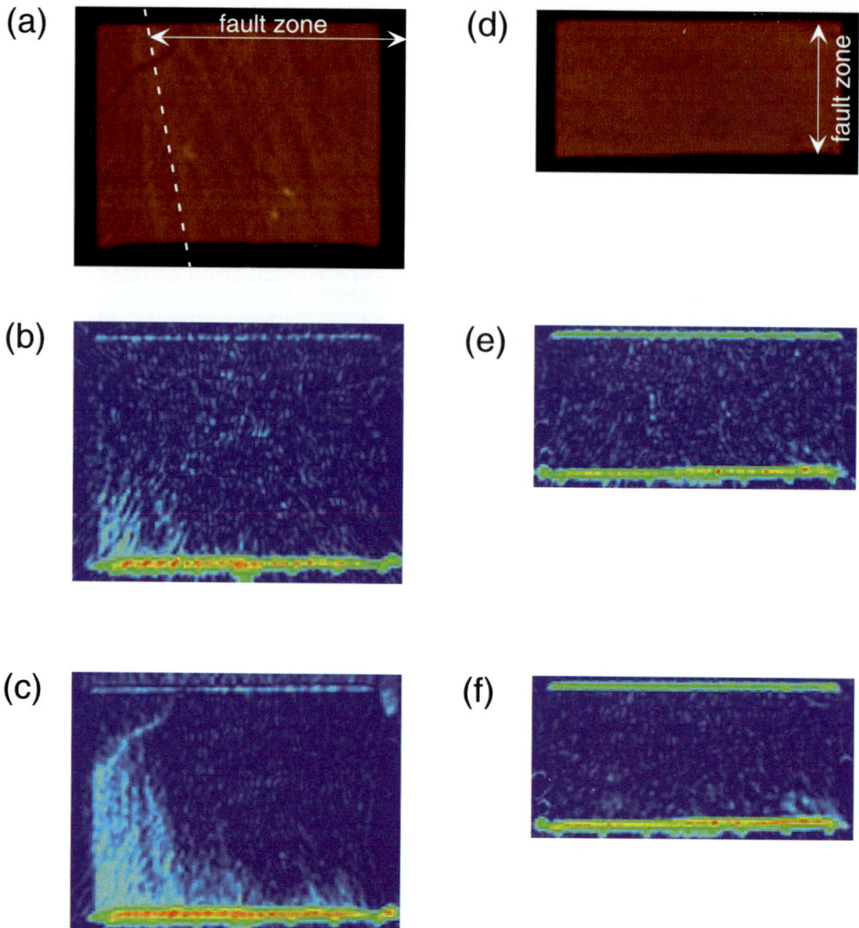

Fig. 7. (**a**) Converted 8-bit CT image of sample CF-P at the initial state. (**b**) Differential image of CT scans of sample CF-P recorded after 0 and 120 minutes. (**c**) Differential image of CT scans of sample CF-P recorded after 0 and 400 minutes. (**d**) Converted 8-bit CT image of sample CF-V at the initial state. (**e**) Differential image of CT scans of sample CF-V recorded after 0 and 180 minutes. (**f**) Differential image of CT scans of sample CF-V recorded after 0 and 540 minutes.

using Microsoft Excel; (3) importing the TEXT-formatted subtracted data into TIFF-formatted image files using NIH-Image. These CT images clearly show that the fluid flow occurs primarily in the fault zone without cataclasis, which plays the role of a fluid conduit.

CT images of fluid flow in sample CF–P, containing some fault zones with cataclasis parallel to the core axis, are shown in Figure 7a. The KI solution appears to flow exclusively within relatively intact parts of the sample in the differential images (Fig. 7b and c). The left part is relatively undeformed and acts as a fluid pathway. The crack in the upper left part of the image, which is not an original texture but a secondary crack due to core drilling, shows a concentration of fluid flow.

The CT images of fluid flow within sample CF–V, containing fault zones with cataclasis perpendicular to the axis of the core, are shown in Figure 7d. The KI solution flows pervasively upward without restriction (Figs 7e and 7f). The permeability is lower than that for CF–P sample (CF–V; 7.57×10^{-17} m^2, CF–P; 1.43×10^{-16} m^2). The cataclastic fault zones appear to seal perpendicular fluid transport. These results suggest that cataclastic fault zones act as barriers to fluid flow in the studied system.

Discussion and conclusions

In this study, X-ray CT imaging was successfully applied for visualization of fluid flow in heterogeneous deformed rocks. High-resolution, three-dimensional fluid flow distributions were measured for fault-related rocks. The results of fluid visualization during permeability testing indicate that fault zones whose deformation mechanism is independent particulate flow acts as a fluid conduit, whereas cataclastic fault zones act rather as barriers. Smith (1980), Pittman (1981) and Seeburger *et al.* (1991) suggested that the permeability perpendicular to a fault zone in sandstone decreases due to granulation and cataclasis. According to Nelson (1985) and Antonellini & Aydin (1994), faults may enhance or decrease permeability depending on whether the faults are characterized by cataclasis or mineral precipitation. They suggested that the intensity of cataclasis strongly reduces permeability. Zhu & Wong (1997) observed that the permeability evolution in porous sandstone depends on effective mean stress and deviatoric stress, and concluded that a drastic decrease in permeability is triggered by the onset of cataclasis. Their interpretation of the behaviour of cataclastic fault zones as barriers is supported by the present experimental results.

References

ANTONELLINI, M.A. & AYDIN, A. 1994. Effect of faulting on fluid flow in porous sandstones: Petrophysical properties. *American Association of Petroleum Geology Bulletin*, **78**, 355–377.

AYDIN, A. & JOHNSON, A.M. 1983. Analysis of faulting in porous sandstones. *Journal of Structural Geology*, **5**, 16–31.

CHU, C.L. & WANG, C.Y. 1988, Permeability and frictional properties of San Andreas fault gouge. *Geophysical Research Letters*, **8**, 565–568.

DARCY, H. 1856. *Les Fontaines Publiques de Dijon*. Victor Dalmont, Paris.

EVANS, J.P., FORSTER, C.B. & GODDARD, J.V. 1997. Permeability of fault-related rocks, and implications for hydraulic structure of fault zones. *Journal of Structural Geology*, **19**, 1393–1404.

HIRONO, T. & OGAWA, Y. 1998. Duplex array and thickening of accretionary prism – An example from Boso Peninsula, Japan. *Geology*, **26**, 779–782.

KNIPE, R.J. 1986. Deformation mechanism path diagrams for sediments undergoing lithification. *Geological Society of America Memoir*, **166**, 151–160.

MORROW, C.A. & BYERLEE, J.D. 1988. Permeability of rock samples from Cajon Pas, California. *Geophysical Research Letters*, **15**, 1033–1036.

MORROW, C.A., SHI, L.Q. & BYERLEE, J.D. 1981. Permeability and strength of San Andreas gouge under high pressure. *Geophysical Research Letters*, **8**, 325–329.

MOZLEY, P.S. & GOODWIN, K.B. 1995. Patterns of cementation along a Cenozoic normal fault: a record of paleoflow orientations. *Geology*, **23**, 539–542.

NELSON, R.A. 1985. *Geologic Analysis of Naturally Fractured Reservoirs*. Gulf Publishing, Houston.

PITTMAN, E.D. 1981. Effect of fault-related granulation on porosity and permeability of quartz sandstones, Simpson Group, Oklahoma. *American Association of Petroleum Geology Bulletin*, **65**, 2381–2387.

RANDOLPH, L. & JOHNSON, B. 1989. Influence of faults of moderate displacement on groundwater flow in the Hickory Sandstone aquifer in central Texas. *Geological Society of America Abstracts*, **21**, 242.

SAWAMURA, K. & NAKAJIMA, T. 1980. Miocene Silicoflagellate zones in the Boso Peninsula. *Bulletin of the Geological Survey of Japan*, **31**, 333–345.

SEEBURGER, D.A., AYDIN, A., WARNER, J.L. & WHITE, R.E. 1991. Structure of fault zones in sandstone and its effect on permeability. *American Association of Petroleum Geology Bulletin*, **75**, 669.

SMITH, D. 1980, Sealing and non-sealing faults in Louisiana Gulf Coast basin. *American Association of Petroleum Geology Bulletin*, **64**, 145–172.

SUZUKI, Y., AKIBA, F. & KAMIYA, M. 1996. Latest Oligocene siliceous microfossils from Hota Group in southern Boso Peninsula, eastern Honshu, Japan. *Journal of the Geological Society of Japan*, **102**, 1068–1071.

ZHU, W. & WONG, T.F. 1997. The transition from brittle faulting to cataclastic flow: permeability evolution. *Journal of Geophysical Research*, **102**, 3027–3041.

Rock drying tests monitored by X-ray computed tomography – the effect of saturation methods on drying behaviour

B. ROUSSET-TOURNIER[1], F. MAZEROLLE[2], Y. GÉRAUD[3] & D. JEANNETTE[3]

[1] *Expert-Center pour la Conservation du Patrimoine Bâti, EPFL MX-G, CH-1015 Lausanne, Switzerland (e-mail: benedicte.rousset@epfl.ch)*
[2] *Laboratoire de Mécanique et d'Acoustique, Chemin Joseph Aiguier, F-13402 Marseille Cedex 20*
[3] *Centre de Géochimie de la Surface – EOST, 1 rue Blessig, F-67084 Strasbourg Cedex*

Abstract: Drying experiments were conducted under controlled conditions (relative humidity, temperature, motionless atmosphere) using Fontainebleau sandstone. X-ray computed tomography images acquired at various stages of the drying show that the location of water in the pore network depends on the method of initial saturation. After capillary absorption, the trapped air, or free porosity, allows the water distribution to be homogeneous and independent of the pore structure. This is because it homogenizes the tensions that apply to the water in the pores, regardless of the pore dimensions. However, this is not the case after a total saturation under vacuum. These differences are visible from the beginning to the end of the drying experiments. These results are important for curators or restorers of the architectural heritage, because they show that before testing a conservation product for stones, the method of saturation of these stones must be considered and attention should be given to the required effect.

The patterns and intensity of alteration depend on the location of salt crystallization. This location is determined by the evaporation kinetics of the solutions circulating through the pore system. The understanding of mechanisms that control drying of stones allows us to explain degradation patterns. This study is an extension of the work of Hammecker (1993), who estimated the effects of external parameters and the capillary properties of the stone on drying kinetics.

In this paper, the influence of the pore structure and the method of initial saturation is studied. The radiological density of geological objects depends on their mineralogical composition, their pore volume and the degree of water saturation (Raynaud *et al.* 1989; Cadoret 1993; Géraud *et al.* 1999). X-ray computed tomography (CT) allows non-destructive localisation of porosity and the visualization of water distribution at different drying stages. The drying behaviour of a sandstone can, therefore, be followed by this method. Before the drying experiments, the pore network of the samples was either partially or totally saturated to study the effect of saturation methods on drying behaviour.

The results show that the method of initial saturation influences the location of the water inside the pore network during the entire drying experiment, which implies that the kinetics are modified from the beginning to the end of the drying process according to the saturation method used.

Material

The Fontainebleau sandstone that is used for this study was formed by hardening of Stampian aeolian sand (Lower Oligocene, 36–60 My) by groundwater processes (Thiry & Bertrand-Ayrault 1988). This rock is monomineralic: the chemical analyses show an SiO_2 content of 98.4%. Variations in radiological density will, therefore, only be determined by variations in porosity or water saturation.

The quartz grains are very clean and well sorted (Fig. 1), with an average grain size of about 300 m. Silicification has strongly reduced the pore volume (11.5% of the bulk volume) and resulted in the development of euhedral shapes (Fig. 1). As a result, the pore network is essentially represented by intergranular polyhedral macropores (mean diameter between 50 and 100 μm), connected by a network around the grains, with a pore diameter that is often smaller

From: MEES, F., SWENNEN, R., VAN GEET, M. & JACOBS, P. (eds) 2003. *Applications of X-ray Computed Tomography in the Geosciences*. Geological Society, London, Special Publications, **215**, 117–125. 0305-8719/03/$15.

Fig. 1. Scanning electron microscope photograph of Fontainebleau sandstone.

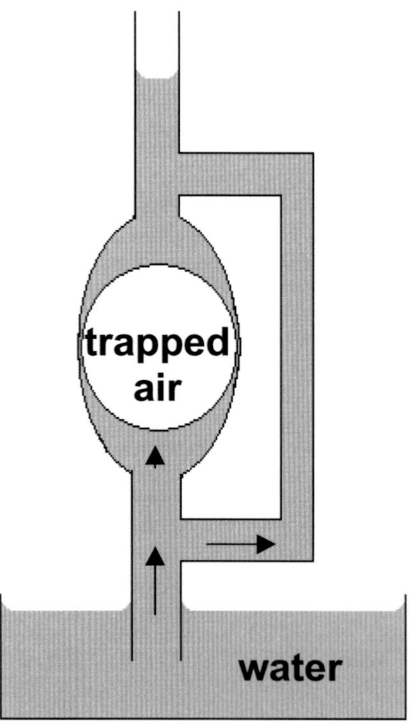

Fig. 2. Capillary absorption: by-pass of the macropores and trapping of air.

than 1 μm. The pore network is, therefore, well connected throughout the sandstone and its pore threshold, as estimated with mercury porosimetry, is 10.5 μm.

Because of these contrasting pore sizes, the Fontainebleau sandstone is very heterogeneous at the pore scale and the capillary properties are very particular: during capillary absorption in the presence of air, when the size difference between a pore and its access is very large, the passage from a constriction to a widening induces a slowing down of the saturation of the pores (Mertz 1991). As the pore network of this sandstone is formed by contrasting but well connected widenings and constrictions, the water always has the possibility of passing more quickly along narrow channels that by-pass the wider pores (Fig. 2). This results in an important volume of air being trapped in the widenings. At the end of capillary absorption, the volume occupied by water represents only 5.6% of the bulk volume (Rousset-Tournier 2001).

Experimental procedure

Specimen preparation and drying conditions

Before being dried, the samples must be saturated with water in a reproducible manner. For a first experiment, the samples were completely saturated under vacuum and dried to complete dryness under controlled conditions. The same samples were then used for a second experiment which involved drying after partial saturation by capillary absorption.

After saturation, the samples were packed in a tight-fitting sleeve so that only the upper flat face acted as transferring surface. In this way, the macroscopic area of the evaporating surface is known ($=\pi r^2$) and comparable for all the samples.

After packing, the samples were dried in a vertical position, in hermetic enclosures without any air movement. The temperature was fixed at 23°C and relative humidity was controlled at 33% with a brine of hexa-hydrated magnesium chloride (standard method NF X 15–119). The stability of these parameters was checked with a thermo-hygrometer. The samples were weighed at regular time intervals during drying, which allowed deriving information of kinetics from experimental curves representing the loss of weight per unit of the macroscopic evaporating surface versus time. As long as water saturation of the pore volume is high enough, capillary transfers wet the evaporating surface and the flux is constant (Fig. 3). When the water saturation becomes lower the critical saturation that is specific to each sample as a function of the experimental conditions (Sc) (Rousset-Tournier 2001), the capillary supply to the surface is no longer sufficient to wet it. The flux then decreases because water diffusion through the pore network becomes gradually predominant, compared to capillary transfers. The establishment of the experimental kinetics for each sample and each experimental condition allowed the determination of the stages of drying in which CT examination could be interesting.

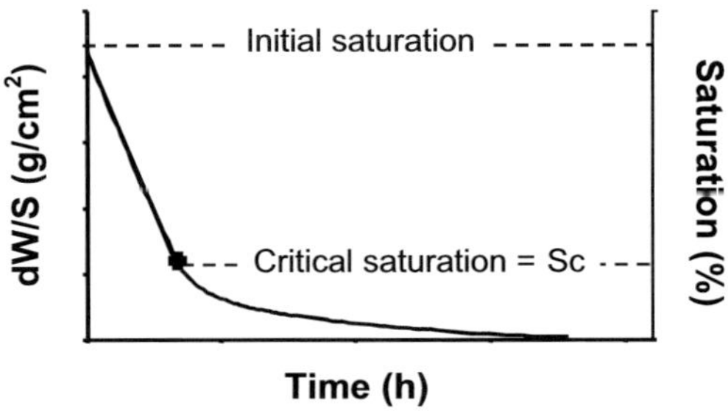

Fig. 3. Standard drying kinetic of a porous medium. dW = weight variation; S = macroscopic evaporating surface; Saturation = percentage of the pore volume filled with water.

CT examinations

A second-generation medical CT scanner (CGR ND 8000) of the Laboratoire de Mécanique et d'Acoustique (Marseille) was used. Acquisition was performed during rotation over 360° with steps of 12°. The acquisition time for a cross-section was 80 seconds and the calculation time was 45 seconds. Voltage and current were 120 kV and 18 mA, respectively. Each slice is 2 mm thick and the voxel dimensions or spatial resolution are, therefore, $2 \times 0.7 \times 0.7$ mm. The density resolution is higher than 0.2% and the precision of the measurements related to the apparent gravimetric density is estimated between 10^{-3} and 10^{-4} (Dou *et al.* 1985). The scale of standard visualization (Dou *et al.* 1985) was modified so that the radiological density (RD) of the air is −1000 UHM (Modified Hounsfield Units) and the RD of water is −600 UHM. This scale increases the contrast for geological materials, to optimize visualization of their structure.

The experiments were carried out on cylindrical samples that were 4 cm in diameter and 8 cm in height. The spatial resolution is low because the voxel volume is approximately 1 mm^3, whereas the pore diameters of the Fontainebleau sandstone are between 300 µm and less than 1 µm. This CT study permits distinguishing zones with high porosity from zones with low porosity, but it does not show the behaviour of the content of individual pores during drying.

Drying of the samples was performed under experimental conditions that were pre-established in order to follow the distribution of water within the pore network by CT. The first examination always corresponds to the maximum saturation (partial or total). During every scan, which took approximately 2 hours (36 sections, at 3 minutes per section), the evaporation of water contained in the sample was stopped with an integral tight sleeve. The images acquired for the same cross-section at the beginning and at the end of a complete scan of the sample show that, at the scale of this scanner, the distribution of water does not change during the examination when carried out on samples in a horizontal position.

Results and discussion

The images in Figure 4 represent RD distributions of cross-sections for each state of saturation examined for sample Ftx2a2. Only 9 of the 36 sections are shown in this figure. The distance between each of the 9 sections is 10 mm.

The rows in Figure 4 show the desaturation after total (Nt) or capillary (Nc) saturation. The

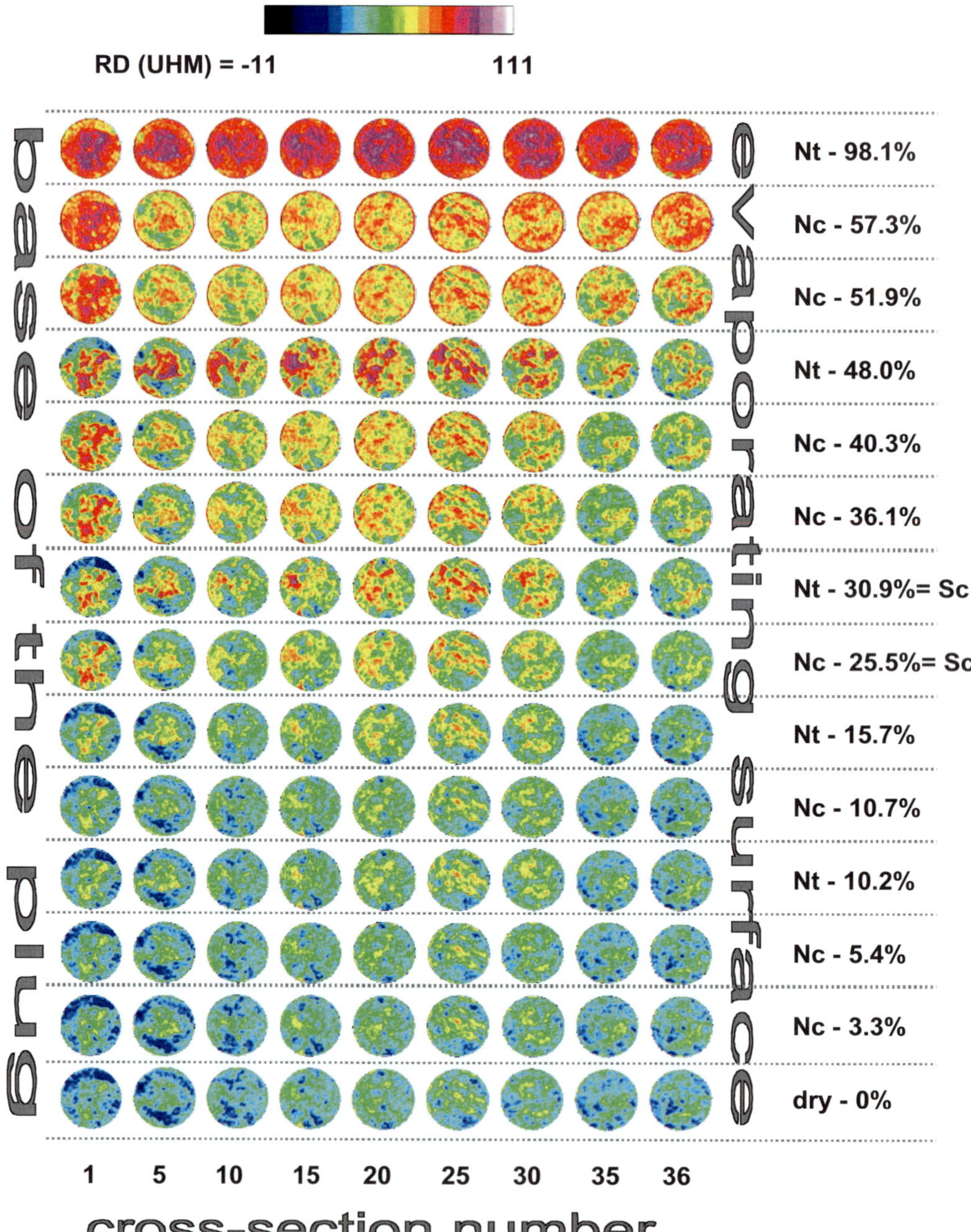

Fig. 4. CT images of the Fontainebleau sandstone sample Ftx2a2. 1 to 36: cross-section number (1 = base of the sample, 36 = evaporating surface); saturation method: N_c = capillary absorption; N_t = total saturation; $X\%$ = saturation at the time of the scan; RD = radiological density in Modified Hounsfield Units.

two upper rows ('Nt-98.1%' and 'Nc-57.3%') correspond to maximum water saturation (98.1% after Nt, 57.3% after Nc) and the bottom row ('dry-0%') corresponds to the dry state (0% water saturation). The maximum saturation observed after total saturation under vacuum was only 98.1% (instead of 100%) because some water was lost during transport between the place of saturation and the place of the scanning, which took 72 hours.

These images represent the matrix (solid phase) plus the water when the sample is wet. As the Fontainebleau sandstone is composed of only one mineral species, the RD variations of a voxel, from the dry state to any state of saturation, can be directly translated to the pore volume saturation of this voxel (Géraud *et al.* 1999).

The images of the dry state ('dry-0%', Fig. 4) confirm that high porosity zones (low RD = dark colours) can be distinguished from low porosity zones (high RD = light colours). Observations made with an optical and electron microscope show that the low porosity zones can be regarded as microporous zones and the high porosity sections as macroporous zones. The results of microfocus CT studies using Fontainebleau sandstone by Auzerais *et al.* (1996) and Coker & Torquato (1996) confirm this hypothesis. Whereas the pore structure of this sandstone is relatively homogeneous on the sample scale, without stratification or variations in structure, CT images obtained with a medical scanner highlights heterogeneities of a few millimetres at the cross-section scale. These heterogeneities are induced by the secondary silicifications related to a more or less compact stacking of the quartz grains before cementation (Thiry & Bertrand-Ayrault 1988).

Initial distribution of water

Figure 5 illustrates the drying kinetics for the experiments: every lozenge corresponds to a CT scan. Figure 6 shows the RD profile of the dry matrix (RD matrix), which gives an image of the porosity (the porosity changes in the different cross-sections because the RD matrix is not flat). The ΔRD profiles in the same figure represent the amount of water contained in every cross-section after total ('Nt-98.1%') and capillary

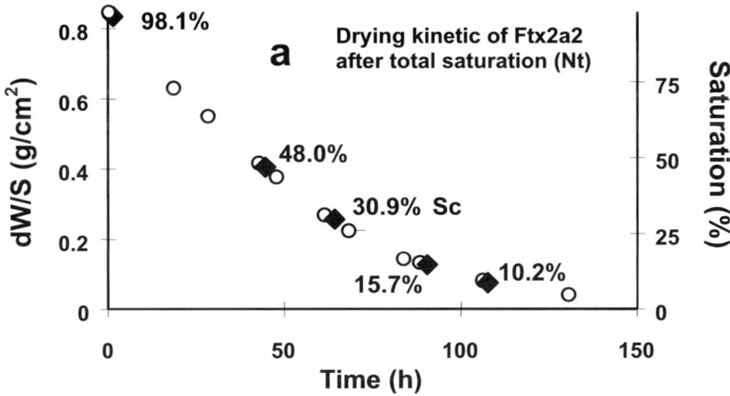

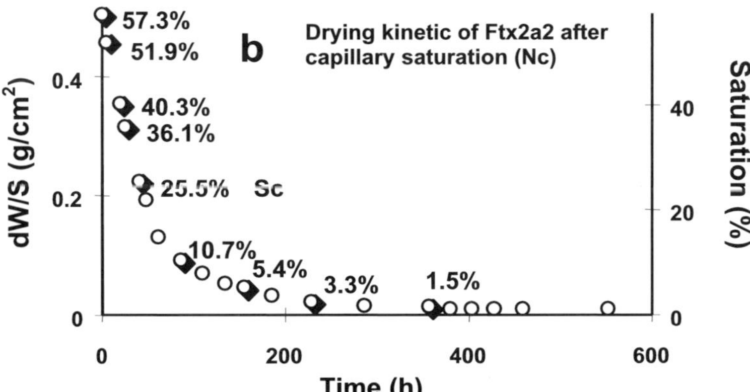

Fig. 5. CT aquisitions plotted on drying curves obtained after total saturation (**a**) and after capillary saturation (**b**). Dots = weight measurements; lozenges = measurements at stages corresponding to the images in Figure 4.

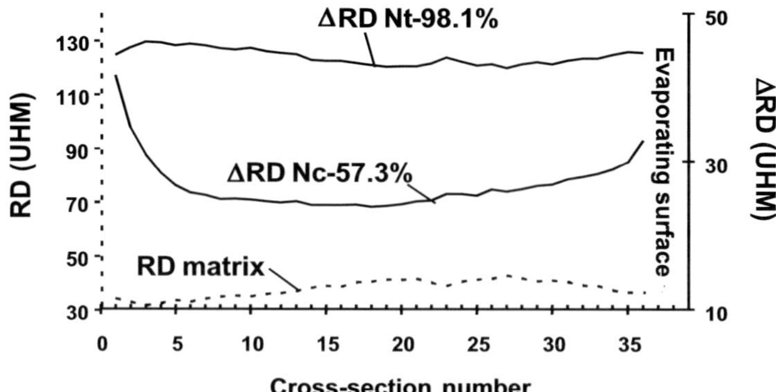

Fig. 6. RD profile of the dry matrix (RD matrix) and ΔRD profiles of the maximum water content after total saturation (Nt-98.1%) and capillary saturation (Nc-57.3%).

('Nc-57.3%') saturation. These ΔRD profiles are RD profiles from which the dry state RD has been subtracted.

Figure 6 shows that after total saturation, by definition, the entire connected pore volume is filled with water. The water distribution is heterogeneous, due to the heterogeneous distribution of porosity within the sample. When the value of the RD matrix profile of a cross-section is high, the pore volume is low and, consequently, the corresponding value of the ΔRD-Nt profile is low as well. In contrast, after capillary saturation the water tends to be distributed in a homogeneous manner, particularly in the centre, whereas saturation at both ends of the sample tends towards total saturation. During capillary absorption the water tends to occupy a part of the pore network whose access has a radius smaller or slightly larger than the pore threshold. The largest pores are trapped. Thus, the part of the network that is freely accessible is relatively homogeneous. This is the case for the centre of sample Ftx2a2 (Fig. 6). Saturation at the base of the sample (on the left-hand side in Figure 6) can be partly explained by the saturation method because the lower 1 or 2 mm are immersed in water during saturation. The high saturation at the top of the sample after capillary absorption (evaporating surface in Figs 6 and 4) is explained by an edge effect. During the experiments of capillary absorption, the capillary fringe reached the top of the sample. Water, therefore, had time to invade part of the trapped pores from which the air is evacuated by diffusion. This 'additional' water is not distributed homogeneously in the sample, unlike the water that penetrates by capillary absorption: the larger the pores and the closer they are to the edges, the easier the diffusion of air will be. The ends of the tested sample happen to be more porous than the centre (RD matrix profile, Fig. 6), thus they can accommodate more water than the centre. Finally, the water that moves from the interior of the pore network to the edges does not only form meniscuses in the cavities open outside, it also creeps and wets external mineral surfaces. This external capillary water explains the saturation corona observed along the edge of the cross-sections after capillary absorption ('Nc-57.3%', Fig. 4).

Distribution of water during desaturation

The profiles in Figures 7 and 8 represent the RD of the matrix and the degree of water saturation for each cross-section (deduced from the ΔRD profiles), for every stage of the drying experiments that are also illustrated in Figure 4. Figure 7 presents bulk saturations greater than or equal to Sc, while Figure 8 only includes bulk saturations lower than Sc.

Figures 4, 7 and 8 show that the spatial distribution of water during desaturation evolves differently when a different method is used to fill the samples with water. These differences are visible from the beginning to the end of the drying experiments, both on the sample scale (Figs 7 and 8) and on the cross-section scale (Fig. 4).

Bulk water saturations greater than Sc. At the beginning of the drying experiments, the macroporous zones are emptied quicker than the rest of the pore network, whatever the initial saturation. In the filled macropores, tensions exerted on the water by the matrix (= matrix potential) are lower than in partially- or totally-filled micropores, or in partially saturated macropores

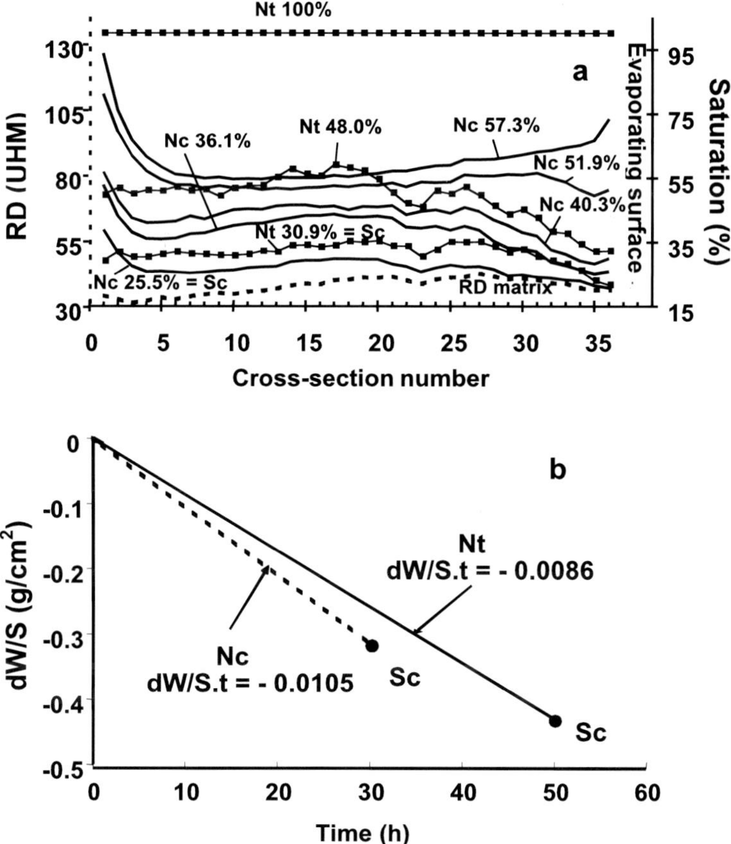

Fig. 7. (**a**) Profiles of the mean saturation for each cross-section at the first drying stage (saturation ≥ *Sc*) and RD of thc dry matrix. (**b**) Drying kinetics of sample Ftx2a2 for saturation ≥ *Sc*.

(Hillel 1971). In saturated macropores, mobilization of the water is therefore easier, which explains why these pores are emptied first.

After capillary saturation, the only filled macropores are located at the top and bottom of the samples. These ends are, therefore, more quickly desaturated (Figs 7 and 4), whereas in the centre of the sample the desaturation tends to be uniform and independent of the pore structure: the same amount of water leaves each section in the same time interval whatever the pore volume. In this part of the sample, the matrix potential is the same in saturated micropores or in partially saturated macropores. The air trapped during capillary absorption allows a uniform distribution of the tensions during drying, which explains the uniform desaturation.

After total saturation all pores are filled with water, whatever their size and their location in the sample. As a result, desaturation is much more heterogeneous than after capillary saturation: the macroporous zones are emptied before the microporous parts. At the start of drying, the water distribution is consequently strongly influenced by the pore structure (Figs 7 and 4).

The drying kinetics obtained after both saturation modes show that the flux measured after capillary saturation is slightly, but systematically, higher than the flux measured after total saturation (Fig. 7). A faster capillary transfer seems to be supported by the more uniform desaturation, which is allowed by the trapped air.

Bulk water saturations equal to Sc. For both drying experiments (Fig. 5) the bulk water saturation Sc is related to the state when the evaporating surface begins to dry. CT observations show that from this stage onward, the water content of the cross-sections nearest to the evaporating surface starts to decrease much more, whatever the initial saturation (Fig. 7).

Although the distribution of water inside the sample tends to be homogenized during the first drying stage, when Sc is reached, this distribution

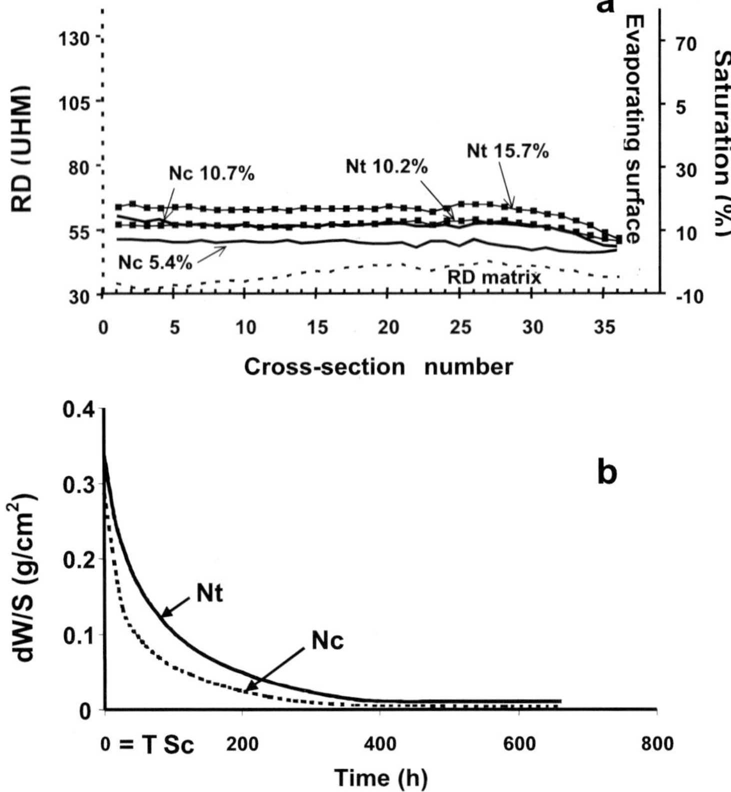

Fig. 8. (**a**) Profiles of the mean saturations for each cross-section at the second drying stage (saturation < Sc) and RD of the dry matrix. (**b**) Drying kinetics of sample Ftx2a2 for saturation < Sc.

is still influenced by the saturation mode (Figs 7 and 4). After capillary absorption the base of the sample (on the left-hand in Figs 7 and 4) remains more saturated than the rest of the sample, where the saturation is uniform and independent of the pore structure, contrary to what is observed after a total saturation: water saturation tends to be high in the microporous zones and low in the macroporous parts (Figs 7 and 4).

This implies that for the same drying conditions and the same sample, the rupture of hydrous continuity does not correspond to a single distribution of water, contrary to what is suggested by Hammecker (1993).

Bulk water saturations lower than Sc. When the evaporating surface is dry, desaturation of the sections nearest to this surface increases slightly and affects an increasing number of slices (Fig. 8). It is probably in this less saturated part of the sample that the capillary transfers, bringing water to the surface, decrease gradually and water vapour diffusion becomes predominant.

The CT scans show that the distribution of water still tends to be homogenized during this last drying stage, but there remains a difference according to the saturation mode almost until the end of the drying experiments. For example, for the same bulk saturation, the most microporous zones in sample Ftx2a2 remain more saturated after total saturation (see for example, slice 25 for bulk saturations 'Nt-10.2%' and 'Nc-10.7%', Fig. 4). For water saturations lower than Sc, this difference in distribution of the water according to the method of initial saturation is still expressed by a difference in drying kinetic: the same value of total saturation is reached more quickly after capillary saturation than after total saturation (Fig. 8). The flux decreases less quickly after capillary saturation, which means that for low bulk saturations the supply of water in the liquid phase to the surface is greater after capillary saturation than after total saturation. In other words, the difference in water distribution implies that, for the same low bulk saturation, hydrous continuity is better after capillary saturation than after total saturation.

Conclusions

The experimental results obtained by CT analysis show a significant effect of the initial saturation method on drying kinetics, from the beginning to the end of drying experiments. Capillary absorption allows trapping of air in the macropores and, consequently, allows the saturation of a 'homogeneous' pore network in which the matrix potential exerted on water is uniform and independent of the pore structure. Total saturation, on the other hand, completely fills small and large pores. In this case the larger the pores, the lower the matrix potential that is induced.

During drying the matrix potentials evolve, but they remain uniform throughout the network after capillary saturation, which is not the case after total saturation. As a result, desaturation is much more uniform and independent of the pore structure after capillary saturation than after total saturation. The drying kinetics are, therefore, faster after capillary absorption than after total saturation.

These results are important for curators or restorers of the architectural heritage because they show that before testing a conservation product for stones the method of saturation of these stones must be selected, taking the required effect into consideration. For example, standard tests of waterproofing products often impose total saturation of the pore network of stones, but the result will obviously be different from what occurs on site, where water saturation is more often closer to capillary absorption than total saturation. The drying of some of the standard tests should be reconsidered.

The authors wish to thank S. Ebersole-Mühlehaus for her comments and assistance in preparing the English manuscript. This is publication number EOST/2002-403-CGS.

References

AUZERAIS, F.M., DUNSMUIR, J., FERREOL, B.B., MARTYS, N., OLSON, J., RAMAKRISHNAN, T.S., ROTHMAN, D.H. & SCHWARTZ, L.M. 1996. Transport in sandstone: a study based on three dimensional microtomography. *Geophysical Research Letters*, **23**, 705–708.

CADORET, T. 1993. *Effet de la Saturation Eau/Gaz sur les Propriétés Acoustiques des Roches – Étude aux Fréquences Sonores et Ultrasonores*. PhD thesis, Université Paris VII.

COKER, D.A. & TORQUATO, S. 1996. Morphology and physical properties of Fontainebleau sandstone via tomographic analysis. *Journal of Geophysical Research*, **101**, 17497–17506.

DOU, H., FLESIA, E., CAGNASSO, A. & LATIERE, H.J. 1985. Étude du charbon de Gardanne par tomodensitométrie. *Analusis*, **13**, 69–75.

GÉRAUD, Y., TOURNIER, B. & MAZEROLLE, F. 1999. Detection of porosity and mineralogical variations in geological materials: radiological density measured by X-ray tomography. *Proceedings of the International Symposium on Imaging Applications in Geology, Géovision 99, Liège, 6–7 May*, 109–112.

HAMMECKER, C. 1993. *Importance des Transferts d'Eau dans la Dégradation des Pierres en Œuvre*. PhD thesis, Université Strasbourg I.

HILLEL, D. 1971. *Soil and Water – Physical Principles and Processes*. Academic Press Inc., Orlando.

MERTZ, J.D. 1991. Structures de porosité et propriétés de transport dans les grès. *Sciences Géologiques, Mémoire*, **90**.

RAYNAUD, S., FABRE, D., MAZEROLLE, F., GÉRAUD, Y. & LATIÈRE, H.J. 1989. Analysis of the internal structure of rocks and characterization of mechanical deformation by a non destructive method: X-ray tomodensitometry. *Tectonophysics*, **159**, 149–159.

ROUSSET-TOURNIER, B. 2001. *Transferts par Capillarité et Evaporation dans des Roches: Rôle des Structures de Porosité*. PhD thesis, Université Strasbourg I.

THIRY, M. & BERTRAND-AYRAULT, M. 1988. Le grès de Fontainebleau: genèse par écoulement de nappes phréatiques lors de l'entaille des vallées durant le Plio-quaternaire et phénomènes connexes. *Bulletin d'Information des Géologues du Bassin de Paris*, **25**, 25–40.

Characterization by X-ray computed tomography of water absorption in a limestone used as building stone in the Oviedo Cathedral (Spain)

V. G. RUIZ DE ARGANDOÑA[1], A. RODRIGUEZ-REY[1], C. CELORIO[2], L. CALLEJA[1] & L. M. SUÁREZ DEL RIO[1]

[1] *Department of Geology, University of Oviedo, Jesus Arias de Velasco s/n, 33005 Oviedo, Spain (e-mail: vgargand@geol.uniovi.es)*

[2] *Radiology Service, Hospital Alvarez Buylla, 33616 Mieres, Spain*

Abstract: Water plays a fundamental role in rock weathering processes. Its penetration and movement inside rocks greatly influences the nature and intensity of damage affecting building stones. X-ray computed tomography (CT) is a useful technique for non-destructive mapping of water penetration. For the reported study, CT was used to investigate the internal structure and water penetration patterns for the Piedramuelle Stone, a Cretaceous limestone used as building material in the Oviedo Cathedral (North of Spain).

X-ray CT provides good images of the internal structure of the samples: the sedimentary layering due to differences in composition and porosity is clearly seen, as well as other textural features. The movement and penetration rate of water was monitored during standard free absorption water tests. The images that were obtained show a clear difference between dry and wet zones in the interior of the sample. Water movement is related to the petrographic characteristics of the rock, mainly to layering which controls the direction of water penetration. Hounsfield Unit numbers provide a quantitative approach for assessing the penetration rate of water.

The behaviour and kinetics of water inside rocks is a field of research that in recent years, has been explored with regard to rock weathering studies. X-ray computed tomography (CT) is a non-destructive technique that has a large potential for this type of research because it provides images of the interior of a sample, allowing monitoring of the evolution of water movement as well as the development of damage caused by ageing cycles, without the need to destroy the studied samples. For technical details of CT, the reader is referred to Wellington & Vinegar (1987), Duliu (1999) and Van Geet *et al.* (2000).

X-ray CT is being increasingly used in rock weathering studies. Jacobs *et al.* (1997) showed the potential of the CT technique in the cultural heritage domain by monitoring various physical and biological weathering processes of calcarenites, marbles and medieval mortars used as natural building stones. Ruiz de Argandoña *et al.* (1999) produced images of the evolution of the internal damage in a dolomitic rock caused during freeze-thaw tests. CT monitoring of water movement has been reported by several researchers. Queisser (1986) compared dry sandstone samples with those partially immersed in water. Mossotti & Castanier (1990) studied the movement of water in limestones in total immersion over time and compared the water distribution in the interior of the samples during drying tests in different conditions. In capillary imbibition tests of granitic rocks, Géraud *et al.* (1998) explored the relationship between porosity, connectivity and permeability. X-ray CT has also extensively been used to monitor the spatial distribution of soil water in the vicinity of plant roots (Hainsworth & Aylmore 1983; Aylmore 1993).

In this paper, we report on a CT study of the internal structure and water movement inside the Piedramuelle Stone, a Cretaceous limestone that has been used as building material in the Gothic Oviedo Cathedral (North of Spain). This rock has mineralogical and textural characteristics that are relevant to the interpretation of water movement in its interior. Its grain size is rather homogeneous but it shows mineralogical and textural layering that conditions the movement of water.

A strategy for studying water imbibition through time in function of variable petrographic characteristics is used in this study to develop a method for the quantification of water penetration rates using CT numbers in Hounsfield Units.

From: MEES, F., SWENNEN, R., VAN GEET, M. & JACOBS, P. (eds) 2003. *Applications of X-ray Computed Tomography in the Geosciences*. Geological Society, London, Special Publications, **215**, 127–134. 0305-8719/03/$15.

Studied material: Piedramuelle Stone

The Piedramuelle Stone represents about 37% of the rocks that were used for the construction of the Oviedo Cathedral (Esbert & Marcos 1983). It has also been widely used for other ancient buildings in the city.

Carbonates (mainly calcite) form the bulk of the material (70–80%), with a smaller percentage of ferrous dolomite and ankerite. Other constituents are quartz (8–15%), phyllosilicates (6–15%) (including muscovite, weathered biotite, chlorite, glauconite, illite and kaolinite), iron oxides (4–6%) (mainly goethite and hematite) and organic matter. It has an average grain size of between 200 and 300 μm. Angular quartz grains range in size between 50 and 100 μm while rounded aggregates of cryptocrystalline glauconite have an average grain size of about 100 μm.

Macroscopically, three different types of layers can be distinguished, based on composition and texture: type A (yellow, with black dots), type B (red) and type C (yellow). Figure 1 shows the macroscopic image of one of the samples.

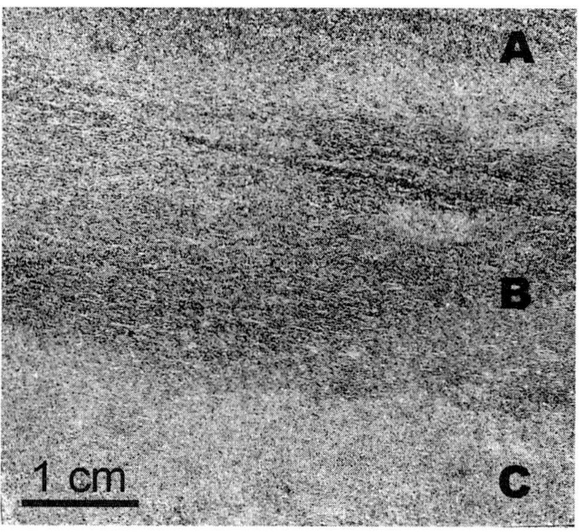

Fig. 1. Macroscopic image of a dry sample showing the three types of layers in the Piedramuelle Stone (A, B and C).

The yellow, black-dotted layers (type A) comprise organic matter that appears as inclusions in the calcite crystals, together with some iron oxides. The reddish layers (type B) are characterized by an abundance of iron oxides. The yellow layer (type C) lacks organic matter and iron oxides and has a larger amount of glauconite grains than the other two types of layers.

The three types of layers have different porosities. Small cores were taken from each of the layer types for porosity measurements by means of mercury porosimetry (Fig. 2). Mean porosity values are 7.8% for type A, 18.6% for type B and 13.7% for type C. The pore size distribution is unimodal for types B and C and bimodal for type A, with mean pore radii of 0.1 and 0.01 μm for type A, 0.5 μm for type B and 0.3 μm for type C.

For this paper, loose masonry blocks taken from the cathedral were cut parallel to the layering to obtain cubic samples approximately 50 mm in length, as required by the free water absorption standard method that was followed. The values of physical properties, measured following the RILEM (1980) standards, are 10.5% open porosity, 2400 kg/m^3 bulk density and 4.3% water saturation content.

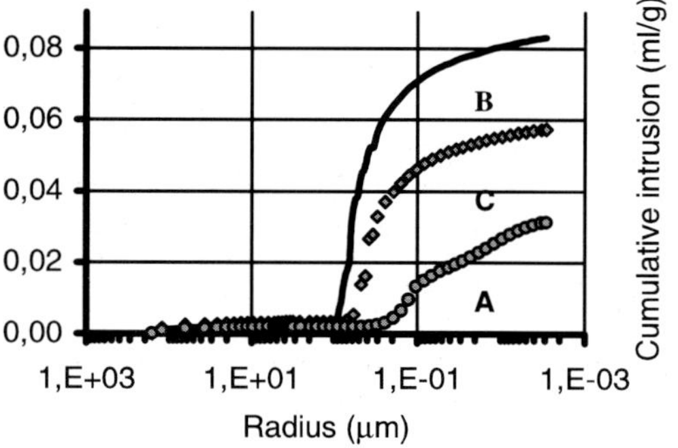

Fig. 2. Mercury porosimetry curves showing differences in pore radius between the three layer types (A, B and C).

Experimental procedure: free water absorption

Free water absorption was measured following the Normal 7/81 standard test (CNR-ICR, Rome). Samples were dried at 60°C until constant weight and subsequently cooled down inside a desiccator. They were then introduced into a methacrylate container that was completely filled with water and the increase in weight, due to free water absorption, was determined at different time intervals. The test was ended when the increment of mass stayed below 0.01% over a 24 hour period.

Figure 3 illustrates the increase in water content (expressed as percentage of the dry weight of the sample) produced over time (expressed as the square root of time in minutes). As in many rocks, the water content increases rapidly at the

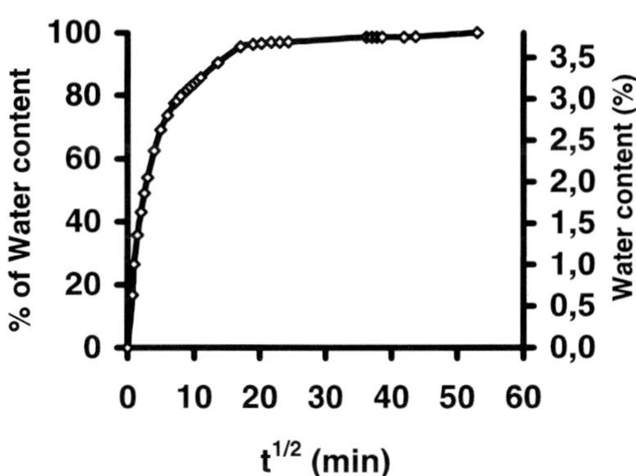

Fig. 3. Free water absorption curve for the Piedramuelle Stone. The Y axis on the left side represents the percentage of the maximum water content.

Fig. 4. Experimental set-up for the study of water absorption during X-ray CT monitoring. The container has four point supports to have a water jacket of 1 cm along the six faces of the sample. The virtual central section studied by CT is indicated in this image.

beginning, then slows down and finally becomes asymptotic. 86% of the water content is absorbed in the first two hours.

The test was then carried out again under X-ray CT monitoring. The cubic samples were placed inside the methacrylate container and scanned under dry conditions. Water was then poured into the container until a water jacket of 1 cm thickness covered all faces of the sample (Fig. 4). Because the water could penetrate under the same conditions through all six faces of the sample, the possibly anisotropic nature of water absorption could not be studied. Some of the air in the pores could get out under these conditions and the capillary behaviour of the sample was not optimal. However, the property of interest here was only the flow of water in free absorption tests.

Images of the same central plane with a thickness of 1 mm were obtained at the same time intervals as those used for the Normal 7/81 standard test (CNR-ICR, Rome). The duration of this test was 133 minutes, covering the most interesting part of the absorption curves of the standard test.

A General Electric medical scanner (SYTEC 3000) was used. The images were obtained with a voltage of 120 kV and a current of 160 mA, with a scan time of three seconds. The Window Width and Window Level values were adjusted to obtain an adequate contrast in the images.

Textural information by X-ray CT

The three layer types (A, B C) with different compositions and porosities, are clearly visible. Figure 5 is an image of the vertical central plane of the cubic sample in Figure 4 recorded under dry conditions.

At the used voltage, the contrast of the image mainly reflects differences in porosity between the three layers: the Compton effect is predominant and the linear attenuation coefficient (μ)

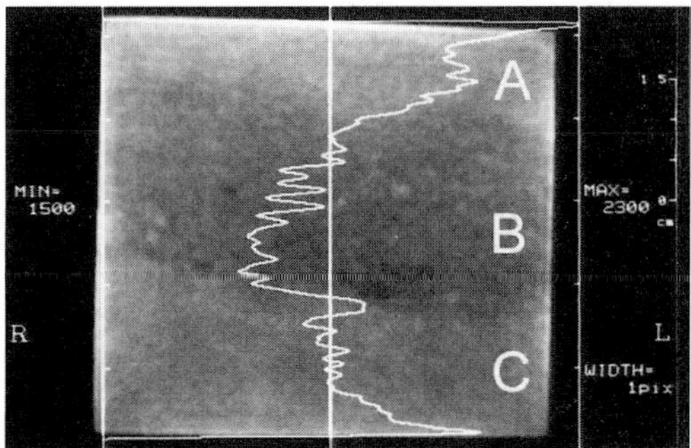

Fig. 5. CT image of the vertical central plane of the sample shown in Figure 4 showing a contrast between the three layers of different types (A, B and C). The variation of CT numbers along the vertical central line of the image is superimposed. Image taken under dry conditions.

mainly depends on the density of the sample rather than on mineralogical composition. Areas of low porosity and high bulk density have greater attenuation coefficient values and show lighter shades in the images. Layer B which has the highest porosity and lowest bulk density, appears darker, whereas layer A with the lowest porosity and the highest density, appears lighter. Layer C with intermediate density values, is characterized by intermediate values.

In Figure 5 the vertical profile of radiological density along the centre of the CT image shows the difference in CT numbers between the three layers. No modified radiological densities for air and water were used, such as those proposed by Raynaud *et al.* (1989) and Géraud *et al.* (1998), because the normal Hounsfield Units seemed to be useful for our studies.

Visualization by X-ray CT of water penetration

Figure 6 shows CT images of the vertical central plane of a sample at different times during the free water absorption test. Consecutive images demonstrate the displacement with time of the water front toward the interior. Wet zones show lighter shades than the dry rocks, allowing an estimation of the position of the water front at each moment. From top to bottom and from left to right, the images become lighter as the sample becomes wet.

It appears that water movement is related to the petrographical characteristics of the rock, as the layered structure conditions the direction of water penetration. The water migrates following the stratification and moves preferentially

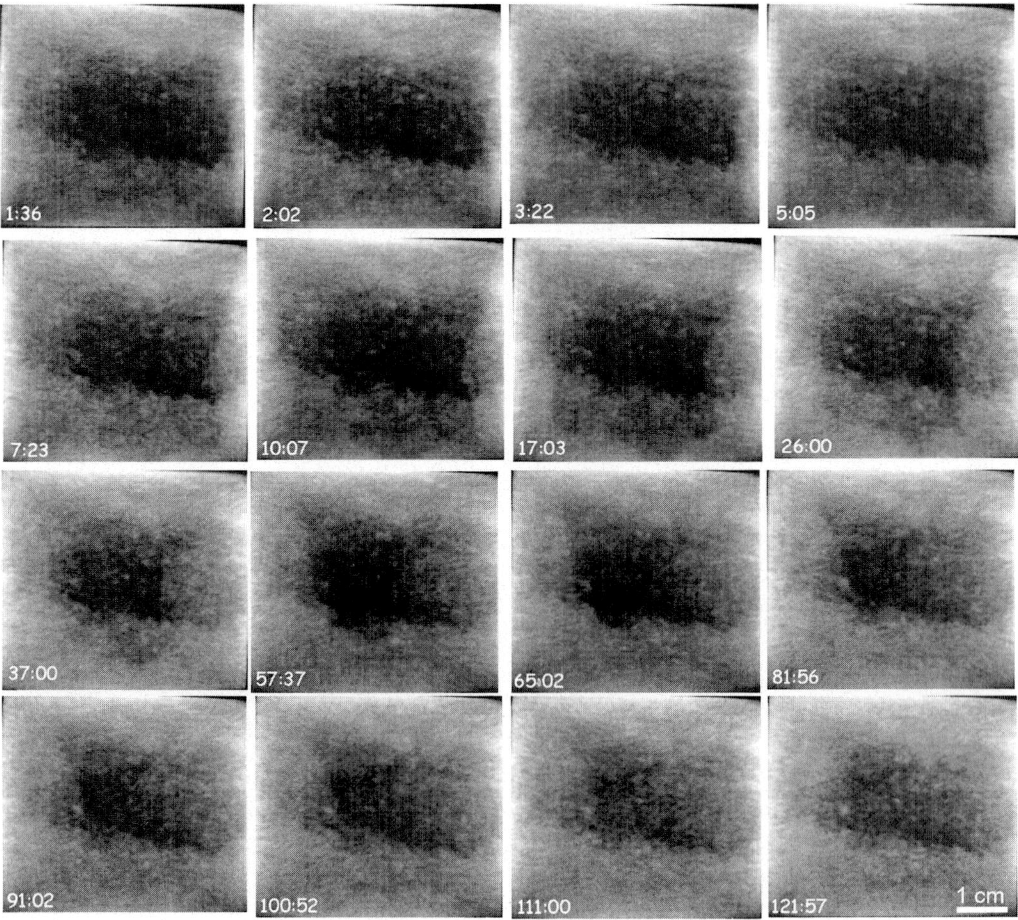

Fig. 6. Evolution over time during the free absorption test of the CT image of the central vertical plane of a sample. Lighter colours correspond to a higher water content. Time in minutes and seconds is indicated at the bottom left of each image.

along the central layer with the highest porosity. Migration in a direction perpendicular to the stratification is less important. This can be seen, for instance, in the image taken after 26 minutes of water absorption. The wet zone on the right is bigger in the central layer than in the top and bottom layers. The pore radius size also conditions the penetration of the water. Layer B which gets soaked more quickly, has a larger mean pore radius (0.5 μm) than layer A the least soaked one (0.1 and 0.01 μm). Layer C with a mean pore radius of 0.3 μm attains intermediate values.

Quantification of water penetration

In an attempt to obtain a quantitative estimate of water penetration inside the rock, 23 to 25 square regions of interest (ROI) were defined for each sample, depending on the sample size, aligned along the central horizontal line (layer B) of the virtual section studied by X-ray CT. Each ROI has $20 \times 20 = 400$ pixels corresponding to $2 \times 2\,\text{mm}^2$. As the thickness of the virtual section is 1 mm, the size of the corresponding voxel is $2 \times 2 \times 1\,\text{mm}^3$. In order to eliminate the artefacts associated to the border effect, a zone of 1 mm in thickness along the rims of the sample was not considered. Figure 7 shows the position of some significant ROI's, whose behaviour is discussed below.

Although the layers are macroscopically homogeneous, the ROIs do not have exactly the same mineralogical composition and porosity and their mean CT numbers under dry conditions are slightly different, varying between 1815 and 2035 HU. Figure 8 shows the porosity, both in percentage and pore size, of two adjacent areas (B1 and B2) in layer B as seen by backscattered scanning electron microscopy. The mean CT numbers measured drops drastically

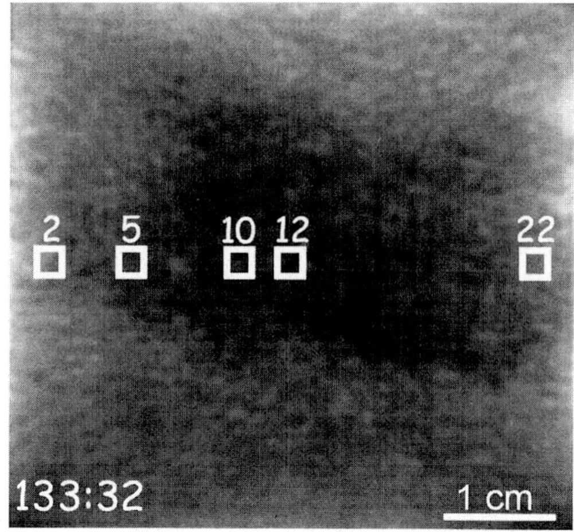

Fig. 7. Location of five ROIs in the central vertical plane imaged by CT after 133 minutes and 32 seconds of water absorption.

from the value for the dry sample when the block is immersed in water, even before water migrates into the sample, due to the influence of the water jacket on X-ray attenuation. This reduction is evident in all ROI's that were studied.

The mean CT number of each ROI was measured at different times during the water absorption test. Table 1 shows the evolution of the mean CT numbers of several ROI's during the test. Figure 9 exemplifies the evolution of the mean CT number of ROI 5 (for location see Fig. 7). Several parts can be distinguished in this curve: (a) the distinct initial drop of the mean CT number, due to the introduction of the water jacket; (b) a first horizontal segment, with statistical variations due to the experimental procedure and equipment; (c) a marked increase in the mean CT number caused by water

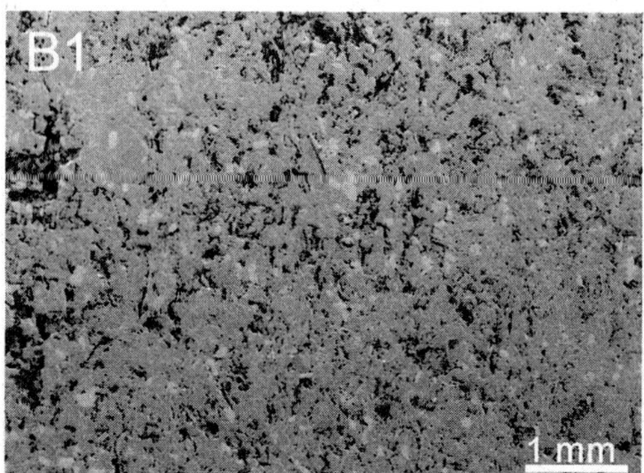

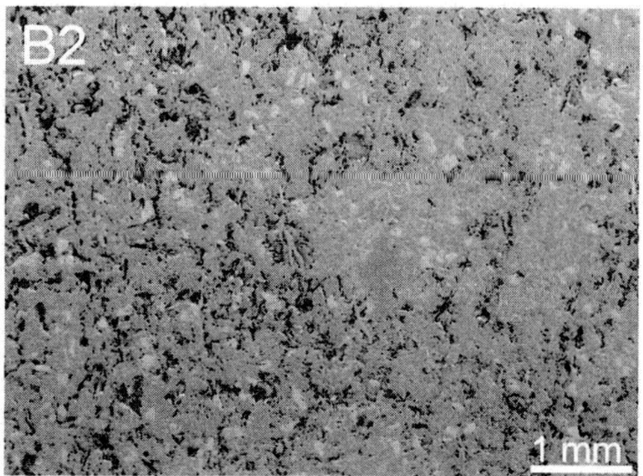

Fig. 8. Scanning electron microscopy images of two adjacent areas (B1 and B2) in the central layer (B).

Table 1. *Evolution of CT numbers (Hounsfield Units) of some ROI's during the absorption test*

Time ($min^{1/2}$)	min:sec	ROI 2	ROI 5	ROI 10	ROI 12	ROI 22
1.3	01:36	1922	1811	1748	1750	1985
1.4	02:02	1908	1806	1749	1747	1974
1.8	03:22	1918	1807	1749	1745	2022
2.3	05:05	1921	1814	1747	1757	2038
2.7	07:23	1936	1801	1739	1749	2056
3.2	10:07	1959	1811	1748	1743	2050
4.1	17:03	1973	1814	1741	1739	2051
5.1	26:00	1982	1819	1737	1752	2047
6.1	37:00	1981	1815	1754	1746	2062
7.6	57:37	2012	1886	1751	1753	2073
8.1	65:02	2007	1888	1750	1763	2067
9.1	81:56	2011	1896	1749	1760	2070
9.5	91:02	2013	1898	1751	1779	2072
10	100:52	2005	1890	1754	1777	2064
10.5	111:00	2008	1892	1752	1783	2074
11	121:57	2010	1887	1753	1790	2070
11.6	133:32	2010	1889	1760	1803	2075

penetrating into ROI 5 (not sharply defined in some other ROIs, due to the time interval between scans and the statistical variation of the CT numbers); and (d) a second horizontal segment after saturation where the mean CT number remains roughly constant, with some statistical variability.

In Figure 10 the initial mean CT number, measured without the water jacket, is not taken into account in order to eliminate the problems related to the introduction of the water jacket. The first value corresponds to the image obtained 96 seconds after the sample container was filled with water.

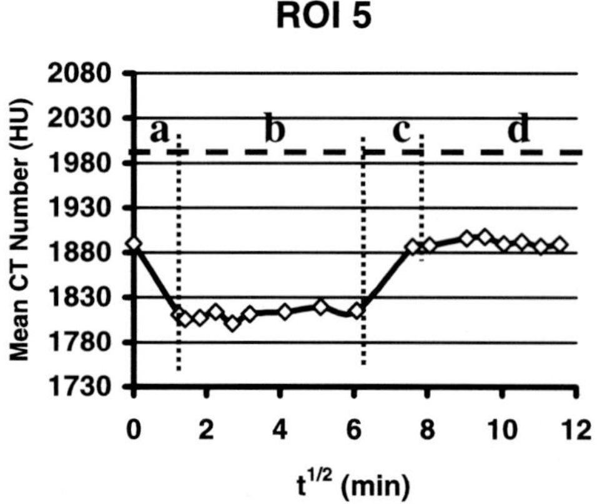

Fig. 9. Evolution of the mean CT number (in Hounsfield Units) for ROI 5 during the water absorption test, showing four different stages (a, b, c and d). The initial CT value was measured under dry conditions.

Water penetrated into ROI 2 between five and seven minutes; saturation was reached after approximately one hour (Fig. 10a). From that moment onwards, this zone does not collect more water, as the later mean CT numbers are within the variation range of the measured values.

The water reaches ROI 22, located in a symmetrical position relative to ROI 2 between two and three minutes: saturation is reached faster than in ROI 2 (Fig. 10b). This suggests that the water penetrates more quickly from the right side of the sample than from the left.

ROI 5 located closer to the centre, was reached between 37 minutes and one hour (Fig. 10c). After one hour, the saturation process is completed. The long scanning interval between the two images (23 minutes) does not allow a higher precision. The time of water penetration into the ROI and the time of saturation are probably superimposed in this interval. By relating these time indications to the distance from the centre of the ROI to the left side of the sample (10 mm) and supposing that water penetration follows the shortest path, a penetration rate between 0.27 and 0.17 mm/min is estimated for water reaching the ROI.

Figure 10d suggests that the water reached ROI 12 (centre of the sample) after approximately one hour and 20 minutes, with complete saturation not being reached during the test. On the other hand, ROI 10, located to the left of the ROI 12, shows no significant increments in mean CT number (Fig. 10e). This might indicate that water reached the centre of the sample (ROI 12) from the right.

By comparing the graphs, it can be seen that there are two types of saturation curves. ROI 22

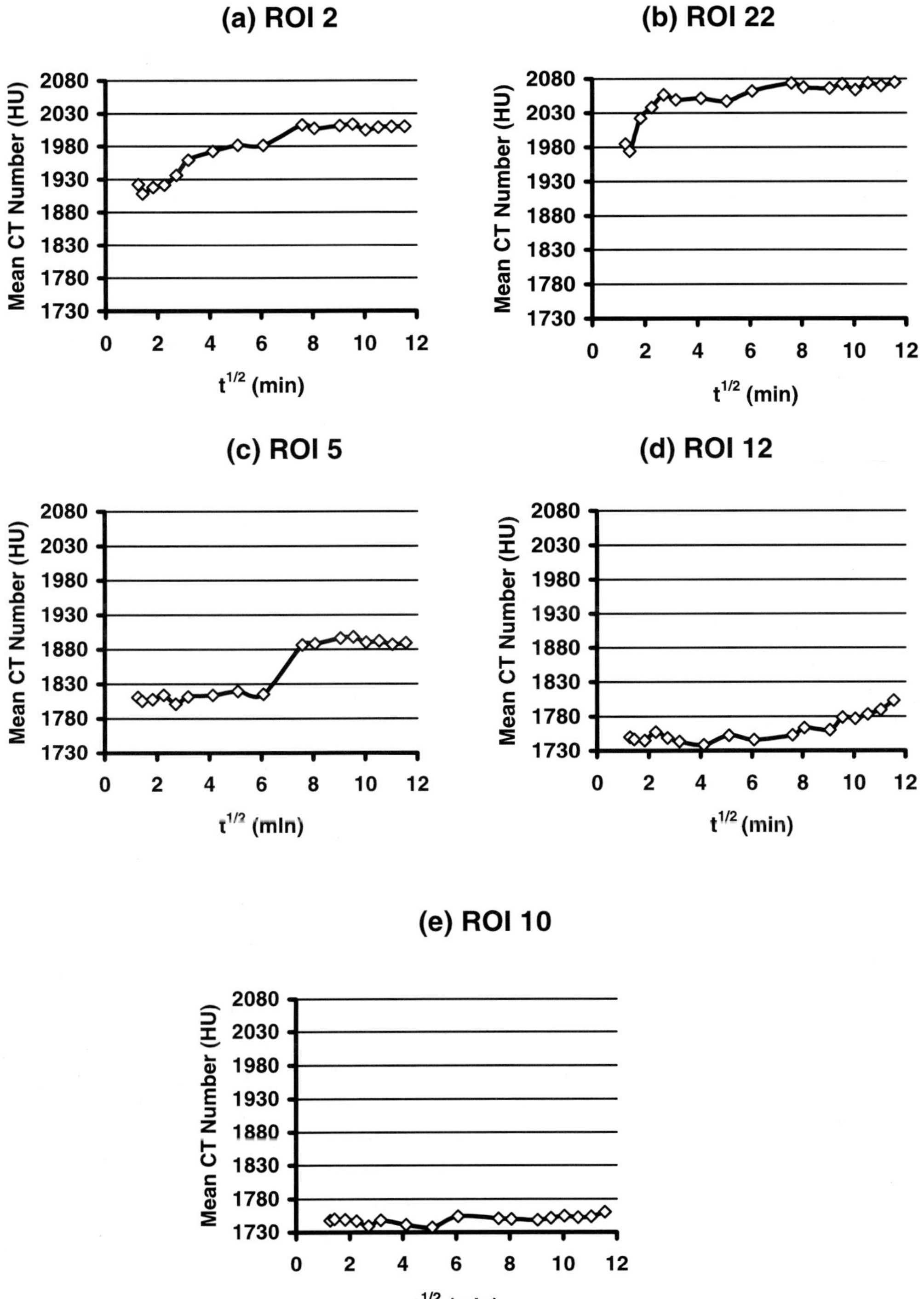

Fig. 10. Evolution of the mean CT number (in Hounsfield Units) for five ROIs during the water absorption test: (**a**) ROI 2, (**b**) ROI 22, (**c**) ROI 5, (**d**) ROI 12, (**e**) ROI 10. Time in square root of minutes.

and ROI 5 show rapid saturation, whereas ROI 2 and ROI 12 show a slow saturation. This different behaviour indicates that there are some differences in total porosity and pore size between the different ROIs, although the studied layer, (B), is macroscopically homogenous (Fig. 8).

These results suggest that the variation of the mean CT numbers of each ROI over time can be considered a proxy for the characterization of the movement and penetration rate of water inside rocks.

Conclusions

This study demonstrates that X-ray CT is useful for the study of the movement of water inside rocks and the study of their texture. This allows a relationship to be established between water movement and the petrographical characteristics of the rock. Under the conditions of the absorption tests performed for this study, some of the air in the pores cannot get out and so the results refer to this type of test and cannot be extrapolated to the capillary behaviour of the samples.

Good quality CT images of the Piedramuelle limestone show its layering to be related to variations in porosity and composition. The images also demonstrate that water movement inside the samples is controlled by the layering and mainly related to pore volume and pore size.

The evolution of the mean CT values of different ROIs of the sample is a good proxy to quantify the kinetics of water inside rocks. For a correct estimation of penetration rates, it is important to achieve the optimized ROI size and the time intervals for CT monitoring.

This methodology, applied to one of the types of layers of the Piedramuelle Stone, shows two types of saturation behaviour in the ROIs (fast and slow). These differences can be due to differences in total porosity and pore size, although all ROIs are located in the layer that appears to be homogeneous at the macroscopic level.

This research was financially supported by the Consejería de Cultura, Principado de Asturias, Spain (Project No. PB-ESE-99-09C).

References

AYLMORE, L.A.G. 1993. Use of computer assisted tomography in studying water movement around plant roots. *Advances in Agronomy*, **49**, 1–54.

DULIU, O.G. 1999. Computer axial tomography in geosciences: an overview. *Earth-Science Reviews*, **48**, 265–281.

ESBERT, R.M. & MARCOS, R.M. 1983. *Las Piedras de la Catedral de Oviedo y su Deterioración.* Colegio Oficial de Aparejadores y Arquitectos Técnicos de Asturias, Oviedo, Spain.

GÉRAUD, Y., MAZEROLLE, F., CARDON, H., VIDAL, G. & CARON, J.M. 1998. Anisotropie de connectivité du réseau poreux d'une bande de cisaillement ductile. *Bulletin de la Société Géologique de France*, **169**, 645–654.

HAINSWORTH, J.M. & AYLMORE, L.A.G. 1983. The use of computer assisted tomography to determine the spatial distribution of soil water content. *Australian Journal of Soil Research*, **21**, 435–443.

JACOBS, P., SEVENS, E., VOSSAERT, P. & KUNNEN, M. 1997. Non-destructive monitoring of interactive physical and biological deterioration of building stones by computerised X-ray tomography. *In*: MARINOS, P.G., KOUKIS, G.C., TSIAMBAOS, G.C. & STOURNARAS, G.C. (eds) *Engineering Geology and the Environment.* Balkema, Rotterdam, 3163–3168.

MOSSOTTI, V.G. & CASTANIER, L.M. 1990. The measurement of water transport in Salem limestone by X-ray computer aided tomography. *In*: MARINOS, P.G. & KOUKIS, G.C. (eds) *Engineering Geology of Ancient Works, Monuments and Historical Sites.* Balkema, Rotterdam, 2079–2082.

QUEISSER, A. 1986. Computerized tomography applied to stone investigation. *In*: ICOMOS (ed.) *Advanced Methods and Techniques for the Study of Stone Decay, Cleaning and Conservation.* ICOMOS, Paris, 125–138.

RILEM 1980. Essais recommendés pour mesurer l'altération des pierres et évaluer l'efficacité des méthodes de traitement. *Materials and Structures*, **13**, 175–252.

RAYNAUD, S., FABRE, D., MAZEROLLE, F., GÉRAUD, Y. & LATIÈRE, H.J. 1989. Analysis of the internal structure of rocks and characterization of the mechanical deformation by a non-destructive method: x-ray tomodensitometry. *Tectonophysics*, **159**, 149–159.

RUIZ DE ARGANDOÑA, V.G., RODRÍGUEZ REY, A., CELORIO, C., SUÁREZ DEL RÍO, L.M., CALLEJA, L. & LLAVONA, J. 1999. Characterization by computed X-ray tomography of the evolution of the pore structure of a dolomitic rock during freeze-thaw cyclic tests. *Physics and Chemistry of the Earth (Part A)*, **24**, 633–637.

VAN GEET, M., SWENNEN, R. & WEVERS, M. 2000. Quantitative analysis of reservoir rocks by microfocus X-ray computerised tomography. *Sedimentary Geology*, **132**, 25–36.

WELLINGTON, S.L. & VINEGAR, H.J. 1987. X-ray computerized tomography. *Journal of Petroleum Technology*, **39**, 885–898.

Estimation of porosity and hydraulic conductivity from X-ray CT-measured solute breakthrough

S. H. ANDERSON[1], H. WANG[2], R. L. PEYTON[3] & C. J. GANTZER[1]

[1] *Department of Soil and Atmospheric Sciences, 302 ABNR Building, University of Missouri, Columbia, Missouri 65211, USA (e-mail: andersons@missouri.edu)*

[2] *Collins Engineers, 300 West Washington Boulevard, Suite 600, Chicago, Illinois 60606, USA*

[3] *Department of Civil and Environmental Engineering, E2509 Eng. Building East, University of Missouri, Columbia, Missouri 65211, USA*

Abstract: Heterogeneities are common in natural porous media and are present on different scales. Use of X-ray computed tomography (CT) may provide a tool for quantifying small-scale heterogeneities in porosity and hydraulic conductivity in porous media. Porosity and saturated hydraulic conductivity distributions were estimated using CT for a series of undisturbed soil core samples taken from a field site. CT measurements were collected during breakthrough experiments using an iodide tracer. Techniques were developed to estimate porosity and hydraulic conductivity from solute breakthrough data. Results were compared with bulk sample measurements. CT-measured porosity compared well with laboratory-measured porosity. Hydraulic conductivity estimated from CT methods slightly over-estimated laboratory-measured values. These techniques provide a method to quantify the spatially variable porosity and hydraulic conductivity on a millimetre scale rather than on a core-averaged scale. Chemical transport through the soil was predicted using a finite element method for each core using the CT-measured soil properties. Comparisons between measured and predicted chemical transport suggest that small-scale heterogeneities cause departures between measured and simulated solute breakthrough curves, and that a smaller grid size may be needed to improve the simulation.

The study of water and solute transport processes in heterogeneous porous media, such as fractured geological formations and soils containing macropores, has long been a field of research interest in the scientific and engineering communities (Grisak & Pickens 1981; van Genuchten *et al.* 1984; Endo *et al.* 1984; Germann *et al.* 1984; Richter & Jury 1986; Smettem 1986; Tsang 1991). Porosity can be highly variable over a distance of only a few millimetres, which causes hydraulic conductivity to vary by orders of magnitude. This spatial heterogeneity makes modelling groundwater flow and chemical transport challenging. Because it is impossible to measure hydraulic conductivity precisely at small scales by conventional methods, averaging is always necessary. When transport processes are strongly dependent on the heterogeneity of the medium, such globally-averaged measurements often provide parameters that are lumped and fail to yield sufficiently accurate information. Under such circumstances, direct chemical transport observation would be a valuable tool for understanding chemical transport processes.

X-ray computed tomography (CT), which has been widely used in diagnostic medicine to observe internal organs and tissue, was not applied to the study of soils until 1982 (Petrovic *et al.* 1982). Application of CT techniques can provide a much finer resolution of soil parameters than previously achieved with bulk sample methods. Crestana *et al.* (1985) used CT to observe the change in the X-ray attenuation coefficient with time at a single point in space within a homogeneous soil core as a wetting front passed. Anderson *et al.* (1988) reported on the methods of calibration for determining mean water contents in homogeneous soil cores for two soils with different iron contents.

CT has also been applied to measuring macropores and macroporosity in soils. Warner *et al.* (1989) distinguished and characterized macropores in soil by manipulation of scanner images. Anderson *et al.* (1990) evaluated natural and

From: MEES, F., SWENNEN, R., VAN GEET, M. & JACOBS, P. (eds) 2003. *Applications of X-ray Computed Tomography in the Geosciences*. Geological Society, London, Special Publications, **215**, 135–149. 0305-8719/03/$15.

constructed soil macropores using CT. Macropore characteristics were also quantified with CT by Peyton *et al.* (1992). Clausnitzer & Hopmans (1999) developed a technique to estimate phase-volume fractions from CT measurements in two-phase systems, which is useful for determining the representative volume for measurements of porosity. Cheng *et al.* (2001) evaluated images of CT density from intact cores with topological analysis and subsequently used fuzzy logic to relate this information to saturated hydraulic conductivity.

Several researchers have developed techniques to use X-ray CT scanners for the study of solute transport processes in undisturbed soil columns. Peyton *et al.* (1990) used CT to study iodide transport in soil cores. Anderson *et al.* (1992) applied CT to measure solute breakthrough curves for selected soil aggregates and determined two-dimensional velocity distributions within the soil cores. Clausnitzer & Hopmans (2000) showed that microfocus CT could be used to characterize the spatial distribution of the hydraulic convective dispersion through a sample core of glass beads. This work was limited to very small cores (<10 mm diameter). Perret *et al.* (2000) conducted tracer breakthrough experiments with an X-ray CT unit to quantify transport in macropore and matrix flow domains.

Application of CT for investigating transport processes has also been made in petroleum engineering and geological sciences. Wang *et al.* (1984) demonstrated that CT can provide fluid saturation data as a function of spatial position and time during dynamic oil displacement experiments. Vinegar & Wellington (1987) and Wellington & Vinegar (1987) discussed methods for quantifying three discrete phases, such as oil, gas and water, using CT. They presented results of carbon dioxide displacement of oil in water saturated sandstone cores, focusing on viscous fingering, gravity segregation, miscibility and mobility control. Withjack (1988) investigated the development and application of CT for the determination of rock properties and the study of miscible displacement in laboratory cores. He showed that saturation and porosity measurements by CT scanning matched well with those determined by conventional methods.

By extending the application of CT to solute transport processes in porous media, the impacts of the small-scale heterogeneity of natural porous media on these processes can be better understood and predicted. The objectives of this study were: (i) to develop methodology to non-destructively quantify macropore-scale porosity distributions using CT-measured solute breakthrough in soil cores; (ii) to estimate the distribution of soil hydraulic conductivity within undisturbed soil cores using an inverse approach with CT-determined porosity distributions; and (iii) to predict three-dimensional chemical transport in soil cores using a finite element approach.

Materials and methods

Experimental details

Undisturbed soil cores were collected from 100 to 200 mm depth at a field site near Rocheport, Missouri. The soil was classified according to US soil taxonomy as Menfro silt loam (fine-silty, mixed, mesic Typic Hapludalf). Cores were collected in 68 mm long Plexiglas cylinders with a 76 mm inside diameter and 3 mm wall thickness. Two cores (Core #1 and Core #2) were removed from soil under forest management. The dominant species of the forest were American elm (*Ulmus americana*), persimmon (*Diospyros virginiana*), black oak (*Quercus velutina*), red oak (*Quercus rubra*) and white oak (*Quercus alba*). The forest had been relatively undisturbed for over 25 years. An additional soil core (Core #3) was removed from a field under tall fescue (*Festuca arundinacea*) or meadow management, which had been untilled for four years. This core was taken approximately 50 m away from the forest sampling site. The physical properties of each soil sample on a bulk core basis are shown in Table 1. Particle size analysis of the soil indicated 0.735 g/g silt and 0.102 g/g clay determined from the pipette method of Gee & Bauder (1986). Soil cores were sealed in plastic bags and stored at 4°C prior to scanning.

To prepare the soil cores for scanning, each core was sealed with Plexiglas end-caps. A schematic of a core prepared for a solute breakthrough experiment is shown in Figure 1a. The end-caps each had a conically shaped reservoir. A Plexiglas plate with a honeycomb pattern of 3.0 mm diameter holes was fitted into each end-cap to distribute and collect flow uniformly. The

Table 1. *Physical properties of undisturbed soil cores determined on a bulk core basis*

Soil core	Land management	Bulk density g/cm^3	Porosity
Core #1	Forest	1.362	0.486
Core #2	Forest	1.391	0.475
Core #3	Meadow	1.471	0.445

Porosity values were estimated from bulk density and particle density (2.65 g cm^{-3}).

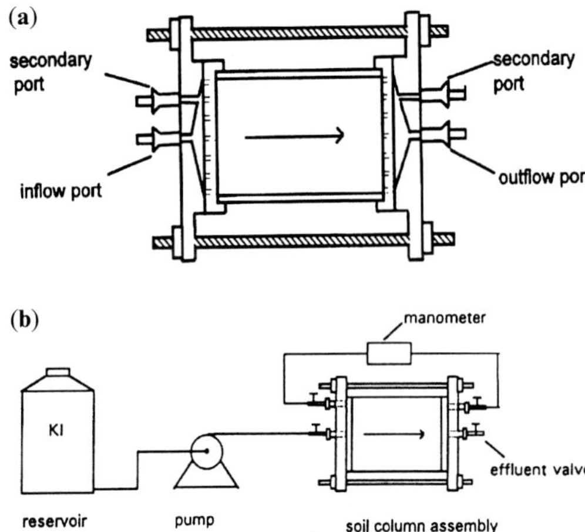

Fig. 1. Schematic of soil column assembly used to saturate and conduct solute breakthrough curves for undisturbed soil cores: (**a**) individual core assembly and (**b**) overall assembly. The core assembly includes end-plates attached with nylon bolts represented on the top and bottom of Figure 1a. The arrow indicates the mean direction of flow.

soil core, with its two end-caps, was held together by three nylon bolts. There were two flow ports with Swagelok connectors in each end-cap. The primary port was located at the centre of the end-cap and a secondary port was located near the outside radius. Primary ports were used to introduce flow at the upstream end and collect flow at the downstream end during breakthrough experiments. Secondary ports were used at the beginning of a breakthrough experiment to flush the water in the upstream reservoir with tracer solution. The secondary port at the downstream end was used to create a small back-pressure gradient during flushing to minimize migration of tracer across the upstream face of the soil. After flushing, these two secondary ports were connected to a piezometer to measure the total hydraulic head drop between the upstream and downstream ends during the breakthrough experiment. During experiments, a plastic tube connected the inflow port to a pump and a bottle of solute solution, while the outflow port was open to the atmosphere (Fig. 1b).

The soil cores were saturated with a background solution of distilled, de-aired water containing 6.06 g/L of $CaCl_2$ and 1.78 g/L of $MgCl_2$ to minimize the dispersion of clay particles (Palmer 1979). Separate breakthrough experiments were conducted using two different solute solutions, 7.5 g/L KI and 50 mg/L 2-chlorophenol. In each case, these solute solutions were prepared using the background solution described above. KI was chosen because the iodide ion is non-adsorbing and non-reactive and has a large X-ray linear attenuation coefficient. Chlorophenol was chosen as the organic chemical because it is a common groundwater contaminant.

A Parker digital positioning rail table system controlled by a Sigma Instruments stepper motor was used to transport soil cores into and out of the CT scan plane. During breakthrough experiments, the soil core was attached to the rail table. The rail table was then placed on the CT scanner table, which remained stationary during the experiment. During scanning, the positioning table was operated electronically, which positioned the soil core within the scan plane to an accuracy of ± 0.02 mm.

The X-ray CT scanner used during the solute breakthrough experiments was a Siemens Somatom DRH third generation full-body scanner with a gantry opening of approximately 0.7 m in diameter. The slice thickness for the scans was 8 mm and the pixel size was 2 mm by 2 mm. The size of the pixel and the thickness of the slice define the size of the volume elements (voxels) within the scanned object.

The saturated soil core assembly was placed on the digital positioning table in the CT unit with the core's longitudinal axis oriented horizontally and perpendicular to the scan plane. During a breakthrough experiment, the soil core was scanned in eight scan planes along the core. The first scan plane was centred 6 mm from the upstream end of the soil surface and the horizontal distance between the centre lines of adjacent scan planes was 8 mm. Therefore, a core length of 64 mm was scanned during a breakthrough experiment. A 2 mm length of the core at each end was not scanned. Prior to the breakthrough experiment, each scan plane was scanned twice to obtain an initial reference CT number with zero concentration of potassium iodide solution.

The KI solution was pumped into the core with the pre-designed flow rate after the flushing of the upstream end-plate, which initiated the breakthrough experiment. The tracer was applied as a step. Experimental flow rates were 3.5 mL/min. Scans were taken continuously at 30 s intervals throughout the breakthrough experiment. The effluent samples were collected from the outlet for later measurement and the outflow rates from the soil core were repeatedly measured. The digital positioning table was moved between each scan. The sequence of positions varied during an experiment depending on the location of the tracer front. The positions were sequenced to obtain the maximum number of data points on the rising portion of the breakthrough curve at each of the eight positions. After the breakthrough experiment, each plane was scanned twice to obtain an ending reference CT number. At the end of this experiment, the soil core was flushed by the background solution for approximately two hours until zero concentration of iodide ion was detected in the outflow. Then, another breakthrough experiment in the same soil core using the 2-chlorophenol solute solution was conducted under the same experimental conditions and procedure but without CT scanning.

Effluent samples collected during the breakthrough were analysed for iodide concentration using an ion selective probe. The 2-chlorophenol samples were analysed by the direct photometric method. Samples collected during breakthrough experiments were stored in a refrigerator at 4°C and analysed within 12 hours of collection.

Soil porosity determination

The linear attenuation coefficient for a given pixel is a function of the density and effective atomic number of the material within the pixel. Typically, the X-ray attenuation coefficient is normalized with respect to water and is referred to as the CT number:

$$CTN(x, y, z) = 1000\frac{\mu(x, y, z) - \mu_w}{\mu_w} \tag{1}$$

where $CTN(x, y, z)$ is the CT number for the pixel centred at position x, y, z expressed in Hounsfield units; $\mu(x, y, z)$ is the linear attenuation coefficient for pixel centred at position x, y, z; and μ_w is the linear attenuation coefficient for water.

To determine the porosity from the CT-measured solute breakthrough experimental data, the relative concentration of each pixel is first calculated. The relative concentration, C^*, of the solute in each local voxel during a breakthrough experiment was evaluated by the following equation:

$$C^*(x, y, z, t) = \frac{CTN(x, y, z, t) - CTN(x, y, z, t = 0)}{CTN(x, y, z, T) - CTN(x, y, z, t = 0)} \tag{2}$$

where $CTN(x, y, z, t)$ is the CT number for the voxel centred at position x, y, z measured at time, t, during the breakthrough experiment; $CTN(x, y, z, t = 0)$ is the CT number for the voxel centred at position x, y, z measured immediately prior to the beginning of the breakthrough experiment ($t = 0$); and $CTN(x, y, z, T)$ is the CT number for the voxel centred at position x, y, z measured at the end of the breakthrough experiment ($t = T$). Equation 2 is only valid for a step application of solute. Scanning a soil core twice, once saturated with water and once saturated with an aqueous solution of KI, allowed the use of the following two equations:

$$\mu_{(p+w)} = \mu_p(1 - f) + \mu_w f \tag{3}$$

$$\mu_{(p+w+c)} = \mu_p(1 - f) + \mu_{(w+c)} f \tag{4}$$

where $\mu_{(p+w)}$ is the linear attenuation coefficient for water-saturated porous media; μ_p is the linear attenuation coefficient for dry porous media; f is porosity; $\mu_{(p+w+c)}$ is the linear attenuation coefficient for the porous media saturated with the aqueous KI solution and $\mu_{(w+c)}$ is the linear attenuation coefficient for the aqueous KI solution. Subtracting Eq. 3 from Eq. 4 gives the following relationship:

$$f = \frac{\mu_{(p+w+c)} - \mu_{(p+w)}}{\mu_{(w+c)} - \mu_w}. \tag{5}$$

Using Eq. 1 with Eq. 5 for a single voxel, the following can be obtained:

$$f(x, y, z) = \frac{CTN(x, y, z)_{(p+w+c)} - CTN(x, y, z)_{(p+w)}}{\overline{CTN}_{(w+c)} - \overline{CTN}_w} \tag{6}$$

where $f(x, y, z)$ is the porosity in the voxel centred at position x, y, z; $CTN(x, y, z)_{(p+w)}$ is the CT number in the voxel centred at position x, y, z containing only water-saturated porous media; $CTN(x, y, z)_{(p+w+c)}$ is the CT number in the voxel centred at position x, y, z containing porous media saturated with aqueous KI solution; $\overline{CTN}_w$ is the mean cross-sectional CT number for the cylinder containing only water; and $\overline{CTN}_{(w+c)}$ is the mean cross-sectional CT number for the cylinder containing only aqueous KI solution. The average CT-determined porosity for each soil core was calculated by averaging the porosity values estimated for the entire soil core. These values were compared with bulk core measured porosity values.

Hydraulic conductivity determination

Three-dimensional, saturated steady flow without a source or sink term is defined by the following equation (Jacob 1950):

$$\frac{\partial}{\partial x}\left(K\frac{\partial h}{\partial x}\right) + \frac{\partial}{\partial y}\left(K\frac{\partial h}{\partial y}\right) + \frac{\partial}{\partial z}\left(K\frac{\partial h}{\partial z}\right) = 0. \tag{7}$$

This equation is subject to the following initial and boundary conditions for the current study:

$$h(x, y, z, t = 0) = h_0(x, y, z) \quad (x, y, z) \in \Omega \tag{8a}$$

$$h(x, y, z, t) = h_1(x, y, z, t) \quad (x, y, z) \in \Gamma_1 \tag{8b}$$

$$\sum_{i=1}^{3}\left[-K\frac{\partial h}{\partial x_i}n_i\right] = q_c(x, y, z, t) \quad (x, y, z) \in \Gamma_2 \tag{8c}$$

where K is the saturated hydraulic conductivity of the porous media and is a function of position (x, y, z); h is hydraulic head; t is time; Ω is the domain of the flow problem; Γ_1 is the portion of the exterior boundary subject to a fixed-head (Dirichlet) boundary condition; Γ_2 is the portion of the exterior boundary subject to a specified-flux (Neumann) boundary condition; h_0 is the initial hydraulic head distribution; h_1 is the hydraulic head specified on the segment Γ_1; q_c is the normal flux of water per unit area of boundary Γ_2; x_i is the coordinate in the i direction; n is the outward unit vector normal to the boundary; n_i is the normal cosine in the i direction; and subscript i denotes the x, y, and z directions, respectively. The governing partial differential equation was solved using the Galerkin finite element technique.

A common use of Eq. 7 is to compute hydraulic head and velocity distributions from a known hydraulic conductivity distribution, which is known as the direct approach. Another use of Eq. 7 is to estimate the hydraulic conductivity distribution from a known hydraulic head or velocity distribution, which is known as the inverse approach. In this study, the inverse approach was used taking advantage of CT-measured porosity and mean velocity values.

The inverse approach was conducted using the following steps. First, data were inputted, which included the following parameters: the CT-measured porosity for each element, a cross-sectional mean velocity measured during breakthrough and the geometry of the flow domain. Secondly, a three-dimensional finite element mesh was generated corresponding to the soil core dimensions. Next, the isotropic saturated hydraulic conductivity for each element was calculated from an empirically measured relationship between bulk porosity and bulk conductivity (conducted with separately packed cores with a range of porosity values). Next, the governing equation was integrated with the finite element method that forms a system of equations. The resulting system of linear equations was solved to obtain the hydraulic head distribution within the soil core. The velocity distribution within the soil core was computed using Darcy's law and a mean velocity was calculated from these estimated velocities. Directional velocities were calculated from the directional hydraulic head gradients, with the CT-measured porosity values assuming isotropic hydraulic conductivity. The estimated mean velocity and measured velocity were compared and the initial hydraulic conductivity estimation was adjusted depending upon the error between the velocity values. Details of this adjustment are explained in the next paragraph. The above steps were repeated until the estimated mean velocity matched the measured mean velocity within a specified error limit.

The relationship between porosity and hydraulic conductivity used for the initial estimates of hydraulic conductivity in each element were taken from empirical formulas. The general form of these formulas is:

$$K = \Psi(\tau, \mu, \gamma, d_e, T)\varphi(f) \tag{9}$$

where K is the saturated hydraulic conductivity of the porous media; Ψ is a function; τ is a coefficient related to temperature; μ is the dynamic fluid viscosity; γ is the fluid specific weight; d_e is the effective grain diameter; T is tortuosity; and $\varphi(f)$ is a function of porosity f. If the temperature and the properties of water remain constant and the variation in effective grain diameter and tortuosity within the soil matrix remain relatively constant, Eq. 9 can be written as:

$$K = A\varphi(f) \tag{10}$$

where A is a coefficient. Equation 10 assumes that the primary factor that controls the spatial variation of hydraulic conductivity within the soil core is the variation in porosity. The Kozeny (Dullien 1992) formula for determination of hydraulic conductivity was used for the initial estimate:

$$K = A\frac{f^3}{(1-f)^2}. \tag{11}$$

This formula was used in the model to determine the initial estimate of the hydraulic conductivity in each volume element from the CT-measured porosity.

The flow domain was a 76-mm-diameter by 68 mm long region with the same geometry as the soil cores. The flow domain was divided into eight slices, the plane of which was normal to the longitudinal axis. The flow region had no-flow boundaries at the wall of the cylindrical flow domain and had specified head boundaries at the up-stream and down-stream faces. The total head drop across the up-stream and down-stream faces measured during the breakthrough experiment was used as the head difference for the simulation of the soil core. The mean velocity measured during the breakthrough experiment was input into the model as the target for model convergence.

The arithmetic average hydraulic conductivity for each scan slice was determined and the mean value for the core was estimated using the following relationship:

$$\bar{K} = \frac{L_t}{\sum_{j=1}^{8} (l_j/K_j)} \tag{12}$$

where $\bar{K}$ is the computed mean saturated hydraulic conductivity; L_t is the length of soil core in the flow direction; l_j is the length of slice j in the flow direction; and K_j is the arithmetic average of the hydraulic conductivity for a scan slice. For purposes of computation, 16 voxels were combined to develop each finite element that had dimensions of $8 \times 8 \times 8$ mm. This resulted in a total of 76 finite elements within each slice.

Simulation of chemical transport

Three-dimensional advective-dispersive solute transport during steady fluid flow in a heterogeneous system may be described by the following equation (Duguid & Reeves 1976):

$$R\frac{\partial C}{\partial t} = \frac{\partial}{\partial X_\alpha}\left(D\frac{\partial C}{\partial X_\alpha}\right) - \frac{\partial (V_\alpha C)}{\partial X_\alpha} \qquad \alpha = 1, 2, 3 \tag{13}$$

where C is the volume-averaged resident concentration of the dissolved chemical species; t is time; V_α is the pore water velocity in the x, y and z directions; D is the hydrodynamic dispersion coefficient; R is the retardation factor; and X_α is the distance in the α direction where the α subscript values of 1, 2 and 3 correspond to the x, y and z directions, respectively. The parameters D, R and V_α are functions of spatial position (x, y, z). This equation is subject to the following initial and boundary conditions for the current study:

$$C(x, y, z, t = 0) = C_i(x, y, z) \qquad (x, y, z) \in \Omega \tag{14a}$$

$$C(x, y, z, t) = C_d(x, y, z, t) \qquad (x, y, z) \in B_1 \tag{14b}$$

$$D\frac{\partial C}{\partial X_\alpha} n_\alpha = q_c(x, y, z, t), \quad \alpha = 1, 2, 3 \qquad (x, y, z) \in B_2 \tag{14c}$$

where $C_i(x, y, z)$ is the initial concentration distribution; $C_d(x, y, z, t)$ is the solute concentration specified on the segment B_1; Ω is the region of interest; B_1 is the portion of the system boundary subject to a prescribed concentration (Dirichlet) boundary condition; B_2 is the portion of the system boundary subject to a specified dispersive-flux (Neumann) boundary condition; $q_c(x, y, z, t)$ is the dispersive flux specified on the segment B_2; n is the outward unit vector normal to the boundary and n_α is the normal cosine in the α direction, where the α subscript values of 1, 2 and 3 correspond to the x, y and z directions, respectively. The Galerkin finite element method was used in a manner similar to the water flow equation to determine the approximate solution for the governing partial differential equation under the prescribed initial and boundary conditions.

Simulations were conducted using the Galerkin finite element solution for the chemical transport equation. Parameter estimates for the simulations were obtained as indicated. Directional pore water velocity values for each element were determined using estimated hydraulic head information obtained during the hydraulic conductivity estimation procedure, CT-measured hydraulic conductivity values and CT-measured porosity values. Hydraulic conductivity was assumed to be isotropic for these estimates. Hydrodynamic dispersion coefficients were assumed to be influenced only by longitudinal dispersion and molecular diffusion was considered negligible. These coefficients were estimated from the pore water velocity estimates and the dispersivity values determined from average CT-measured solute KI breakthrough curves in each scan plane. Dispersivity was defined as the longitudinal dispersion coefficient divided by the pore water velocity. The retardation factor for chlorophenol was calculated using laboratory-measured 2-chlorophenol adsorption data on bulk samples and from the CT-measured porosity for each element. Thus, dispersivity values were obtained for each scan plane and retardation data were calculated for each element in each slice.

Results and discussion

Results for a typical solute breakthrough through an undisturbed soil core showing CT-measured and effluent-measured solute concentrations are illustrated in Figure 2. The CT measurements were taken near the down-stream end of the soil core. These results show disagreement between the CT-measured solute concentrations averaged

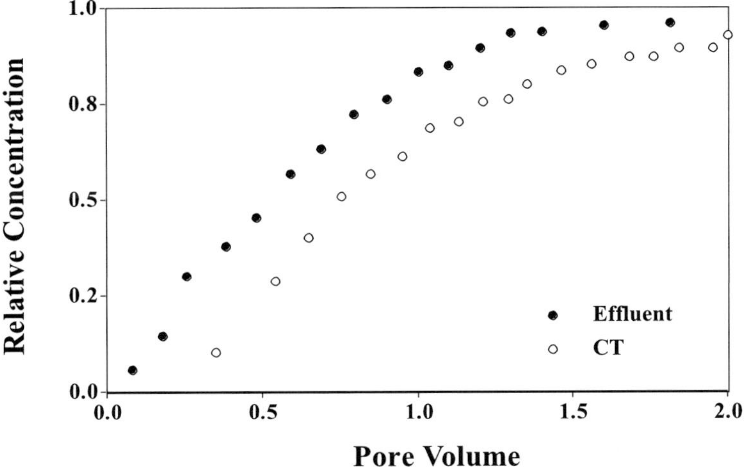

Fig. 2. Relative iodide concentration versus pore volume for solute breakthrough using both effluent samples and CT measurements. This core sample was taken near the sampling location of Core #3.

across a scan plane near the down-stream end of the core and the effluent sample concentrations. One of the main reasons for differences between these two methods is probably due to CT measurements being resident or volume-averaged concentrations, while effluent measurements are flux-averaged concentrations. Effluent-measured or flux-averaged concentrations are primarily measures of arrival time in the preferential flow channels, whereas CT-measured or volume-averaged concentrations give equal weight to solute transport through the entire cross-sectional area. Thus, effluent-measured concentrations produce the faster breakthrough or higher concentrations as a function of pore volume compared to the CT-measured concentrations, as shown in Figure 2. Another reason for differences is also due to the additional solute dispersion occurring through the downstream end-plate prior to sample collection. The advantage of the CT method, in addition to the macropore scale of the measurements, is that sample concentrations

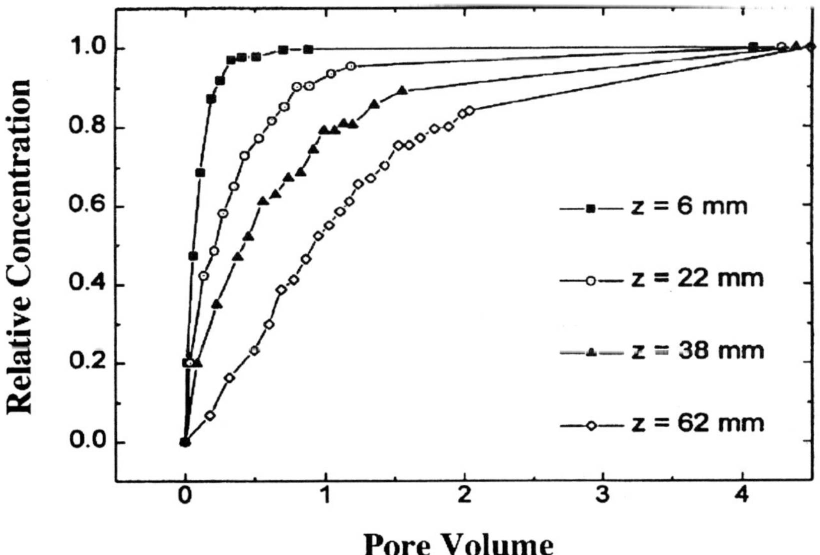

Fig. 3. Relative solute concentration (C^*) versus cumulative outflow measured using CT at four selected positions along the soil core for Core #3.

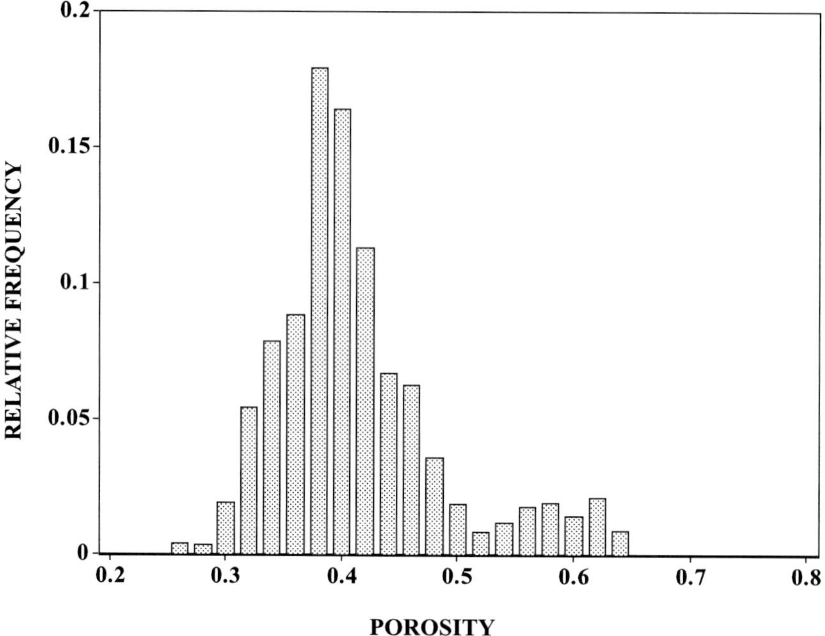

Fig. 4. Frequency distribution of CT-measured porosity for Core #3. Measurements were made in eight scan planes.

can be measured within the soil prior to movement through an end-plate. Another advantage is that spatial concentrations for each measuring plane (and in the ideal case for the whole core) are available. Breakthrough curves are shown in Figure 3 for the soil core taken from the site under meadow (Core #3) as a function of position along the core length. The breakthrough curves illustrate the additional solute dispersion, which is occurring as the longitudinal length is increased along the soil core.

Soil porosity

Results of the frequency distribution of CT-measured porosity [$f(x, y, z)$] for Core #3 are shown in Figure 4. CT-measured porosity for this core ranged from 0.3 to 0.7. These results show that the porosity measured for Core #3 has a slight bi-modal distribution with a dominant peak around 0.4 and a secondary peak around 0.6, which indicates that the porosity distribution is positively skewed. This positively skewed distribution indicates the presence of macropore features. The other two cores had similar frequency distributions.

Results from the comparison of laboratory-measured mean porosity and CT-measured mean porosity for the three soil cores are illustrated in Table 2. The differences between the lab-measured and CT-measured porosity were small, ranging from −4.5 to 4.2% with a mean of −0.9% indicating good agreement for this silt loam soil under two types of field management. The approach for determining porosity illustrated in this paper has advantages over another technique commonly used where f is expressed as

$$f = \frac{\mu_p - \mu_{(p+w)}}{\mu_p - \mu_w}. \quad (15)$$

Use of Eq. 15 requires the core to be scanned first dry and then, without removing the core from the CT scanner, scanned saturated with water. The core cannot be removed because the location of each volume element in the dry scan must be

Table 2. *Average porosity for undisturbed soil cores determined using CT-measured solute breakthrough methods*

Soil core	Lab-measured porosity	CT-measured porosity	Difference %
Core #1	0.486	0.474	−2.5
Core #2	0.475	0.495	4.2
Core #3	0.445	0.425	−4.5

Lab-measured porosity values were estimated from bulk density and particle density (2.65 g cm^{-3}).

exactly identical to the wet scan. At the small scales involved in CT measurements, this is difficult to achieve if the core is removed. In addition, air is easily entrapped during saturation and the presence of air invalidates Eq. 15. Also, the use of techniques that eliminate air entrapment generally involve the risk of moving the core, which would introduce large errors in the numerator of Eq. 15. Finally, the transition from dry to wet generally involves swelling of the clays present in the silt loam soil, which will also introduce large errors in the numerator of Eq. 15.

Hydraulic conductivity

Results of the CT-determined saturated hydraulic conductivity for the soil cores are presented in this section. In order to estimate the hydraulic conductivity on a voxel basis, estimates of the fluid velocity measured by CT are needed. Estimates of the fluid velocity were determined for each of the soil cores. Results of the frequency distribution of fluid velocity for Core #3 are shown in Figure 5. Velocity in the x-direction is shown in Figure 5a, in the y-direction in Figure 5b and in the z-direction in Figure 5c. The z direction is along the soil column, while the x direction is the horizontal direction normal to the flow direction and the y direction is the vertical direction normal to the flow direction. These distributions represent the computed velocities in all elements for the eight slices scanned for the core. Results for the x- and y-directions indicate that the average fluid velocity was close to zero, which was expected. Small deviations from zero are due to the numerical approximation. The distribution in the z-direction indicated that the highest frequency of values occurred slightly higher than zero, with a generally decreasing frequency at higher velocities. All values were lower than 2.5 mm/min for Core #3.

Model simulations were performed on the three undisturbed soil cores, which were scanned by CT during breakthrough experiments. The laboratory-measured values for mean saturated hydraulic conductivity, as well as the experimental flow rate and hydraulic gradient for each of the three cores are listed in Table 3. The controlled fluid flow rates were uniform for the three soil cores. However, the hydraulic head drop across the length of the soil column ranged from 170 to 188 mm for the three cores.

The resulting hydraulic conductivity distribution for Core #3 is presented in Figure 6 (computed average hydraulic conductivity values are listed in Table 4). This distribution represents the

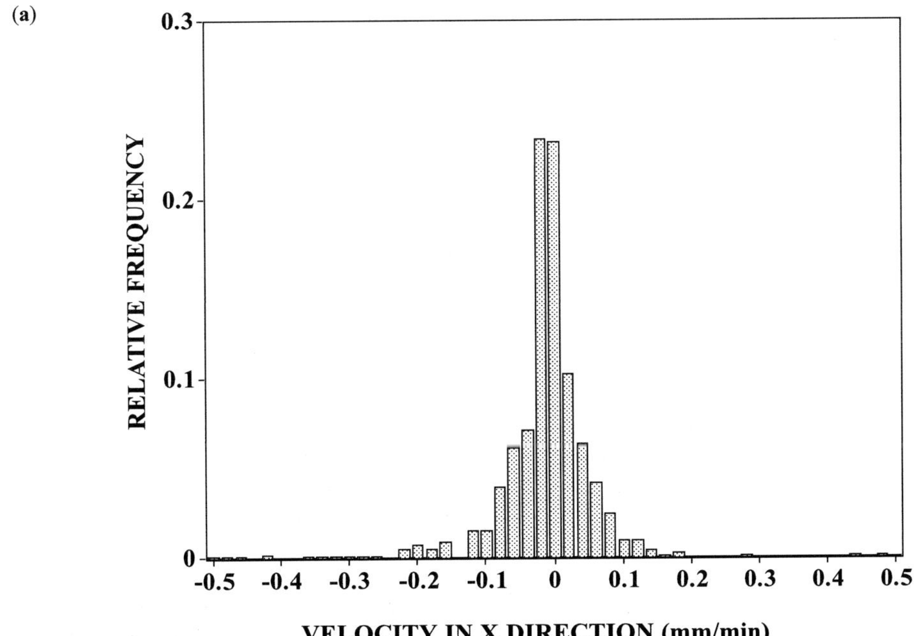

Fig. 5. Frequency distributions of CT-measured solute velocity for Core #3: (**a**) velocity in x direction; (**b**) velocity in y direction; (**c**) velocity in z direction.

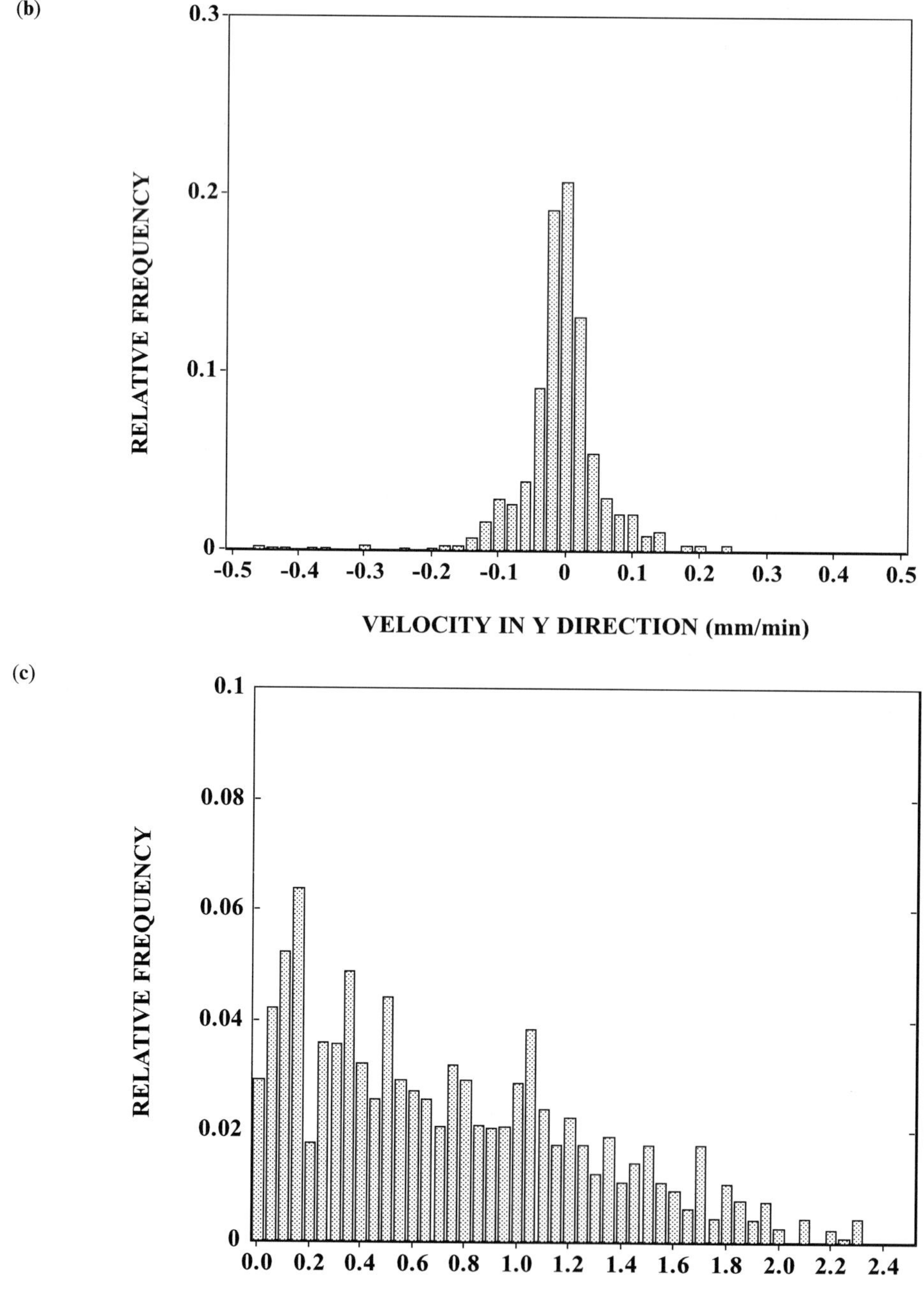

Fig. 5. (*continued*).

computed hydraulic conductivity values in all 608 elements of this core. As expected, it is apparent that the distribution is positively skewed, due to the presence of macropores that were also present in the CT-measured porosity (Fig. 4). Most of the values of hydraulic conductivity ranged from 0 to 0.5 mm/min with a few values ranging from 0.5 to 1.5 mm/min. Similar distributions were found for the other two soil cores.

Table 3. *Hydraulic conditions for soil cores during the solute breakthrough experiments*

Soil core	Hydraulic head drop mm	Flow rate mL/min	Average K_{sat} mm/min
Core #1	173	3.5	0.303
Core #2	170	3.5	0.309
Core #3	188	3.5	0.279

Table 4. *Average hydraulic conductivity for undisturbed soil cores determined using CT-measured solute breakthrough methods*

Soil core	Lab-measured K_{sat} mm/min	CT-measured K_{sat} mm/min	Difference %
Core #1	0.303	0.344	13.5
Core #2	0.309	0.375	21.4
Core #3	0.279	0.316	13.3

To examine the accuracy of the estimated results, the mean hydraulic conductivity of each soil core was calculated and compared with the laboratory-measured mean hydraulic conductivity. The mean hydraulic conductivity was determined using Eq. 12. The basic assumption of Eq. 12 is that the hydraulic head drop over all the elements in a slice should be the same. It is clear that the equation is not valid for heterogeneous porous media. Because no other analytical solution was available for evaluating the mean hydraulic conductivity in a heterogeneous porous media, Eq. 12 was used to gain a quantitative idea of the value for the entire soil core.

Table 4 presents the mean hydraulic conductivity values computed by the estimation procedure for the three soil cores. These values can be compared to the laboratory-measured mean hydraulic conductivities. The calculated values ranged from 13 to 21% higher than the measured values.

The mean fluid velocity values for the three soil cores (Core #1 – 0.77 mm/min, Core #2 – 0.80 mm/min, Core #3 – 0.77 mm/min) were used also to compute a mean hydraulic conductivity for the soil cores using Darcy's equation. These values were much closer to the laboratory-measured hydraulic conductivity values (Core #1 – 0.303 mm/min, Core #2 – 0.322 mm/min, Core #3 – 0.279 mm/min). These computed mean values exactly matched the laboratory-measured values for Cores #1 and #3, while the value for Core #2 was 4.2% higher. Use of the mean fluid velocity for the three soils cores to

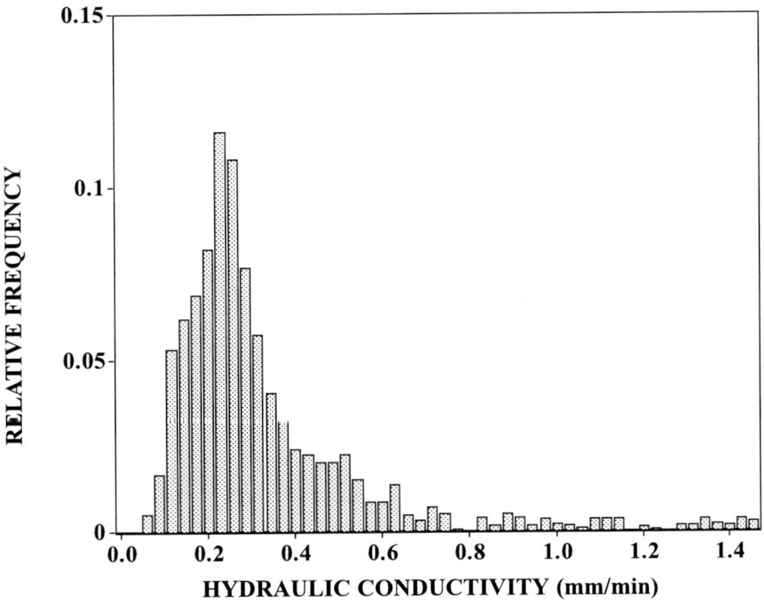

Fig. 6. Frequency distribution of hydraulic conductivity estimated using the inverse approach from CT-measured porosity and velocity distributions for Core #3.

Table 5. *Calculated dispersivities and retardation coefficients as a function of scan plane for undisturbed soil cores*

	Longitudinal distance from upstream face mm	Dispersivity mm	Retardation coefficient	
			mean	std. dev
Core #1	6	0.004	2.38	0.65
	14	0.455	2.59	0.46
	22	1.991	2.57	0.44
	30	4.701	2.64	0.55
	38	8.001	2.67	0.60
	46	12.983	2.68	0.71
	54	16.491	2.67	0.75
	62	22.505	2.73	0.70
Core #2	6	0.002	2.45	0.30
	14	0.050	2.41	0.43
	22	0.162	2.50	0.56
	30	0.821	2.48	0.59
	38	1.154	2.58	0.55
	46	2.910	2.54	0.54
	54	3.881	2.70	0.55
	62	6.164	2.69	0.55
Core #3	6	0.027	3.19	0.35
	14	0.472	3.32	0.35
	22	1.012	3.39	0.33
	30	1.854	3.26	0.36
	38	2.590	3.19	0.33
	46	5.401	2.03	0.27
	54	10.541	2.77	0.30
	62	18.617	2.65	0.23

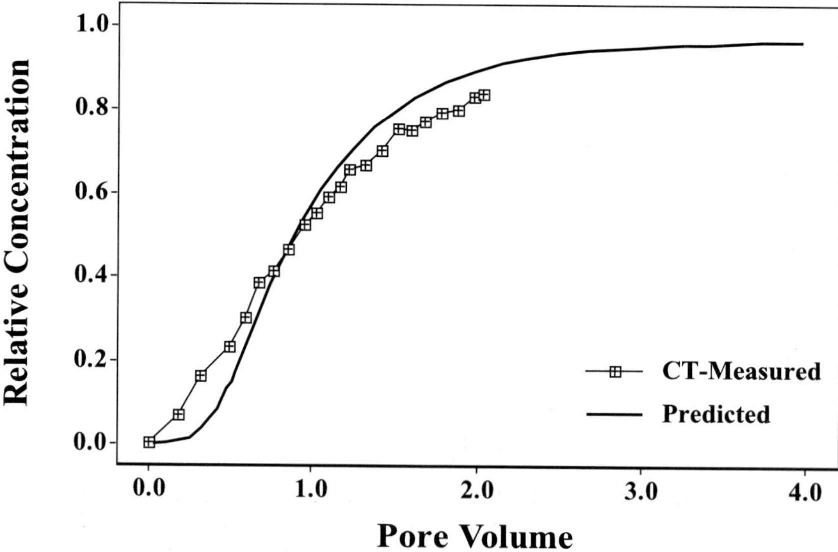

Fig. 7. Relative concentration (C^*) of iodide versus cumulative outflow for both CT-measured and predicted solute breakthrough for Core #3.

compute the hydraulic conductivity resulted in a much closer approximation to the bulk core laboratory measurements.

Chemical transport simulation

Results of the CT-measured dispersivitiy values for each scan plane for each core are presented in Table 5. For each soil core, the dispersivity increased as a function of distance along the soil core as would be expected. Higher dispersivity values occurred in Cores #1 and #3 compared to Core #2. Calculated retardation coefficients for 2-chlorophenol for each scan plane are also shown in Table 5. The variability within each scan plane is due to variations in the porosity, a parameter used to calculate the retardation coefficient. There were no general trends in the retardation coefficient as a function of distance along the soil core.

Results of the chemical transport simulation for iodide for Core #3 are shown in Figure 7, along with the CT-measured solute concentrations. It is apparent from the simulated results that the prediction underestimated the initial breakthrough prior to one pore volume and overestimated the iodide breakthrough after one pore volume. This was attributed to the large grid size used for the finite element simulation model (8 × 8 × 8 mm). It is recommended that future work should concentrate on using smaller volume elements for the simulation to hopefully provide a more accurate prediction.

Results of the simulation for chlorophenol for Core #3 are shown in Figure 8, along with effluent-measured concentrations. These results show a similar pattern as with the iodide simulation: initial underestimation of the solute concentration prior to two pore volumes and overestimation subsequent to two pore volumes. The higher pore volume was due to the adsorption of chlorophenol to soil particles.

Conclusions

This study developed an approach to estimate porosity from CT-measured solute breakthrough data. CT-estimated porosity values were found to be similar to laboratory-measured values with differences ranging between −4.5 to 4.2%. An inverse procedure was developed to estimate hydraulic conductivity using CT-estimated porosity and fluid velocity distributions. CT-determined hydraulic conductivity values overestimated laboratory-measured values from 13 to 21%. Chemical transport simulations using a finite element solution underestimated the early rise in the breakthrough for both potassium iodide and chlorophenol. This was attributed to the coarse grid size used for estimating the transport properties. Future studies could utilize the technology of microfocus X-ray CT (Van Geet *et al.* 2000; Van Geet & Swennen 2001) to improve the scanning resolution. This approach may improve the estimate of the parameters and accuracy of the simulation.

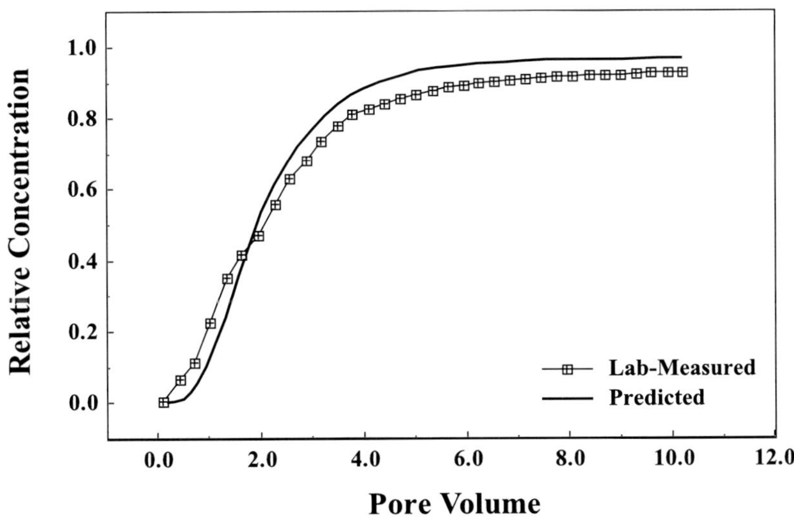

Fig. 8. Relative concentration (C^*) of chlorophenol versus cumulative outflow for both laboratory-measured and predicted solute breakthrough for Core #3.

This study dealt with the determination of the spatial distribution of saturated hydraulic conductivity values on a macropore-scale rather than a core-averaged scale. By extending the application of CT to solute transport processes in porous media, the impacts of the small-scale heterogeneity of natural porous media on these processes can be better understood and predicted.

References

ANDERSON, S.H., GANTZER, C.J., BOONE, J.M. & TULLY, R.J. 1988. Rapid nondestructive bulk density and soil-water content determinations by computed tomography. *Soil Science Society of America Journal*, **52**, 35–40.

ANDERSON, S.H., PEYTON, R.L. & GANTZER, C.J. 1990. Evaluation of constructed and natural soil macropores using x-ray computed tomography. *Geoderma*, **46**, 13–29.

ANDERSON, S.H., PEYTON, R.L., WIGGER, J.W. & GANTZER, C.J. 1992. Influence of aggregate size on solute transport as measured using x-ray computed tomography. *Geoderma*, **53**, 387–398.

CHENG, Z., ANDERSON, S.H., GANTZER, C.J. & CHU, Y. 2001. Fuzzy logic for predicting soil hydraulic conductivity using CT images. *In*: DAGLI, C.H., BUCZAK, A.L., GHOSH, J., EMBRECHTS, M.J., ERSOY, O. & KERCEL, S. (eds) *Intelligent Engineering Systems Through Artificial Neural Networks, Volume 11*. ASME Press, New York, 307–312.

CLAUSNITZER, V. & HOPMANS, J.W. 1999. Determination of phase-volume fractions from tomographic measurements in two-phase systems. *Advances in Water Resources*, **22**, 577–584.

CLAUSNITZER, V. & HOPMANS, J.W. 2000. Pore-scale measurements of solute breakthrough using microfocus X-ray computed tomography. *Water Resources Resources*, **36**, 2067–2079.

CRESTANA, S., MASCARENHAS, S. & POZZI-MUCELLI, R.S. 1985. Static and dynamic three-dimensional studies of water in soil using computed tomographic scanning. *Soil Science*, **140**, 326–332.

DUGUID, J.O. & REEVES, M. 1976. *Material Transport in Porous Media: a Finite-element Galerkin Model*. Oak Ridge National Laboratory. Oak Ridge, Tennessee, **ORNL-4928**.

DULLIEN, F.A.L. 1992. *Porous Media: Fluid Transport and Pore Structure*. Academic Press, San Diego.

ENDO, H.K., LONG, J.C.S., WILSON, C.R. & WITHERSPOON, P.A. 1984. A model for investigating mechanical transport in fracture networks. *Water Resources Research*, **20**, 1390–1400.

GEE, G.W. & BAUDER, J.W. 1986. Particle size analysis. *In*: KLUTE, A. (ed.) *Methods of Soil Analysis, Part 1: Physical and Mineralogical Methods*. Monograph, No. 9. American Society of Agronomy, Madison, Wisconsin, 383–411.

GERMANN, P.F., EDWARDS, W.M. & OWENS, L.B. 1984. Profiles of bromide and increased soil moisture after infiltration into soils with macropores. *Soil Science Society of America Journal*, **48**, 237–244.

GRISAK, G.E. & PICKENS, J.F. 1981. An analytical solution for solute transport through fractured media with matrix diffusion. *Journal of Hydrology*, **52**, 47–57.

JACOB, C.E. 1950. Flow of groundwater. *In*: ROUSE, H. (ed.) *Engineering Hydraulics*. John Wiley and Sons, New York, 321–386.

PALMER, C.J. 1979. *Hydraulic and electrical properties in sodium amended soils*. PhD Thesis, University of Missouri, Columbia.

PERRET, J., PRASHER, S.O., KANTZAS, A. & LANGFORD, C. 2000. A two-domain approach using CAT scanning to model solute transport in soil. *Journal of Environmental Quality*, **29**, 995–1010.

PETROVIC, A.M., SIEBERT, J.E. & RIEKE, P.E. 1982. Soil bulk density analysis in three dimensions by computed tomographic scanning. *Soil Science Society of America Journal*, **46**, 445–450.

PEYTON, R.L., HAEFFNER, B.A., ANDERSON, S.H. & GANTZER, C.J. 1992. Applying x-ray CT to measure macropore diameters in undisturbed soil cores. *Geoderma*, **53**, 329–340.

PEYTON, R.L., ANDERSON, S.H. & GANTZER, C.J. 1990. *Measurement of Soil Structure, Water Movement and Solute Transport Using X-ray Computed Tomography*. Report to US Geological Survey, Water Resources Research Program, Grant No. 14-09-001-G1643.

RICHTER, G. & JURY, W.A. 1986. A microlysimeter field study of solute transport through a structured sandy loam soil. *Soil Science Society of America Journal*, **50**, 863–868.

SMETTEM, K.R.J. 1986. Analysis of water flow from cylindrical macropores. *Soil Science Society of America Journal*, **50**, 1139–1142.

TSANG, C.F. 1991. Coupled hydromechanical-thermochemical processes in rock fractures. *Reviews of Geophysics*, **29**, 537–551.

VAN GEET, M. & SWENNEN, R. 2001. Quantitative 3D-fracture analysis by means of microfocus X-ray computer tomography (μCT): an example from coal. *Geophysical Research Letters*, **28**, 3333–3336.

VAN GEET, M., SWENNEN, R. & WEVERS, M. 2000. Quantitative analysis of reservoir rocks by microfocus X-ray computerized tomography. *Sedimentary Geology*, **132**, 25–36.

VAN GENUCHTEN, M.T., TANG, O.H. & GUENNELON, R. 1984. Some exact solutions for solute transport through soils containing large cylindrical macropores. *Water Resources Research*, **20**, 335–346.

VINEGAR, H.J. & WELLINGTON, S.L. 1987. Tomographic imaging of three-phase flow experiments. *Review of Scientific Instruments*, **58**, 96–107.

WANG, S.Y., AYRAL, S., CASTELLANA, F.S. & GRYTE, C.C. 1984. Reconstruction of oil saturation distribution histories during immiscible liquid-liquid displacement by computer-assisted tomography.

American Institute of Chemical Engineering Journal, **30**, 642–646.

Warner, G.S., Nieber, J.L., Moore, I.D. & Geise, R.A. 1989. Characterizing macropores in soil by computed tomography. *Soil Science Society of America Journal*, **53**, 653–660.

Wellington, S.L. & Vinegar, H.J. 1987. X-ray computerized tomography. *Journal of Petroleum Technology*, **39**, 885–898.

Withjack, E.M. 1988. Computed tomography for rock-property determination and fluid-flow visualization. *Society of Petroleum Engineers Formation Evaluation*, **3**, 696–704.

Assessment of soil structure using X-ray computed tomography

H. ROGASIK[1], I. ONASCH[1], J. BRUNOTTE[2], D. JEGOU[3] & O. WENDROTH[1]

[1] *Centre for Agricultural Landscape and Land Use Research Müncheberg, Eberswalder Strasse 84, D-15374 Müncheberg, Germany (e-mail: hrogasik@zalf.de)*
[2] *Federal Agricultural Research Centre Braunschweig-Völkenrode, Bundesallee 50, D-38116 Braunschweig, Germany*
[3] *Université de Rennes 1 Station Biologique, F-35380 Paimpont, France*

Abstract: Assessment of soil structure, characterized by complex morphological and functional properties, is difficult because most conventional soil physical investigations are destructive and variable in spatial resolution. The use of X-ray computed tomography, as a non-destructive technique, presents significant progress. It can be used to study soil structure at the millimetre scale, e.g. with a resolution of 0.25 mm in the horizontal direction and 1 mm in the vertical direction for the reported study. The measured Hounsfield Unit (HU) values characterize X-ray attenuation for each volume element of the soil core samples. From HU values, soil physical properties of soil cores or their subunits can be derived. They enable: (i) visual assessment of the soil structural condition through inspection of the X-ray CT images; (ii) 3D visualization of air-filled macropores; and (iii) calculation of the mean dry bulk density and standard deviation of voxel-related HU values for successive slices of soil cores. The degradation of structure of loamy and silty soils by tillage could be assessed by CT through quantification of decreased air-filled porosity, destroyed macropore connectivity, increased dry bulk density and decreased standard deviation of HU values in horizontal slices. Small-scale compactions near earthworm burrows could also be detected.

A significant limitation in soil science is the lack of knowledge concerning the effect of soil structure on functional processes. X-ray computed tomography (CT) is a non-destructive imaging technique that allows 3D morphological characterization of soil structure at high resolution and offers the possibility of quantifying the soil structure by derived soil physical parameters.

CT has previously been used to obtain non-destructive measurements of water content, dry bulk density and macroporosity (Brown *et al.* 1987; Jenssen & Heyerdahl 1988; Tollner & Ramseur 1988; Grevers *et al.* 1989; Warner *et al.* 1989; Anderson *et al.* 1990; Peyton *et al.* 1992; Hopmans *et al.* 1994). It allows an investigation of the morphological properties of earthworm burrows (Joschko *et al.* 1993; Capowiez *et al.* 1998; Jégou *et al.* 1998), characterization of pore continuity and quantification of macropore networks (Grevers & de Jong 1994; Perret *et al.* 1999; Pierret *et al.* 2002), identification of water flow patterns in macroporous soils (Heijs *et al.* 1996), characterization and modelling of preferential solute flow in soil columns (Perret *et al.* 1998, 2000*a*, *b*), and measuring pore-scale solute breakthrough (Clausnitzer & Hopmans 2000). CT scanning is also a suitable tool for characterizing soil tillage effects (Vaz *et al.* 1989; Olsen & Borresen 1997; Gantzer & Andersen 2002), for detecting water drawdowns by individual plant roots (Hainsworth & Aylmore 1986; Hamza *et al.* 2001), for documenting plant root development and for imaging plant roots in three dimensions (Tollner *et al.* 1994; Heeraman *et al.* 1997), and for predicting fungal growth in the rhizosphere (Grose *et al.* 1996). Computer tomography has also been used to investigate soil crusting and sealing at the micrometre scale (Macedo *et al.* 1998).

A strong linear bivariate relationship exists between Hounsfield Units (*HU*) and dry bulk density ($HU = f(\rho_b)$) and between *HU* and water volume fraction ($HU = f(WV)$), with coefficients of determination r^2 close to unity (Petrovic *et al.* 1982; Crestana *et al.* 1986; Tollner & Verma 1989; Hopmans *et al.* 1992; Vaz *et al.* 1989). However, *HU* is a measure of the wet bulk density of soil, i.e. it depends on both dry bulk density and gravimetric water content. Separate bivariate relations cannot, therefore, represent the measurements.

From: MEES, F., SWENNEN, R., VAN GEET, M. & JACOBS, P. (eds) 2003. *Applications of X-ray Computed Tomography in the Geosciences*. Geological Society, London, Special Publications, **215**, 151–165. 0305-8719/03/$15.

In this contribution we summarize results obtained with a medical scanner for analysis of narrow-spaced heterogeneity, which is the spatial heterogeneity in soil phase composition at the mm-scale. The assessment of soil structure was conducted for undisturbed soil columns of $785\,cm^3$ in volume (diameter and height equal to 100 mm). As known from the literature (Phogat *et al.* 1991; DiCarlo *et al.* 1997; Garnier *et al.* 1998; Rogasik *et al.* 1999), the three-phase composition of subunits of soil cores can be calculated on the basis of dual energy CT scanning. However, dual energy CT scanning is a rather expensive and labour-intensive technique. Comparable results can be obtained by using single energy CT scanning, only if: (i) volume elements are composed of only two phases, and (ii) soil samples with a three-phase composition are scanned twice – first in their present moisture state and then after converting into a two-phase composition, i.e. completely dry or fully saturated (Hainsworth & Aylmore 1983, 1986; Heijs *et al.* 1996; Grose *et al.* 1996). The last possibility mentioned assumes that variations in water content do not cause soil structural changes, which can generally not be anticipated for any soil, especially for those containing clay minerals.

This study aims to show that single energy CT scanning can deliver detailed information about the structural condition of soil samples. The objective of this study is to present possible uses of *single* energy CT scanning such as: (i) 3D visualization of macropores; (ii) calculation of bulk density distribution within soil samples; (iii) calculation of standard deviation of HU values within horizontal slices; and (iv) assessment of dry bulk density of the compressed soil around earthworm burrows.

Theory

The volumetric fractions of all three soil components – solid, water and air – specifically contribute to the attenuation of X-rays for any soil volume. A linear relationship exists between the linear attenuation coefficient (μ), as measured for defined parts of a soil with a three-phase composition (Hainsworth & Aylmore 1983):

$$\mu = SV\mu_{solid} + WV\mu_{water} \tag{1}$$

where μ is the linear X-ray attenuation coefficient of the soil (cm^{-1}), SV is the solid volume fraction ($m^3\,m^{-3}$), μ_{solid} is the linear X-ray attenuation coefficient of the solid phase of the soil (cm^{-1}), WV is the water volume fraction ($m^3\,m^{-3}$), μ_{water} is the linear X-ray attenuation coefficient of the water phase of the soil (cm^{-1}), AV is the air volume fraction ($m^3\,m^{-3}$) and μ_{air} is the linear X-ray attenuation coefficient of the air phase of the soil (cm^{-1}). The third term $AV\mu_{air}$ does not appear in Eq. 1 because μ_{air} is zero.

Often, the linear attenuation coefficient μ is converted into Hounsfield Unit (HU) values according to Eqs 2a to 2d:

$$HU = [(\mu - \mu_{water})/\mu_{water}]1000 \tag{2a}$$

$$HU_{solid} = [(\mu_{solid} - \mu_{water})/\mu_{water}]1000 \tag{2b}$$

$$HU_{water} = [(\mu_{water} - \mu_{water})/\mu_{water}]1000 = 0 \tag{2c}$$

$$HU_{air} = [(\mu_{air} - \mu_{water})/\mu_{water}]1000 = -1000. \tag{2d}$$

The conversion has the advantage that HU values of the water phase (HU_{water}) and the air phase (HU_{air}) are known and defined.

The following linear relationship exists between HU values and soil phase composition (Rogasik *et al.* 1999):

$$HU = \frac{SV\,HU_{solid} + WV\,HU_{water} + AV\,HU_{air}}{SV + WV + AV}. \tag{3a}$$

Since $HU_{water} = 0$ (Eq. 2c), $HU_{air} = -1000$ (Eq. 2d) and $(SV + WV + AV) = 1$, Eq. 3a may be written as:

$$HU = SV\,HU_{solid} - 1000AV. \tag{3b}$$

In order to derive soil physical parameters from HU values it is necessary to calculate the HU_{solid} values, which reflect the X-ray attenuation as influenced by the solid component of the soil. For calculating HU_{solid}, the mean HU value (HU_{core}), gravimetric water content (w_{core}) and dry bulk density (ρ_{bcore}) of the entire soil core have to be estimated. The volumetric fractions of its soil components can be derived according to Eqs 4 to 6.

$$SV_{core} = \rho_{bcore}/\rho_s \tag{4}$$

$$WV_{core} = \rho_{bcore}w_{core}/\rho_w \tag{5}$$

$$AV_{core} = 1 - SV_{core} - WV_{core} \tag{6}$$

where ρ_s is the particle density ($Mg\,m^{-3}$) and ρ_w is the density of water ($Mg\,m^{-3}$).

From these equations, HU_{solid} can be calculated using the modified Eq. 3b:

$$HU_{solid} = (HU_{core} + 1000AV_{core})/SV_{core}.$$

The HU_{solid} values can be considered as fixed for any investigated soil, depending on particle size distribution and organic matter content. As a general rule, HU_{solid} increases with clay content and decreases with organic matter content for a particular X-ray energy level and sample diameter.

By replacing SV with the ratio of dry bulk density and particle density (Eq. 4) in Eq. 3b, the relationship between HU and dry bulk density in a three-phase system can be expressed as:

$$\rho_b = \rho_s(HU + 1000AV)/HU_{solid}. \quad (7a)$$

Replacing AV with $(1 - SV - WV)$ and WV with $(\rho_b w/\rho_w)$ allows the following equation to be derived (Rogasik *et al.* 1999):

$$\rho_b = \frac{\rho_s\rho_w(HU + 1000)}{[\rho_w(HU_{solid} + 1000) + 1000w\rho_s]}. \quad (7b)$$

Equation 7b allows calculating the soil dry bulk density for any voxel within a sample, as well as for any horizontal slice or for the entire soil core, provided HU_{solid} and the HU values and gravimetric water contents of the corresponding soil volumes are known.

Materials and methods

Soils

Undisturbed soil core samples were taken from A-horizons of loamy and silty soils, following the method described by Rogasik *et al.* (1997). This sampling procedure starts with excavating a soil column corresponding to the geometry of the soil core sample, though larger in diameter and height. A frame is centred above the soil column. The frame enables vertical pushing of a plexiglass cylinder, with a ring knife attached, into the soil column. The soil mantle outside the cylinder is allowed to fall freely. Once the plexiglass cylinder is pushed to the intended depth, the undisturbed soil core can be excavated by using a spade. The height as well as the inside diameter of the cylindrical cores were 100 mm.

Information about the experimental sites is given in Table 1. Soil properties measured for the selected soils are shown in Table 2.

Scanning system

The investigations were carried out with a Siemens Somatom Plus-CT scanner, operated at 120 kV and 165 mA. The scanning time was 2 seconds per slice (multiscan technique). The reconstruction matrix consisted of 512 by 512 pixels. The scanner consists of a rotating fan-beam system, including an X-ray source with a 0.8×0.9 mm spot size and a detector made of 768 multi-used chambers. The apparatus allows a minimum slice thickness of 1 mm. Slice thickness and spacing can vary between 1 2, 5 and 10 mm.

Scanning procedure

Soil samples were scanned horizontally with both a slice thickness and a spacing of 1 mm. The pixel size was 0.25×0.25 mm. It was calculated as the inner diameter of the core divided by the number of pixels of the reconstruction matrix. In addition, 5 mm thick horizontal slices of the soil cores were scanned and characterized by the mean HU and the standard deviation of pixel-related HU values obtained by region of interest (ROI) measurements.

The selected Window and Centre adjustments of 3000 and 500 HU fixed the visible grey-scale range in the CT images in such a way that pure air-filled regions ($HU = -1000$) appeared as black and dense regions ($HU \geq 2000$) appeared as white.

Table 1. *Sampling sites*

Experimental site	Region	Horizon	Comments	Soil type
Adenstedt, A1	Lower Saxony	Ap	Sampling before sugar beet harvest, October	Orthic Luvisol
Adenstedt, A2	Lower Saxony	Ap	Sampling after sugar beet harvest, October	Orthic Luvisol
Adenstedt, A3	Lower Saxony	Ap/Ah	Sampling in winter wheat, April	Orthic Luvisol
Adenstedt, A4	Lower Saxony	Ap/Ah	Sampling in winter wheat, April	Orthic Luvisol
Brandis, B1	Saxony	Ah	Sampling after wheat harvest, September	Calcic Chernozem
Brandis, B2	Saxony	Ah	Sampling after wheat harvest, September	Calcic Chernozem
Brandis, B3	Saxony	Ap	Sampling after wheat harvest, September	Gleyic Luvisol
Paimpont, F	France	A	Disturbed soil column	Gleyic Podzoluvisol

Table 2. *Properties of soils and horizons investigated*

Site	Horizon	Depth	Sampling location	Particle size distribution			Textural class (USDA)	Water content (at scanning)	Organic matter	Dry bulk density (ρ_b)
				>63 µm	2–63 µm	< 2 µm				
		cm		$g\,kg^{-1}$	$g\,kg^{-1}$	$g\,kg^{-1}$		$kg\,kg^{-1}$	$g\,kg^{-1}$	$Mg\,m^{-3}$
A1	Ap	15–25	Field	33	812	155	SiL	0.257	21.3	1.319
A2	Ap	15–25	Field	33	861	106	SiL	0.263	20.4	1.451
A3	Ap/Ah	20–30	Field	27	819	154	SiL	0.234	19.3	1.405
A4	Ap/Ah	20–30	Field	24	821	155	SiL	0.233	17.4	1.212
B1	Ah	31–41	Lysimeter	83	716	201	SiL	0.225	25.4	1.285
B2	Ah	31–41	Field	27	744	229	SiL	0.204	24.0	1.456
B3	Ap	5–15	Lysimeter	532	381	87	SL	0.177	23.9	1.212
F	A	0–28	Laboratory	105	703	192	SiL	0.287	21.7	1.468

Visual characterization of soil structure

Differences in attenuation of X-rays, caused by narrow-spaced heterogeneity in soil phase composition, are documented by differences in grey-scale levels between the extremes for air-filled regions and high-density objects, such as stones or pebbles. Grey-scale levels increase with increasing bulk density of soil fragments. The two-dimensional CT images, available for each horizontal slice, give an impression of the soil structural state. They allow the identification of soil structural elements, such as macropores and aggregates.

Data processing

For assessment of soil structure, e.g. the morphological investigations and calculations of soil physical parameters, the measured *HU* values (at different resolutions) were generally aggregated.

(i) The calculation of *HU* values for the entire soil core sample (HU_{core}) was conducted by averaging the mean *HU* values of 5 mm thick horizontal slices (HU_s), obtained by ROI-measurements of the total cross-section of the soil core.

(ii) 5 mm thick squared horizontal slices consisting of 14×14 voxels, $5 \times 5 \times 5$ mm in size, were formed by aggregating voxels of $0.25 \times 0.25 \times 1$ mm, located inside the largest possible inscribed reference cube with edge lengths of $70 \times 70 \times 90$ mm (Fig. 1). The estimated *HU* values of the aggregated voxels were then used to calculate the standard deviation SD_{src} and the arithmetic mean HU_{src} of the corresponding squared horizontal slices (the subscript 'src' stands for 'slice reference cube'). In this context the SD_{src} values are a measure for the internal heterogeneity of the horizontal slices, whereas the HU_{src} values are a precondition for the calculation of the dry bulk density of squared horizontal slices (ρ_{bsrc}).

(iii) For characterization of dry bulk density distribution around earthworm burrows, mean *HU* values were calculated for soil cylinders of different sizes (HU_c) and for neighbouring cylinder mantles (HU_{cm}) (Fig. 2b).

3D visualization of air-filled macropores

Using CT images with a pixel size of 0.25×0.25 mm, it was only possible to identify macropores ≥ 0.5 mm in diameter, because an object can only be identified when it is at least twice the pixel dimensions (Aylmore 1994).

Three-dimensional reconstructions of the macropores were created according to a method described in detail by Joschko *et al.* (1991). This method is based on data conversion, using the digitized picture matrix, described by *HU* values, linear attenuation coefficients or grey levels.

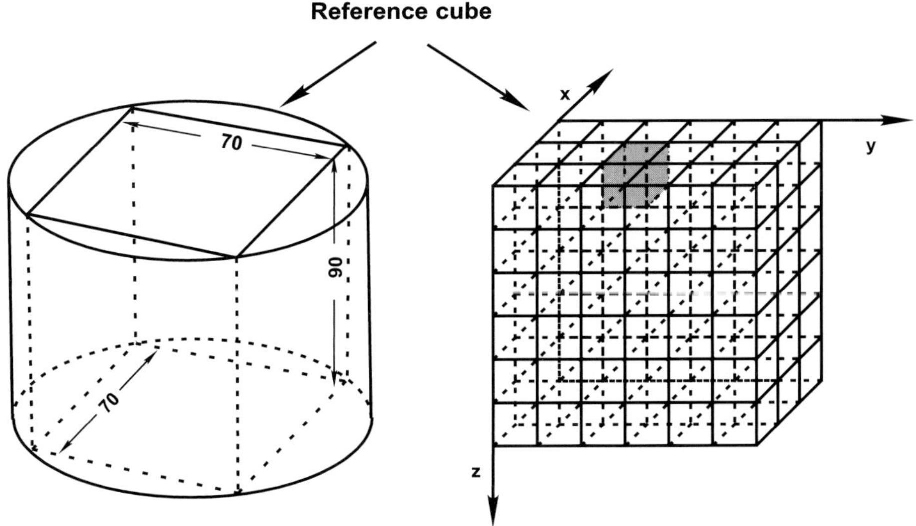

Fig. 1. Scheme of the computationally inscribed largest possible reference cube in the soil core samples.

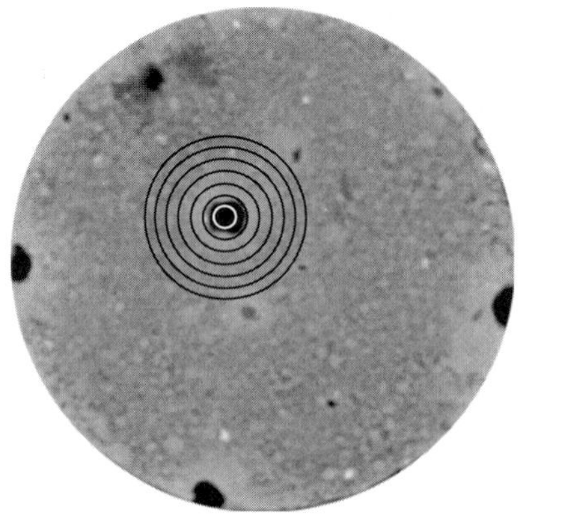

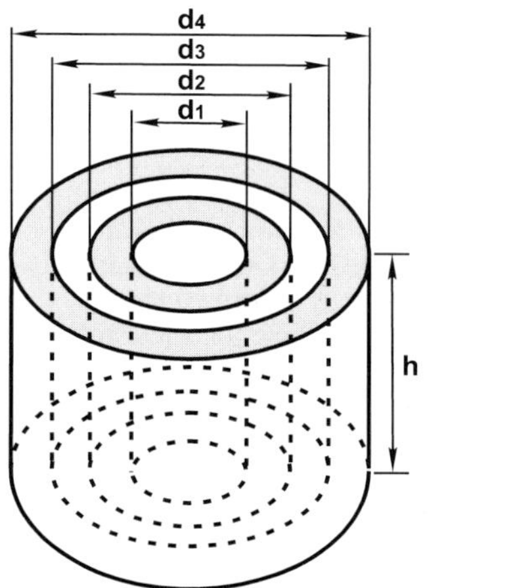

Fig. 2. Scheme of the measurement of Hounsfield Unit values around an earthworm burrow by stepwise increasing ROI cylinders (**a**) and by cylinder mantles (**b**).

From a visual analysis of the two-dimensional slices, the macropores were found to be characterized by grey levels ≤ 10, which could be transformed into a threshold value of $-875\,HU$. This threshold value was used to digitize the voxel data by setting the HU values ≤ -875 to 1 for the air-filled macroporosity (dry bulk density equal to zero) and by setting HU values > -875 to 0 for the undifferentiated soil matrix as a background. In this way, the air-filled macroporosity was separated from the soil matrix. Since all voxels were defined by their x, y and z coordinates, the macropores could be represented in three-dimensional projections for soil core samples in vertical position, in so-called pseudo-3D visualizations.

Calculation of dry bulk density distribution for horizontal slices in soil core samples

After CT scanning, the soil core was pushed out of the plexiglass cylinder using a piston. It was cut with a thin wire into slices of 1 or 2 cm thickness, depending on the stability of the soil structure. For each of the horizontal slices, the mean gravimetric water content (w_s) was estimated, assigned to the corresponding inscribed 5 mm thick horizontal slices and then used for calculation of their dry bulk density according to Eq. 7b.

The calculated dry bulk densities averaged over all horizontal slices are in close agreement with the bulk density of the entire soil core sample (ρ_{bcore}), determined gravimetrically after drying. Differences in gravimetric water content between horizontal slices led to maximum deviations in the range of $\pm 0.001\,\mathrm{Mg\,m^{-3}}$ between both bulk densities, which are negligible.

Additional calculations of standard deviation (SD_{src}) of HU values of aggregated voxels within inscribed horizontal slices of the reference cube (Fig. 1) allowed assessment of the ‘internal heterogeneity’ of the horizontal slices. The highest SD_{src} values can be found in loose or well structured, segregated soil with high percentages of macropores and aggregates. Compaction zones are characterized by high bulk densities and low standard deviations, whereas non-compacted soils have a lower bulk density and higher standard deviation. A combination of high bulk density and high standard deviation can, for example, be observed for compacted zones penetrated by biopores, e.g. earthworm burrows or root channels.

Calculation of dry bulk density distribution around pressed earthworm burrows

The calculation of dry bulk density distributions around earthworm burrows was conducted using 5 mm thick horizontal CT images. It was based on measured mean HU values for cylindrical ROIs of increasing diameter (Table 3, Fig. 2a) and calculated HU values of the neighbouring cylinder mantles (Fig. 2b) according to Eq. 8:

$$HU_{cmij} = \frac{V_{cj}HU_{cj} - V_{ci}HU_{ci}}{V_{cj} - V_{ci}} \qquad (8)$$

where V_{ci} is the volume of the smaller soil cylinder i ($h_s \pi d_i^2/4$), V_{cj} is the volume of the larger soil cylinder j ($h_s \pi d_j^2/4$), d is the diameter of soil cylinder, h_s is the thickness of horizontal slice, HU_{ci} is the mean HU value of the smaller

Table 3. *Circle areas, diameters, mean Hounsfield Unit values and mean dry bulk densities of the ROI cylinders investigated*

ROI No.	Circle area cm^2	Diameter cm	Hounsfield Unit *HU*	Dry bulk density $Mg\,m^{-3}$
1	0.53	0.82	−770.8	0.179
2	1.15	1.21	−182.8	0.639
3	2.01	1.60	363.0	1.065
4	3.13	2.00	610.6	1.259
5	5.37	2.62	762.1	1.377
6	8.24	3.24	816.4	1.420
7	10.34	3.63	832.0	1.432
8	15.27	4.41	846.5	1.443
9	21.82	5.27	855.2	1.450

soil cylinder *i*, HU_{cj} is the mean *HU* value of the larger soil cylinder *j* and HU_{cmij} is the mean *HU* value of the cylinder mantle *ij*.

Finally, Eq. 7b was used for calculation of dry bulk density of cylinders and cylinder mantles under the basic assumption that water is homogeneously distributed around the macropores.

Functional investigations of soil structure

Near water saturation, the unsaturated hydraulic conductivity (K) was measured by means of a disc-infiltrometer, developed by Perroux & White (1988). This was incorporated in a steady state percolation device, similar to the one described by Booltink *et al.* (1991) and modified by Wendroth & Simunek (1999). Negative pressure heads ($h = -10, -5, -1$ cm) were applied successively to assess flow through three macropore classes of equivalent diameters ≤ 0.3 mm, ≤ 0.6 mm and ≤ 3.0 mm respectively. Finally, we calculated the unsaturated hydraulic conductivity for macropores with equivalent diameters in the range between 0.3 and 3.0 mm, which is the difference $K(-1\,\text{cm}) - K(-10\,\text{cm})$.

Results and discussion

Assessment of soil compaction, caused by a high axle load sugar beet harvester

Immediately before (site A1) and after (site A2) the sugar beet harvest with a high axle load harvester (Hollmer, 6-rowed machine, maximum axle load 12 *t*), undisturbed soil core samples were taken from 15–25 cm depth. The soil structure before harvesting was characterized by a distinct initial macroporosity with high continuity and connectivity, as can be seen in the 3D visualization (Fig. 3 left). The dry bulk density distribution before harvesting was relatively homogeneous within the sample (Fig. 3 centre). The CT images of three sample depths all show a mixture of dense fragments, more or less loose soil matrix, an interaggregate pore space and parts of macropores created by roots or earthworms (Fig. 3 right). After compaction, due to harvesting, most of the macropores were destroyed, their vertical continuity was reduced and macroporosity was restricted to isolated regions (Fig. 4 left). Compaction led to an increase in dry bulk density from 1.319 to 1.451 $Mg\,m^{-3}$, homogeneously distributed from top to bottom and a decrease in standard deviation of voxel values in horizontal slices from 200 to 100 *HU* (Fig. 4 centre). For the compacted soil, it is difficult to distinguish between soil structural elements in the CT images (Fig. 4 right). Correspondingly, the unsaturated hydraulic conductivity (K) of macropores with 0.3–3.0 mm equivalent diameters was reduced from 71.1 to 4.3 $cm\,d^{-1}$ (Table 4).

Comparison of degraded and well segregated soil structures

At site A3, long-term conventional tillage led to compaction of soil by induced compactive forces through tillage, vehicle traffic and machine harvest processes. The compaction results from relatively high inflation pressure of tractor tyres and relatively high axle loads of the machinery. The pressures led to the formation of a compacted plow sole or traffic pan at depths of 23–27 cm, which is characterized by a disruption in continuity and connectivity of macropores (Fig. 5 left), locally higher dry bulk densities ($\rho_b > 1.50\,Mg\,m^{-3}$) and lower standard deviation of voxel related *HU* (Fig. 5 centre), as well as by a nearly undifferentiated structure in the CT images, representing a uniformly compacted soil

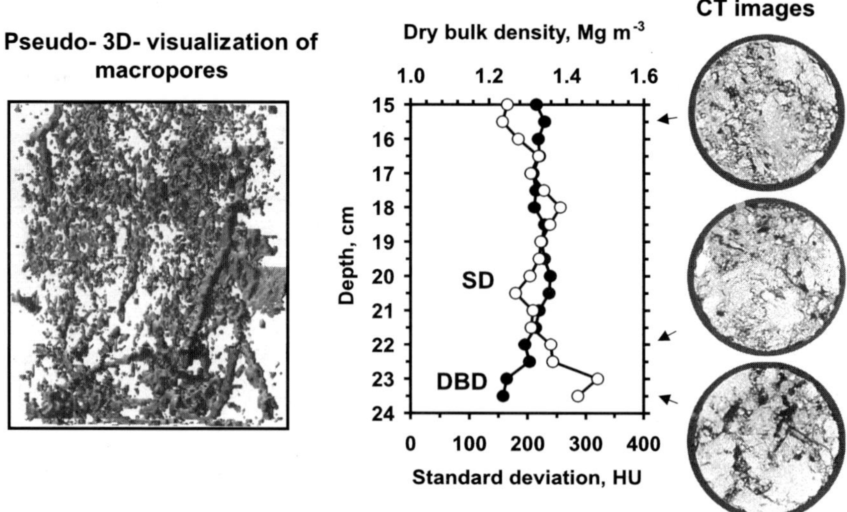

Fig. 3. Pseudo-3D visualization of macropores (left), dry bulk density and standard deviation distribution of horizontal slices over depth (centre) and selected CT images (right) for a soil core sample from site A1 (Ap-horizon, loess soil, 15–25 cm depth). Arrows indicate depths for which horizontal slices are shown.

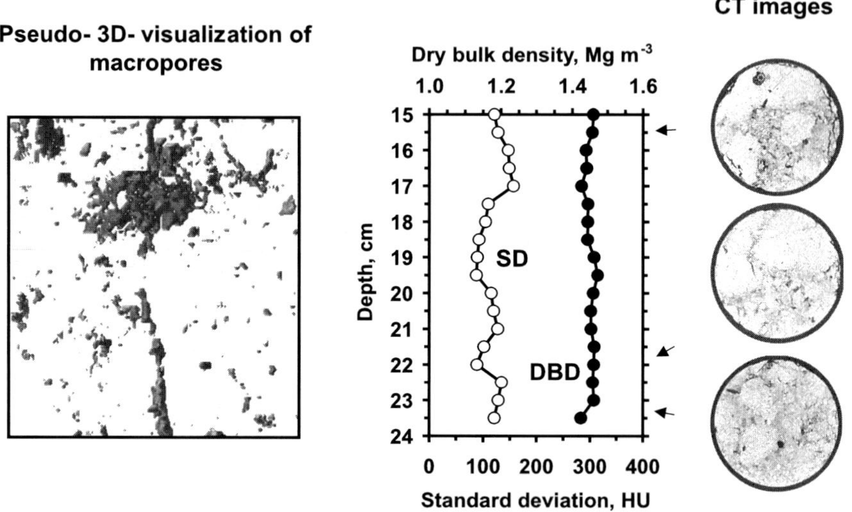

Fig. 4. Pseudo-3D visualization of macropores (left), dry bulk density and standard deviation distribution of horizontal slices over depth (centre) and selected CT images (right) for a soil core sample from site A2 (Ap-horizon, loess soil, 15–25 cm depth). Arrows indicate depths for which horizontal slices are shown.

Table 4. *Unsaturated hydraulic conductivity (K) of the soils and horizons investigated, measured for negative pressure heads of −10, −5 and −1 cm; the last column gives K values for macropores with an equivalent diameter between 0.3 and 3 mm*

Experimental site	Depth (cm)	Sampling location	Hydraulic conductivity K(cm d^{-1})			
			$K(-10)$	$K(-5)$	$K(-1)$	$K(-1)$–$K(-10)$
A1	15–25	field	0.4	1.3	71.5	71.1
A2	15–25	field	0.3	0.4	4.6	4.3
B1	31–41	lysimeter	1.9	3.7	137.0	135.1
B2	31–41	field	1.0	2.0	2.6	1.6

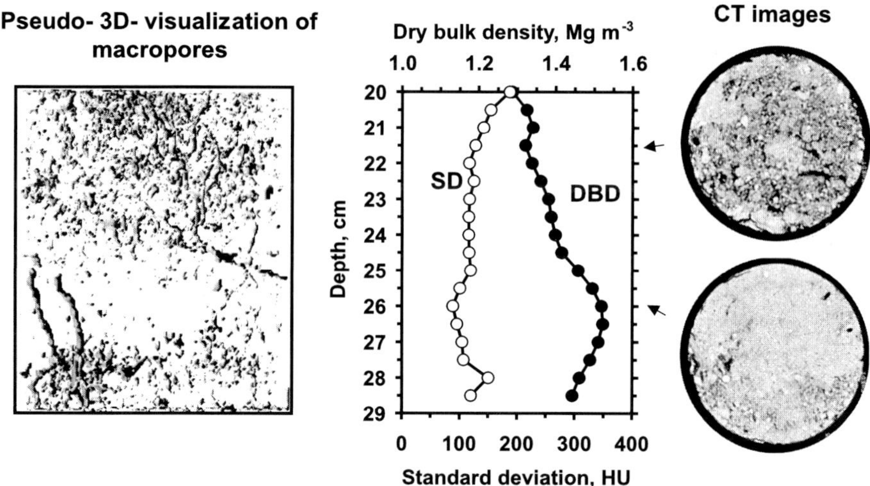

Fig. 5. Pseudo-3D visualization of macropores (left), dry bulk density and standard deviation distribution of horizontal slices over depth (centre) and selected CT images (right) for a soil core sample from site A3 (Ap-horizon, loess soil, 20–30 cm depth). Arrows indicate depth for which horizontal slices are shown.

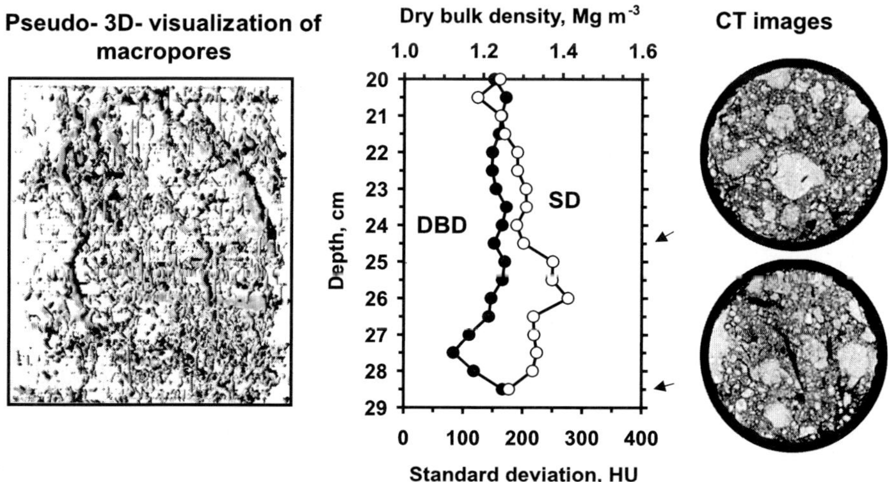

Fig. 6. Pseudo-3D visualization of macropores (left), dry bulk density and standard deviation distribution of horizontal slices over depth (centre) and selected CT images (right) for a soil core sample from site A4 (Ap-horizon, loess soil, 20–30 cm depth). Arrows indicate depths for which horizontal slices are shown.

state (Fig. 5 right). It is important to recognize those locally existing relatively homogeneous compacted zones because they can restrict peneration by roots or earthworms, especially under dry soil conditions. Within compaction zones, roots or earthworms can hardly find any zones of weakness (zones of lower bulk density), which they preferably use to penetrate this layer. The ecological consequence of compaction zones can be that plants are unable to use the water and nutrients stored in the subsoil. The effect of compaction zones as barriers depends on soil conditions like dry bulk density, gravimetric water content and penetrometer resistance. The latter gives an indication of the mechanical forces that roots and earthworms have to exert (Bengough & Mullins 1991; Kretzschmar 1991). Reasons for different root penetration depth in soils with compacted pans can be related to differences in rooting potential of plants (Materechera *et al.* 1991; Dannowski 1992). The values of maximal axial and radial growth pressures, which plant roots can exert (Misra *et al.* 1986), depend on the health of the plant, as well as on climatic conditions, which can cause stress for plants due to high demands for transpiration.

At site A4 characterized by a relatively lower mechanical stress by machinery, the loess soil is well structured (Fig. 6). In comparison with site A3 (Fig. 5), the macropore system is characterized by high continuity and connectivity (Fig. 6, left). The mean dry bulk density is lower ($\rho_b = 1.21\,\mathrm{Mg\,m^3}$), without any kind of differentiation from top to bottom, and the standard deviation is higher (Fig. 6 centre). The size and shape of aggregates can be clearly recognized in the two CT images (Fig. 6 right). It is possible to identify the interaggregate pore space as well as parts of vertically or horizontally oriented macropores. In this case, the soil structure and the corresponding functional properties are favourable for plant growth. Aggregates are responsible for the mechanical stability of a soil (Horn *et al.* 1995). Macropores are the paths available for gas exchange between soil and atmosphere. They play an important role in preferential water and solute transport through soil (Beven & Germann 1982; White 1985; Booltink & Bouma 1991; Gerke & van Genuchten 1993; Flury *et al.* 1994; Bosch & King 2000) and they can also be used by roots to penetrate the compaction zones towards the subsoil (Dexter 1986; Bennie 1991; Whalley & Dexter 1994; Stewart *et al.* 1999). The interaggregate pore space determines at the same time the gas, water and solute transport conditions in the soil through transfer from and to the soil matrix and inner aggregate porosity.

Differentiation in the structure of a loess soil caused by land use – a comparison between different structural development in lysimeters and in the field

At the experimental site Brandis, the comparison between a naturally layered soil in a lysimeter and a soil in the field shows significant soil structural differences in the lower topsoil. The Ah-horizon of the lysimeter profile (site B1) has a well developed segregation structure, characterized by a narrowly spaced change in occurrence of aggregates (crumbs) and interaggregate pores, as well as macropores with high continuity and connectivity (Fig. 7 left and right). The mean dry bulk density is $1.28\,\mathrm{Mg\,m^{-3}}$. At 35–38 cm depth, a zone of lower bulk density occurs together with a higher standard deviation of the *HU* values of the corresponding horizontal slice (Fig. 7 centre). The well-structured state of the Ah-horizon is due to the absence of any kind of mechanically induced stress, as the topsoil is generally only cultivated manually with a spade. In the field, on the other hand, at site B2 all sources of mechanically induced stress (tillage, vehicle traffic and machine harvest processes) acted on the soil and led to a degradation of soil structure, even in the

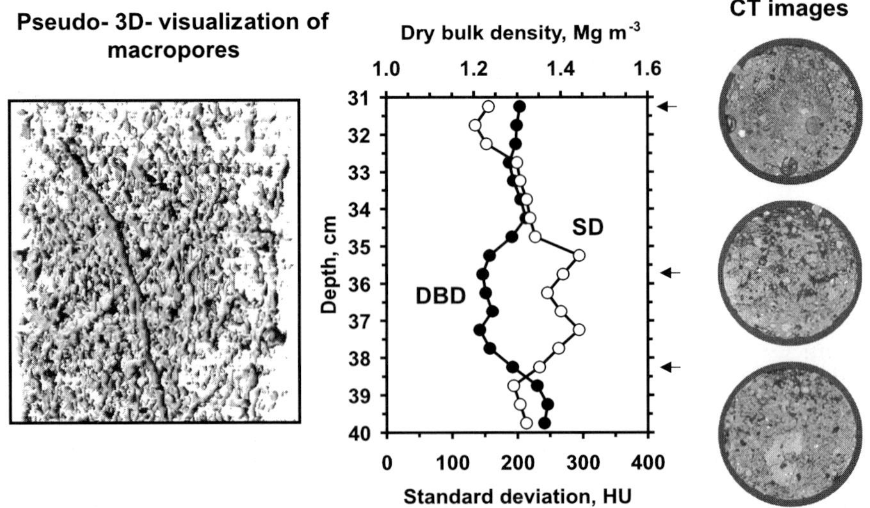

Fig. 7. Pseudo-3D visualization of macropores (left), dry bulk density and standard deviation distribution of horizontal slices over depth (centre) and selected CT images (right) for a soil core sample from site B1 (Ah-horizon, loess soil, 31–41 cm depth). Arrows indicate depths for which horizontal slices are shown.

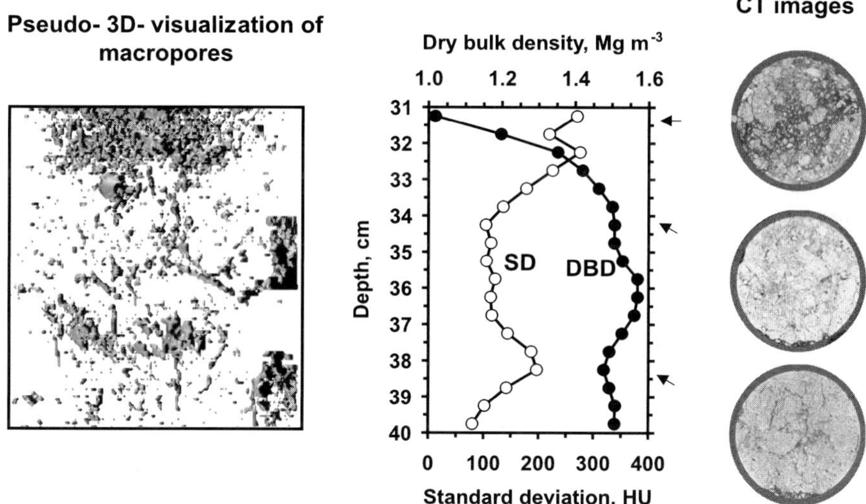

Fig. 8. Pseudo-3D visualization of macropores (left), dry bulk density and standard deviation distribution of horizontal slices over depth (centre) and selected CT images (right) for a soil core sample from site B2 (Ah-horizon, loess soil, 31–41 cm depth). Arrows indicate depths for which horizontal slices are shown.

Ah-horizon. The continuity and connectivity of air-filled macropores and the interaggregate pore system were destroyed, resulting in a higher amount of dead-end macropores within the soil core sample. The pore sizes were reduced and the size and shape of individual aggregates could not be identified anymore (Fig. 8 left and right). Compaction led to an increase in mean dry bulk density to 1.456 Mg m^{-3} and a decrease in standard deviation of the *HU* values of the slices (Fig. 8 centre). The unsaturated hydraulic conductivity of the macropores with 0.3–3.0 mm equivalent diameters was lower compared to site B1 by 1 to 2 orders of magnitude (from 135.1 to 1.6 cm d^{-1}, see Table 4).

Spatial heterogeneity in soil structure

In addition to the analytical opportunities of CT shown above, scanning of soil core samples also allows characterization of specific details of the structural state of the soil. In the Ap-horizon of the soil at experimental site B3 the skeleton content (particles >2 mm in diameter) increases in a vertical direction, accompanied by an increase in macropore volume. This differentiation can be clearly recognized from the CT images (Fig. 9 right). The bulk density distribution of horizontal slices gives no clear indication about differentiation in skeleton content or macroporosity. However, the permanent increase in standard deviation of HU values of horizontal slices characterizes this specific soil structural state and would allow quantifying these structural changes over depth (Fig. 9 left).

Dry bulk density distribution around an earthworm burrow

Results for site F were obtained for soil biological investigations, described by Jégou *et al.*

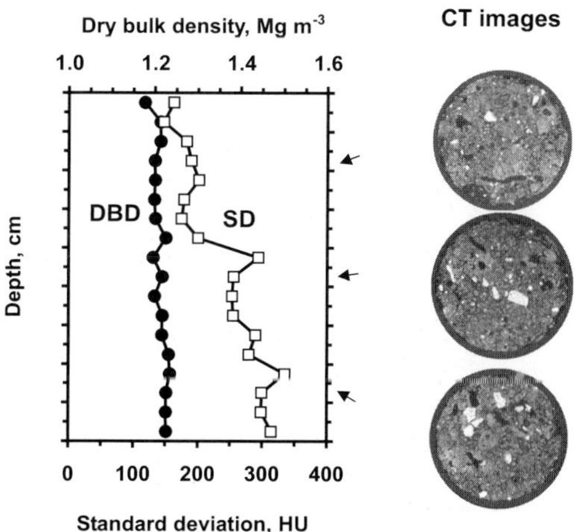

Fig. 9. Dry bulk density and standard deviation distribution for horizontal slices over depth (left) and selected CT images (right) for a soil core sample from site B3 (Ap-horizon, loamy soil, 5–15 cm depth). Arrows indicate depths for which horizontal slices are shown.

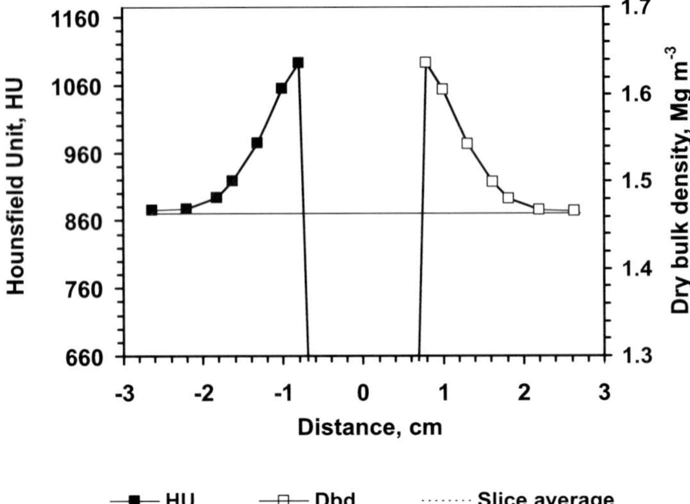

Fig. 10. Horizontal distribution of earthworm-induced variations in Hounsfield Unit values and dry bulk densities, as a function of the distance from the centre of the burrow, compared with the slice averages.

(2002). Disturbed soil core samples with defined dry bulk density (Table 2) were created and inoculated with the earthworm species *Lumbricus terrestris*. Individuals of this earthworm species penetrated the soil and created burrows. During its burrowing activity, the earthworm exerted axial pressures on the soil and displaced soil particles. The worm could not use zones of weakness, which occur in naturally layered soils. As a result, the soil surrounding the burrow was compacted, as shown by the CT images, the measured *HU* values and calculated bulk densities (Fig. 10, Fig. 11). The maximum *HU* values and bulk densities were reached near the wall of the burrow. Here, the earthworm increased the dry bulk density by $0.17\,\mathrm{Mg\,m^{-3}}$. The compactive effects decreased with increasing distance from the central axis of the earthworm burrow. The bulk density decreases from a maximum of $1.638\,\mathrm{Mg\,m^{-3}}$ at the burrow wall to 1.469 $\mathrm{Mg\,m^{-3}}$ at a distance of 27 mm (Fig. 10). This cylindrical volume with a 54 mm diameter can be considered as the limit for the soil volume that is influenced by forces exerted by the earthworm. The mean dry bulk density of the horizontal slice is $1.466\,\mathrm{Mg\,m^{-3}}$.

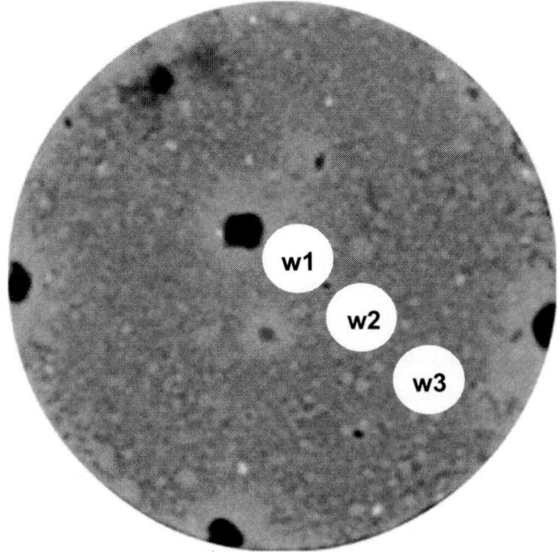

Fig. 11. Local gravimetric water content estimations at different distances from the earthworm burrow.

Several narrow spaced gravimetric water content measurements at different positions within the horizontal slices (Fig. 11) confirm the assumption that the water content is similar at different distances from the earthworm burrow. The differences are statistically not significant

Table 5. *Gravimetric water content near earthworm burrow, site F*

Distance from earthworm burrow[1]	Water content[2] $\mathrm{kg\,kg^{-1}}$
low (w1)	0.289
medium (w2)	0.299
high (w3)	0.291

[1] See Figure 11.
[2] Four replicates; LSD at the $P = 0.05$ level: $0.025\,\mathrm{kg\,kg^{-1}}$.

(Table 5). Nevertheless, a narrowly spaced variation in soil water content could be a source of error. Water content distribution can only be determined precisely by dual energy CT scanning.

Conclusions

This paper assesses the soil structural phenomena of selected undisturbed soil core samples of loamy and silty soils by using single energy level CT scanning. The non-destructive investigations characterize narrowly spaced variations in soil structure and provide detailed knowledge of soil structural state and development. Investigations using X-ray computed tomography help to explain and broaden understanding of the development of soil structure and its functional properties such as hydraulic conductivity. Suitable tools are CT images, 3D visualization of macropores for continuity and connectivity assessment, the calculation of dry bulk density of horizontal slices and the estimation of standard deviation of voxel-related Hounsfield Unit values within the slices. The results indicate that a characterization of soil structure cannot only be based on sample-average soil physical parameters. Identification of causal relationships between morphological and functional soil properties requires the investigation of soil core samples at a high spatial resolution. CT scanning can help to assess effects of soil tillage and land use systems and to elaborate guidelines for management parameters, such as allowed tractor wheel inflation pressure or axle load of agricultural machinery.

The authors wish to thank A.-J. Lemke and J. Döring for X-ray computed tomography measurements in the Virchow hospital in Berlin, and U. Haferkorn as well as S. Knappe for the possibility to investigate the soil core samples from the Brandis site.

References

ANDERSON, S.H., PEYTON, R.L. & GANTZER, C.J. 1990. Evaluation of constructed and natural soil macropores using X-ray computed tomography. *Geoderma*, **46**, 13–29.

AYLMORE, L.A.G. 1994. Application of computer assisted tomography to soil-plant-water studies: an overview. *In*: ANDERSON, S.H. & HOPMANS, J.W. (eds) *Tomography of Soil-Water-Root Processes*. Soil Science Society of America, Special Publications, **36**, 7–15.

BENGOUGH, A.G. & MULLINS, C.E. 1991. Penetrometer resistance, root penetration and root elongation rate in two sandy loam soils. *Plant and Soil*, **131**, 59–66

BENNIE, A.T. 1991. Growth and mechanical impedance. *In*: WAISEL, Y., ESHEL, A. & KAFKAFI, U. (eds) *Plant Roots: The Hidden Half*. Marcel Dekker Inc., New York, Basel, Hong Kong, 393–414.

BEVEN, K. & GERMANN, P. 1982. Macropores and water flow in soils. *Water Resources Research*, **18**, 1311–1325.

BOOLTINK, H.W.G. & BOUMA, J. 1991. Physical and morphological characterization of bypass flow in a well structured clay soil. *Soil Science Society of America Journal*, **55**, 1249–1254.

BOOLTINK, H.W.G., BOUMA, J. & GIMENEZ, D. 1991. Suction crust infiltrometer for measuring hydraulic conductivity of unsaturated soil near saturation. *Soil Science Society of America Journal*, **55**, 566–568.

BOSCH, D. & KING, K. 2000. *Preferential Flow: Water Movement and Chemical Transport in the Environment. Proceedings of the 2nd International Symposium on Preferential flow, January 3–5, 2001, Honolulu, Hawaii.* American Society of Agricultural Engineers, St. Joseph, Michigan.

BROWN, J.M., FONTENO, W.C., CASSEL, D.K. & JOHNSON, G.A. 1987. Computed tomographic analyses of water distribution in three porous foam media. *Soil Science Society of America Journal*, **51**, 1121–1125.

CAPOWIEZ, Y., PIERRET, A., DANIEL, O., MONESTIEZ, P. & KRETZSCHMAR, A. 1998. 3D skeleton reconstructions of natural earthworm burrow systems using CAT scan images of soil cores. *Biology & Fertility of Soils*, **27**, 51–59.

CLAUSNITZER, V. & HOPMANS, J.W. 2000. Pore-scale measurements of solute breakthrough using microfocus X-ray computed tomography. *Water Resources Research*, **36**, 2067–2079.

CRESTANA, S., CESAREO, R. & MASCARENHAS, S. 1986. Using a computed tomography miniscanner in soil science. *Soil Science*, **142**, 56–61.

DANNOWSKI, M. 1992. Das Penetrationsvermögen von Wurzeln unterschiedlicher Roggen- und Triticalegenotypen in Abhängigkeit von der Bodenlagerungsdichte. *Journal of Agronomy and Crop Science*, **168**, 169–180.

DEXTER, A.R. 1986. Model experiments on the behaviour of roots at the interface between a tilled seedbed and a compacted subsoil. Entry of pea and wheat roots into cylindrical biopores. *Plant and Soil*, **95**, 149–161.

DICARLO, D.A., BAUTERS, T.W.J., STEENHUIS, T.S., PARLANGE, J.Y. & BIERCK, B.R. 1997. High-speed measurements of three-phase flow using synchroton X-rays. *Water Resources Research*, **33**, 569–576.

FLURY, M., FLÜHLER, H., JURY, W.A. & LEUENBERGER, J. 1994. Susceptibility of soils to preferential flow of water: a field study. *Water Resources Research*, **30** 1945–1954.

GANTZER, C.J. & ANDERSON, S.H. 2002. Computed tomographic measurement of macroporosity in chisel-disk and no-tillage seedbeeds. *Soil & Tillage Research*, **64**, 101–111.

GARNIER, P., ANGULO-JARAMILLO, R., DICARLO, D.A., BAUTERS, T.W.J., DARNAULT, C.J.G.,

STEENHUIS, T.S., PARLANGE, J.-Y. & BAVEYE, P. 1998. Dual-energy synchroton X-ray measurements of rapid soil density and water content changes in swelling soils during infiltration. *Water Resources Research*, **34**, 2837–2842.

GERKE, H.H. & VAN GENUCHTEN, M.T. 1993. A dual-porosity model for simulating the preferential movement of water and solutes in structured porous media. *Water Resources Research*, **29**, 305–319.

GREVERS, M.C.J., DE JONG, E. & ARNAUD, R.J.ST. 1989. The characterization of soil macroporosity with CT scanning. *Canadian Journal of Soil Science*, **69**, 629–637.

GREVERS, M.C.J. & DE JONG, E. 1994. Evaluation of soil pore continuity using geostatistical analysis on macroporosity in serial sections obtained by computed tomography scanning. *In*: ANDERSON, S.H. & HOPMANS, J.W. (eds) *Tomography of Soil-Water-Root Processes*. Soil Science Society of America, Special Publication, **36**, 73–86.

GROSE, M.J., GILLIGAN, C.A., SPENCER, D. & GODDARD, B.V.D. 1996. Spatial heterogeneity of soil water around single roots: use of CT-scanning to predict fungal growth in the rhizosphere. *New Phytologist*, **133**, 261–272.

HAINSWORTH, J.M. & AYLMORE, L.A.G. 1983. The use of computer assisted tomography to determine spatial distribution of soil water content. *Australian Journal of Soil Research*, **21**, 435–443.

HAINSWORTH, J.M. & AYLMORE, L.A.G. 1986. Water extraction by single plant roots. *Soil Science Society of America Journal*, **50**, 841–848.

HAMZA, M.A., ANDERSON, S.H. & AYLMORE, L.A.G. 2001. Studies of soil water drawdowns by single radish roots at decreasing soil water content using computer-assisted tomography. *Australian Journal of Soil Research*, **39**, 1387–1396.

HEERAMAN, D.A., HOPMANS, J.W. & CLAUSNITZER, V. 1997. Three dimensional imaging of plant roots in situ with X-ray computed tomography. *Plant and Soil*, **189**, 167–179.

HEIJS, A.W.J., RITSEMA, C.J. & DEKKER, L.W. 1996. Three-dimensional visualization of preferential flow patterns in two soils. *Geoderma*, **70**, 101–116.

HOPMANS, J.W., VOGEL, T. & KOBLIK, P.D. 1992. X-ray tomography of soil water distribution in one-step outflow experiments. *Soil Science Society of America Journal*, **56**, 355–362.

HOPMANS, J.W., CISLEROVA, M. & VOGEL, T. 1994. X-ray tomography of soil properties. *In*: ANDERSON, S.H. & HOPMANS, J.W. (eds) *Tomography of Soil-Water-Root Processes*. Soil Science Society of America, Minneapolis, Special Publication, **36**, 17–28.

HORN, R., BAUMGARTL, T., KAYSER, R. & BAASCH, S. 1995. Effect of aggregate strength on strength and stress distribution in structured soils. *In*: HARTGE, K.H. & STEWART, B.A. (eds) *Soil Structure: Its Development and Function*. CRC Press, Boca Raton, Fl., 31–52.

JÉGOU, D., CLUZEAU, D., WOLF, H.J., GANDON, Y. & TREHEN, P. 1998. Assessment of the burrow system of *Lumbricus terrestris, Aporrectodea giardi,* and *Aporrectodea caliginosa* using X-ray computed tomography. *Biology & Fertility of Soils*, **26**, 116–121.

JÉGOU, D., BRUNOTTE, J., ROGASIK, H., CAPOWIEZ, Y., DIESTEL, H., SCHRADER, S. & CLUZEAU, D. 2002. Impact of soil compaction on earthworm burrow systems using X-ray computed tomography – preliminary study. *European Journal of Soil Biology*, **38**, 329–336.

JENSSEN, P.D. & HEYERDAHL, P.H. 1988. Soil column descriptions from X-ray computed tomography density images. *Soil Science*, **146**, 102–107.

JOSCHKO, M., GRAFF, O., MÜLLER, P.C., KOTZKE, K., LINDNER, P., PRETSCHNER, D.P. & LARINK, O. 1991. A non-destructive method for the morphological assessment of earthworm burrow systems in three dimensions by X-ray computed tomography. *Biology & Fertility of Soils*, **11**, 88–92.

JOSCHKO, M., MÜLLER, P.C., KOTZKE, K., DÖHRING, W. & LARINK, O. 1993. Earthworm burrow system development assessed by means of X-ray computed tomography. *Geoderma*, **56**, 209–221.

KRETZSCHMAR, A. 1991. Burrowing ability of the earthworm *Aporrectodea longa* limited by soil compaction and water potential. *Biology & Fertility of Soils*, **11**, 48–51.

MACEDO, A., CRESTANA, S. & VAZ, C.M.P. 1998. X-ray microtomography to investigate thin layers of soil clod. *Soil & Tillage Research*, **49**, 249–253.

MATERECHERA, S.A., DEXTER, A.R. & ALSTON, A.M. 1991. Penetration of very strong soil by roots of different plant species. *Plant and Soil*, **135**, 31–41.

MISRA, R.K., DEXTER, A.R. & ALSTON, A.M. 1986. Maximum axial and radial growth pressures of plant roots. *Plant and Soil*, **95**, 315–326.

OLSEN, P.A. & BORRESEN, T. 1997. Measuring differences in soil properties in soils with different cultivation practices using computer tomography. *Soil & Tillage Research*, **44**, 1–12.

PERRET, J., PRASHER, S.O., KANTZAS, A. & LANGFORD, C. 1998. Characterization of solute breakthrough and preferential flow in intact soil columns using X-ray CAT scanning. *In*: *Proceedings of the 7th International Drainage Symposium, 8–10 March, Orlando, Florida*, 629–636.

PERRET, J.P., PRASHAR, S.O., KANZAS, A. & LANGFORD, C. 1999. Three-dimensional quantification of macropore networks in undisturbed soil cores. *Soil Science Society of America Journal*, **63**, 1530–1543.

PERRET, J.P., PRASHAR, S.O., KANZAS, A., HAMILTON, K. & LANGFORD, C. 2000*a*. Preferential solute flow in intact soil columns measured by SPECT scanning. *Soil Science Society of America Journal*, **64**, 469–477.

PERRET, J.P., PRASHAR, S.O., KANZAS, A. & LANGFORD, C. 2000*b*. A two-domain approach using CAT scanning to model solute transport in soil. *Journal of Environmental Quality*, **29**, 995–1010.

PETROVIC, A.M., SIEBERT, J.E. & RIEKE, P.E. 1982. Soil bulk density analysis in three dimensions by computed tomographic scanning. *Soil Science Society of America Journal*, **46**, 445–450.

Perroux, K.M. & White, I. 1988. Designs for disc permeameters. *Soil Science Society of America Journal*, **52**, 1205–1215.

Peyton, R.L., Haeffner, B.A., Anderson, S.H. & Gantzer, C.J. 1992. Applying X-ray CT to measure macropore diameters in undisturbed soil cores. *Geoderma*, **53**, 329–340.

Phogat, V.K., Aylmore, L.A.G. & Schuller, R.D. 1991. Simultaneous measurement of the spatial distribution of soil water content and bulk density. *Soil Science Society of America Journal*, **55**, 908–915.

Pierret, A., Capowiez, Y., Belzunces, L. & Moran, C.J. 2002. 3D reconstruction and quantification of macropores using X-ray computed tomography and image analysis. *Geoderma*, **106**, 247–271.

Rogasik, H., Weinkauf, H. & Seyfarth, M. 1997. Methodik und Technologie zur Entnahme ungestörter Bodenproben. *Archives of Agronomy and Soil Science*, **41** 199–207.

Rogasik, H., Crawford, J.W., Wendroth, O., Young, I.M., Joschko, M. & Ritz, K. 1999. Discrimination of soil phases by dual energy X-ray tomography. *Soil Science Society of America Journal*, **63**, 741–751.

Stewart, J.B., Moran, C.J. & Wood, J.T. 1999. Macropore sheath: quantification of plant root and soil macropore association. *Plant and Soil*, **211**, 59–67.

Tollner, E.W. & Ramseur, L.E. 1988. Using computed tomography to measure soil moisture and bulk density in the presence of a growing plant. *International Winter Meeting of the American Society of Agricultural Engineers, St. Joseph.* Paper **88–1625**.

Tollner, E.W. & Verma, B.P. 1989. X-ray CT for quantifying water content at points within a soil body. *Transactions of the American Society of Agricultural Engineers*, **32**, 901–905.

Tollner, E.W., Ramseur, L.E. & Murphy, C. 1994. Techniques and approaches for documenting plant root development with X-ray computed tomography. *In*: Anderson, S.H. & Hopmans, J.W. (eds) *Tomography of Soil-Water-Root Processes.* Soil Science Society of America, Special Publication, **36**, 115–133.

Vaz, C.M.P., Crestana, S., Mascarenhas, S., Cruvinel, P.E., Reichardt, K. & Stolf, R. 1989. Using a computed tomography miniscanner for studying tillage induced soil compaction. *Soil Technology*, **2**, 313–321.

Warner, G.S., Nieber, J.L., Moore, I.D. & Geise, R.A. 1989. Characterizing Macropores in Soil by Computed Tomography. *Soil Science Society of America Journal*, **53**, 653–660.

Wendroth, O. & Simunek, J. 1999. Soil hydraulic properties determined from evaporation and tension infiltration experiments and their use for modeling field moisture status. *In*: van Genuchten, M.Th., Leij, F.J. & Wu, L. (eds) *Proceedings of the International Workshop on Characterization and Measurement of the Hydraulic Properties of Unsaturated Porous Media.* Riverside, California, Volume 1, 737–748.

Whalley, W.R. & Dexter, A.R. 1994. Root development and earthworm movement in relation to soil strength and structure. *Archives of Agronomy and Soil Science*, **38**, 1–40.

White, R.E. 1985. The influence of macropores on the transport of dissolved and suspended matter through soil. *Advances in Soil Science*, **3**, 94–120.

3D soil image characterization applied to hydraulic properties computation

J. F. DELERUE[1,2], E. PERRIER[1], A. TIMMERMAN[3] & R. SWENNEN[4]

[1] *UR Geodes, IRD, 32 avenue Henri Varagnat, 93143 Bondy Cedex, France (e-mail: delerue@bondy.ird.fr)*

[2] *Laboratory of Building Physics, K.U. Leuven, Celestijnenlaan 131, B-3001 Heverlee, Belgium*

[3] *Institute for Land and Water Management, K.U. Leuven, Vital Decosterstraat 102, B-3000 Leuven, Belgium*

[4] *Physico-chemical Geology, K.U. Leuven, Celestijnenlaan 200C, B-3001 Heverlee, Belgium*

Abstract: We propose a novel method to characterize the fluid-filled (usually air or water) space in images of porous media at the pore scale. First, an aperture map is created based on a skeleton process, to describe all local sizes in the pore space. Then the pore space is segmented in pores, defined as elementary objects that compose the pore space. Using this segmented image, a pore network is created, which is a graphic representation of the pore space that includes local sizes and direct information about connectivity at the pore scale. As an application of this method for pore space modelling, the equivalent hydraulic conductivity or permeability for a soil sample is computed.

Image acquisition and analysis has been used in soil science for a long time to get information about soil composition and organization. At different scales, ranging from that of human vision using photography to smaller scales using different types of microscopic observation devices, large sets of data obtained by bi-dimensional imaging are available. Measurements can be performed in 2D using classical image analysis software to get global information such as soil porosity or solid space volume, as well as local information about size shape and distributions of solids and voids (Cousin *et al.* 1996; Ringrose-Voase & Bullock 1984). However, mere 2D information is not enough, especially when topological properties are considered, particularly with regards to the connectivity of the pore space, which affects fluid dynamics (Vogel 1997). Measurement devices are, therefore, being developed to directly address 3D issues.

Development of X-ray computed tomography (CT) provides 3D images of soil samples with good accuracy. Analysis of 3D images is of increasing importance, concentrating on aspects such as aperture characterization and bottleneck localization (Lindquist & Venkataranga 1999; Lindquist *et al.* 1999; Lin & Miller 2000; Vogel & Roth 2001). However, the need for not only geometrical but also functional information is still strong. Functional information can be given by using lattice-Boltzmann simulations at a very low scale. Such application needs very high computational power (Miller *et al.* 1998) and cannot be used easily on large amounts of data. Another way is to use a Pore Network model (Constantinides & Payatakes 1989; Ioannidis & Chatzis 1993). There are numerous studies using Pore Network models, which mainly consist of building Pore Network models on square or cubic grids using statistical information such as pore size distribution or co-ordination number (Fisher & Celia 1999; Man & Jing 1999; Lin & Miller 2000; Dixit *et al.* 2000; Peat *et al.* 2000; Tsakiroglou & Payatakes 2000; Knackstedt *et al.* 2001).

The aim of this study is to build a Pore Network directly from soil images, which integrates both size parameters and connectivity and that represents exactly the geometry of a representative sample. This does not exclude of course the necessity of a statistical approach using several realisations. To proceed, first the pore space needs to be divided in elementary objects that are called pores. The concept of a pore is intuitive and well known in geosciences, but no rigorous definition has been actually given in the context of the complex geometry of a real 3D image of a porous medium. A pore will be defined here as a part of the void space surrounded by the solid

From: MEES, F., SWENNEN, R., VAN GEET, M. & JACOBS, P. (eds) 2003. *Applications of X-ray Computed Tomography in the Geosciences.* Geological Society, London, Special Publications, **215**, 167–176. 0305-8719/03/$15.

space, with a homogeneous local aperture. This definition of a pore is driven by the will to use a Poiseuille integrated simple form (Eq. 6) of the Navier–Stokes equations to compute local conductivity in pores.

Once the pore space is divided into pores, each pore is individually analysed to extract its geometrical properties and relationship with its neighbours. A network is then created using previously extracted information and a local hydraulic conductance is associated to each extracted pore based on the aperture or width of the pore, its length and the Poiseuille equation. The pressure, hydraulic conductance and the volumetric flow are respectively analogous to the potential, electrical conductance and the current flow in electric circuits. Kirchoff's rules formulate that the sum of the potential is zero for every closed loop of conductors and that the algebraic sum of the currents flowing into each node is also zero. From the application of these rules to the pore network, a system of linear equations for the potential at each node is obtained. This system is solved using an iterative Gauss–Seidel method. Substituting the volumetric flow rate, q, at the inlet of the network and the pressure drop, ΔP, along the network in Darcy's law (Eq. 1), the absolute conductivity k is obtained:

$$q = k\,\Delta P. \tag{1}$$

In the first part, we describe in detail the different parts of our method: the estimation of local aperture, the segmentation of the pore space, the definition of the pore network and the computation of the hydraulic conductivity. Then this method is applied to a 2D synthetic image. In the second part, the method is applied to a 3D image obtained by X-ray CT analysis. Local geometric and hydraulic properties are calculated and hydraulic conductivity is compared to the hydraulic conductivity measured experimentally. In the third part, conclusions are formulated and plans for further work are presented.

Theory and methods

Estimation of the hydraulic conductivity is based on a Pore Network model. The Pore Network model can be built on a partition of the pore space in 'pore objects'. In our model, a pore is a part of the pore space, defined both by size and geometric consideration. Before defining the segmentation process, which as a result gives the definition of a pore object, first the notion of size in the pore space as the local aperture size is introduced (Delerue *et al.* 1999*b*).

Local pore aperture

Definition of the local aperture size: the local aperture size, $Ap(p)$, defined for each point, p, of the pore space, P, is equal to the diameter of the maximum ball of centre, C, and diameter, d, included in the pore space and including the point, p:

$$Ap(p) = \max\{r \mid B(C,d) \subset \mathsf{P}, p \in B(C,d)\}. \tag{2}$$

Using this definition of the local aperture size, an aperture map, M, can be constructed which gives for each point, P, of the pore space, P, the local aperture:

$$\mathsf{M} = \{P, Ap(P) | P \in \mathsf{P}\}. \tag{3}$$

Aperture maps can be computed in an efficient way, as described by Delerue *et al.* (1999*a*). First the skeleton of the pore space needs to be computed. Such a skeleton can be based on Voronoi diagrams (Ogniewicz *et al.* 1993; Delerue *et al.* 1999*b*), distance transformations (Rosenfeld & Pfaltz 1966; Gagvani & Silver 1997) or thinning processes (Lee *et al.* 1994; Marion-Poty 1994). The skeleton of the object must be computed and the distance to the border for each point of the skeleton must be kept. To compute the aperture map, the following definition of the skeleton is used: the skeleton is defined as the locus of all inscribed maximal spheres of the object where these spheres touch the object boundary at more than one point (Lam *et al.* 1992). Thus, the aperture map can be reconstructed by drawing a ball centred on each point of the skeleton with a radius equal to the distance to the border. The value associated with each ball is the distance to the border. When two or more balls overlap in one point, the highest value for that point is kept.

Pore segmentation

A pore is an abstract object used to describe the pore space. Different definitions of the pore object have been given. For Marsily (1986), a pore object is a part of the pore space limited by the matrix and bottlenecks. However, this definition of the pore object is not suitable for use in this study, in which the definition of a pore object has to be compatible with the definition of the Poiseuille Law for fracture and cylinder, i.e. the geometry of elementary pore objects has to be simple enough to be compared with cylinders or fractures and pore objects have to be surrounded with solid parts and other pore objects (see Fig. 1).

Thus, a pore is defined as a connected cluster of voxels, limited by at least two boundaries at

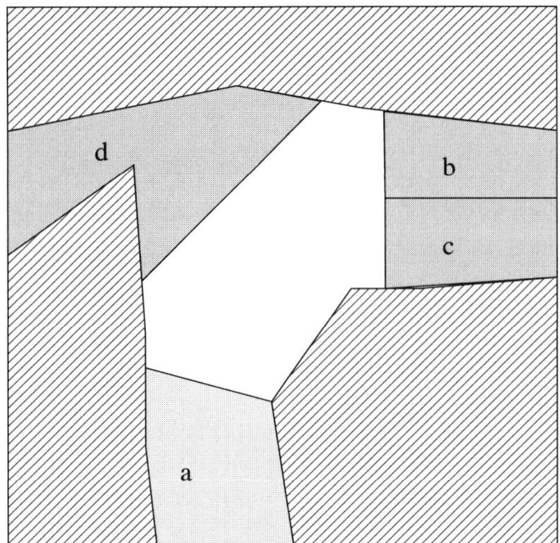

Fig. 1. Definition of a pore object. On this image, region *a* can be considered as a pore, but regions *b*, *c* and *d* cannot: the borders of regions *b* and *c* do not touch opposite borders and the aperture in region *d* is not homogeneous.

the border of the solid space and with all voxels belonging to the same class of local aperture size.

Region growing algorithm. The algorithm used to segment the pore space with pore objects is now described. The main constraints to define pore objects are that: (i) the object must touch boundaries in at least two different points opposite from the centre of the object; and (ii) the local aperture in the object must be bounded.

The first constraint is solved using maximum balls. Maximum balls in an object touch the border of the object in points that are 'opposite' with regard to the centre of the balls. Thus, if pores are derived from maximum balls, they will inherit that property. The second constraint is, in general, impossible to satisfy: because of the very irregular shape of the interface between pore and matrix (Fig. 2), local apertures near a pore interface vary and reach very small values. This problem would not occur if the pore interface was very smooth. All small irregularities have to be neglected at a given space resolution, otherwise no simple model can be used.

Maximum unoverlapping balls are placed in the pore space. Different strategies can be used to dispose those balls and results can vary greatly. depending on the strategy that is used. The objective is to fill the maximum space in the pore space with unoverlapping maximum balls. First, the pore space is filled with the biggest balls and then, step after step, is filled with smaller maximum balls. This operation is done efficiently using the skeleton required for the computation of the local aperture map in section two. A unique colour is associated with each ball (Fig. 3).

Now those balls in pores need to expand simultaneously. This is done with a growing region algorithm. An empty list of points is created. Every point located on the border of a ball is

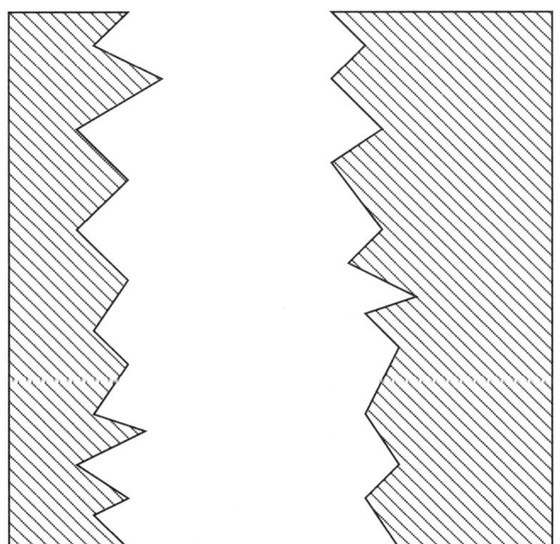

Fig. 2. In this image, all pore space belongs to the same pore. However, due to the complexity of the void-solid interface, regions of the pore space with a very small aperture have to be added to the pore.

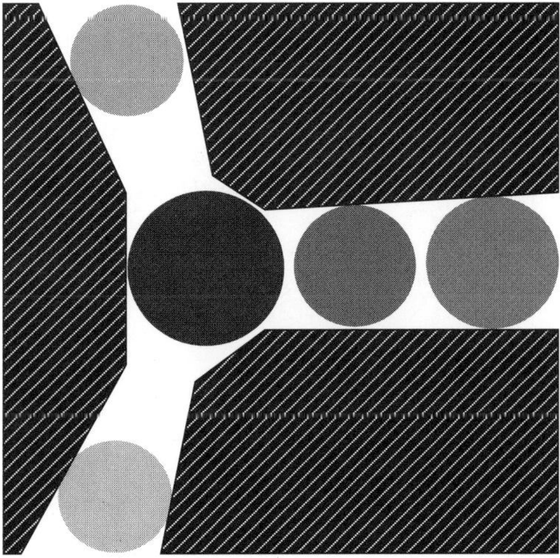

Fig. 3. Seed map: maximum balls (or seeds) have been set in the pore space. Two seeds cannot overlap and must touch opposite borders of the pore space. After the growing region process, each seed will become a pore object.

Fig. 4. Pore map: a region growing process has been applied to the seed map (Fig. 3). Each region in the pore space corresponds to a distinct pore object from which aperture, length and also connectivity with its neighbours can be computed.

added to the list. Then the first point of the list is removed. All of its neighbours that have not yet been associated with a colour are appended to the list and the colour of the removed point is associated with the points that have been appended to the list. This is repeated until the list is empty. At the end of the process, every point of the pore space has been filled with a colour. The pore space has been split into regions of identical colour. A pore is defined as a region with a unique colour (Fig. 4). The aperture of each pore is defined as the maximum aperture size in the pore. This value corresponds to the diameter of the maximum ball from which the region results. The result of this algorithm is a pore map, which is used to define a pore network.

Definition of a pore network

To compute the hydraulic conductivity of the soil sample, a pore network model is used. A pore network model consists of a set of pore objects and links between those pores to describe connectivity. A local property is associated to each pore, namely its hydraulic conductance. This local conductance depends on the geometry, shape, aperture and the size of the pore and can be computed using Navier–Stokes equations or Poiseuille's Law for simple forms. Then, as well as for the Kirchoff network, the macroscopic hydraulic conductivity is calculated by a numerical integration of local fluxes in the graph. It is assumed that the pore space is filled with a fluid, water for example, and that this fluid is flowing between an inlet and an outlet associated with two disjointed sets of pores. Pores located in the first set are called entrance pores and pores located in the second set are called exit pores. The flux of fluid flowing between the entrance and exit pores if a gradient of pressure $P_2 - P_1$ is applied to those pores needs to be ascertained. Pressure in the entrance pores is set to P_2 and pressure in the exit pores is set to P_1. Based on the fact that the local sum of fluid flux must be equal to zero at each node in the network (for example, for every amount of fluid which enters a pore full of fluid, an equal amount of fluid has to go out of the pore), it is possible to

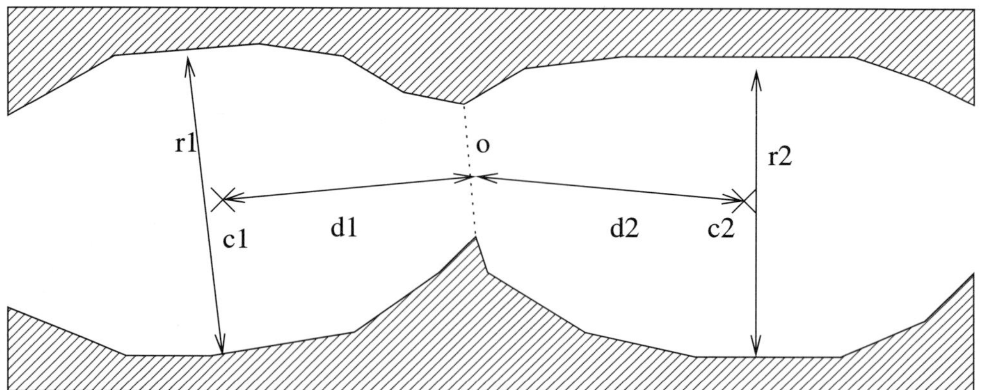

Fig. 5. Construction of a link c_1c_2 in the network from two neighbouring pores p_1 and p_2. The conductivity of the link is computed from the combination of hydraulic conductance of the two half pores between the pore centre c_1 and c_2. Each hydraulic conductance is computed using Poiseuille's law and the combination is done using a harmonic mean.

write for each node the following linear equation, whereby q_u are positive or negative fluxes at each node i:

$$\sum_u q_u = \sum_u k_u(P_i - P_u) = 0. \quad (4)$$

In this case, the set of all equations is a system of linear equations in which the unknowns are local pressures, P_i, at each node. To be solvable, at least one path must exist between entrance and exit pores and all dead branches must be removed. This system of linear equations has a unique solution that can be expressed using the Gauss algorithm.

From a practical point of view, it must be decided where exactly the nodes are in the network and, thus, also where the links of the bonds connecting these nodes are located. In some theoretical approaches, the nodes are situated at the intersection between the pores, in other ones the nodes are the so-called pore bodies and the links are the throats or constrictions between pores. In this case, based on real images, a slightly different approach is needed: the nodes are here the centre of the pore objects, which have been previously defined and the links go from the centre of the pore to the centre of each neighbouring pore, going through their interface (Fig. 5).

Then a local conductance is calculated for each link using a serial composition law, which is in this case a harmonic mean between the conductance, k_1 and k_2, of the two pores connected by this link:

$$\frac{2}{k} = \frac{1}{k_1} + \frac{1}{k_2}.$$

Thus, the conductance of a sub parts path is calculated with the following equation:

$$k(p_1, p_2) = \frac{2}{\dfrac{1}{k(r_1, d(c_1, 0))} + \dfrac{1}{k(r_2, d(c_2, 0))}}. \quad (5)$$

It is assumed that each part or sub pore can be compared to a cylinder of length, L, and radius, r, and that its conductivity can be computed in a first approximation using Poiseuille's Law:

$$q = \frac{\pi r^4 \, \Delta P}{8 \mu L} = k(r, L) \, \Delta P \quad (6)$$

where ΔP is the gradient of pressure applied to the boundaries of the sub pore.

Thus, as for traditional pore network models, it is possible to write an equation for each connection between pores. Again, the set of all those equations is a system of linear equations where unknowns are pressures at each connection. The solution of this system using the Gauss algorithm gives local pressure at each connection and we can deduce the global flux of fluid between entrance and exit pores.

Results on 2D synthetic data

Displaying results on 3D soil images is difficult because of the combination of colours and luminosity effects in 3D rendered images. Thus, we first want to show the application on a synthetic 2D image representing a porous media constructed with a fractal model (Perrier *et al.* 1995).

The image in Figure 6a represents a porous medium. The black colour represents the solid part and the white colour represents the voids. Figure 6b shows the aperture map. Each grey shade represents a different aperture. Figure 6c shows the pore map. Each region represents a pore. Grey intensity represents aperture in the pore and corresponds to the greatest local aperture in the pore. From that image, a pore network can be constructed and local pore sizes can be computed. This network is shown in Figure 6d: each grey ball represents the centre of a pore and the links represent connections between pores.

Then the pore network is used to compute hydraulic properties. An imaginary gradient of pressure is applied between the upper and lower face, with a greater pressure at the lower face than at the upper face. Then the hydraulic conductivity for the image is calculated. Figure 6e shows the gradient of pressure: balls are located on pore centres and their size indicates the local pressure. Figure 6f shows the local flux direction and intensity. The flux intensity is computed using Poiseuille's equation and flux vectors are oriented toward lower pressures. In the following section, the results on a 3D image of the pore system in a soil sample are presented.

Applications

Data acquisition

A soil core (9.94 cm diameter, 15 cm length) was taken from the top layer of a well-drained macroporous sandy loam soil (Eurtic Regosol/Udifluvent) at Bekkevoort (Belgium) by gently pushing down a PVC cylinder into the soil. Saturated hydraulic conductivity, K_{sat}, was measured using the constant-head permeameter method (Klute 1965).

After determination of hydraulic conductivity at saturation, the soil core was dried and

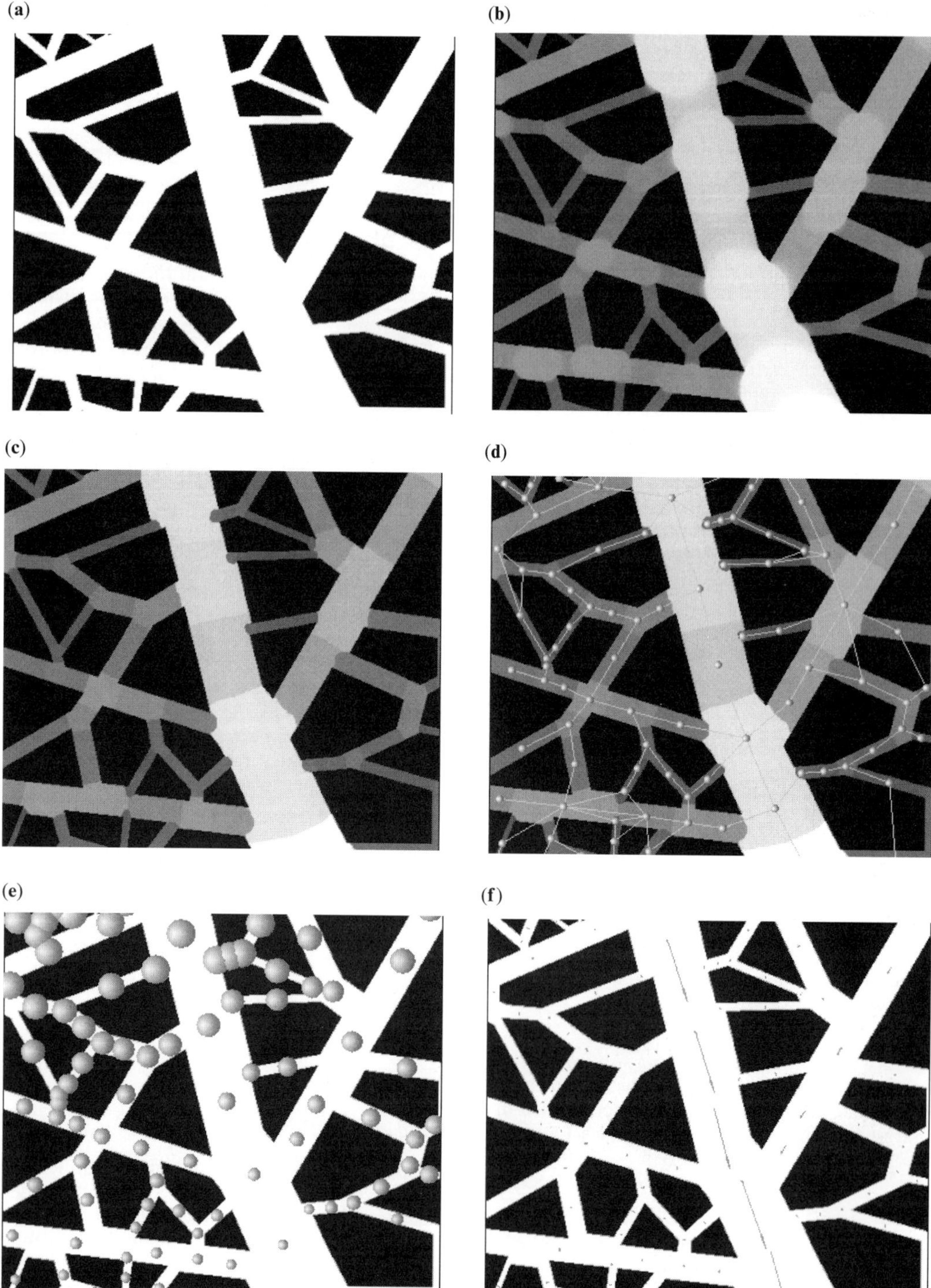

Fig. 6. (**a**) 2D original image – black represents solid space and white represents void space; (**b**) aperture map – brighter colours represent bigger apertures; (**c**) pore map – brighter colours represent bigger apertures; (**d**) pore network representation – spheres represent centres of pores and links represent connections; (**e**) pressure gradient – bigger spheres represent higher pressures; (**f**) lux direction and intensity in the pore network.

transported to the Gasthuisberg University Hospital in Leuven. The soil core was scanned with the spiral scanning method using a fourth generation scanner (Siemens Somatom Plus S). The voltage for the X-ray source was 137 kV and the source current was 180 mA.

The collected raw data were reconstructed with an ultra-high resolution algorithm (CH40)

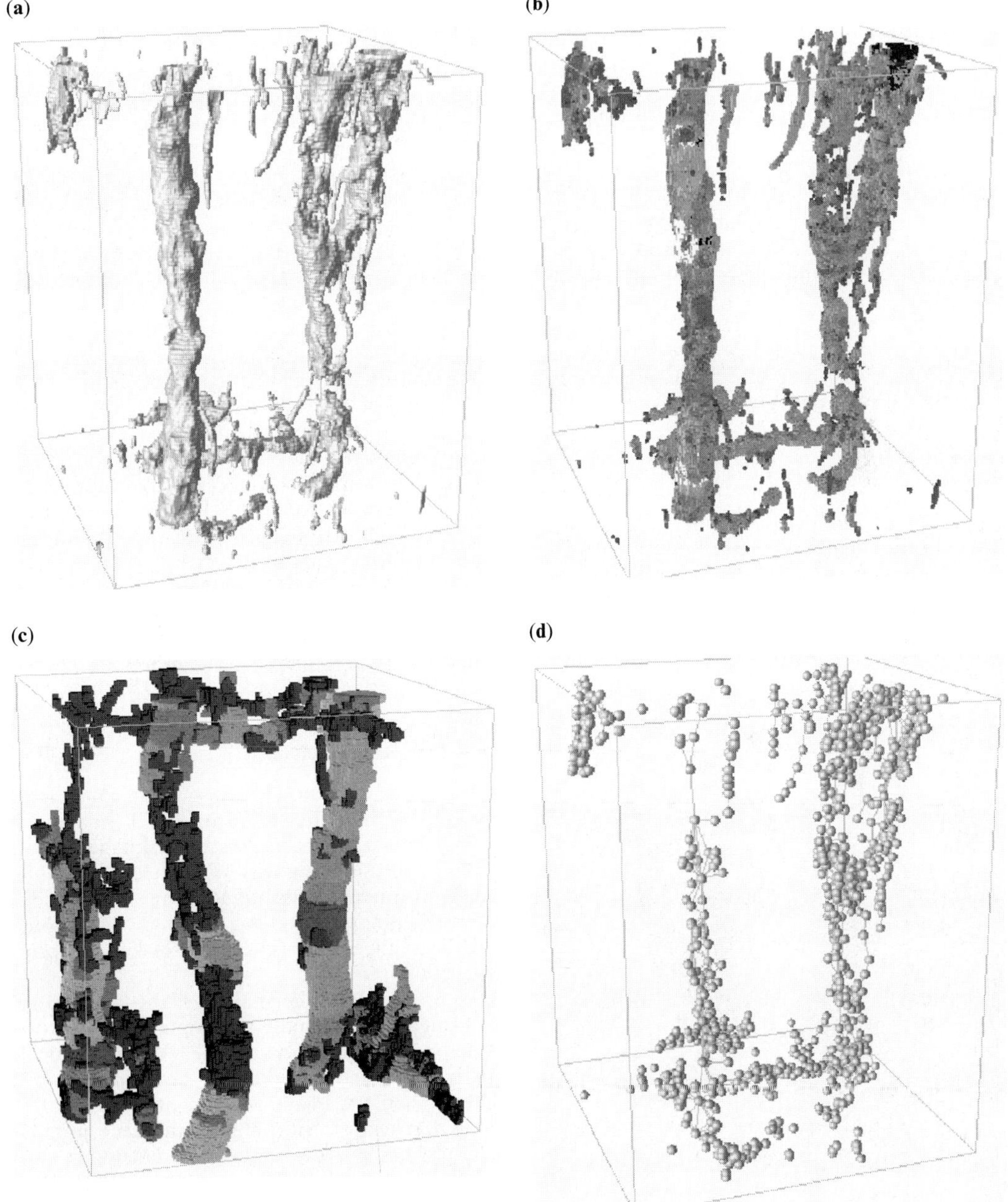

Fig. 7. (**a**) 3D soil image – grey represents the void space; (**b**) aperture map – each grey shade represents a different aperture size; (**c**) pore map – pore space has been divided into pores, each shade represents a different aperture; (**d**) graph network – each sphere localizes the centre of a pore, whereas links represent connectivity between pores; (**e**) local hydraulic properties – spheres are located on centres of pores and their diameter shows local pressures; arrows show intensity and direction of flux.

to derive images with good contrast between air-filled pores and soil matrix. The minimal diameter of the X-ray beam is 1 mm (slice thickness) and this determines the resolution of the images in the Z-direction. With spiral CT, however, over laps of images can be obtained to improve the vertical resolution. In our case, slices with an interspace of 0.75 mm were obtained. The final 3D image consists of $203 \times 203 \times 180$ voxels, with a voxel size of $0.5 \times 0.5 \times 0.5$ mm.

The linear attenuation coefficients obtained during scanning are transformed to standardized CT measuring units (Hounsfield Units, HU), with a total range of −1024 to 3071 HU. To distinguish the soil matrix and air-filled pores, a cut-off value (CTC) of 300 HU was used, based on earlier experience and calibrated relationships between CTC values and air opening diameters (Timmerman *et al.* 1999). The CT images were transformed to images with 256 grey levels, whereby the CTC value of 300 HU corresponds to a value of 80 on this grey level scale.

Results

The results of the algorithms presented in this paper on the image described in the previous part are now given. A 3D rendering of the pore space is presented in Figure 7a. In all images, only objects connected at the given resolution to either the upper or lower face are displayed in order to simplify the clarity of the images.

(e)

Fig. 7. (*continued*)

Figure 7b displays the aperture map: each aperture is shown with a different grey intensity. On Figure 7c, the pore space has been divided into pore objects. Each pore is represented with a grey level that depends on its aperture. Figure 7c has been used to obtain the pore network presented in Figure 7d. Balls represent centres of pores and links represent connections.

As before, a gradient of pressure was applied between the upper face and the lower face, with the greatest pressure on the upper face, to compute an equivalent hydraulic conductivity. Local pressure and fluid flow were computed and are shown in Figure 7e. Balls are centred on pores and their diameter depends on local pressure. Local fluxes are shown using arrows. They are locally oriented in the direction of lower pressure and their length depends on local flux intensity computed with Poiseuille's equation.

The image clearly shows that, at that scale and resolution, only one path connects the upper and lower section of the soil image and all fluxes follow that path. The numerical hydraulic conductivity computed for this image is 9.6 cm/hr, which is in close agreement with the experimentally measured value of 9.5 cm/hr.

Algorithms were implemented as modules in OpenDX visualization software. The complete run, for an image with a size of $203 \times 203 \times 180$ voxels, takes 3 minutes on a Pentium III 1 GHz computer.

Conclusions

In this paper a method to compute an equivalent hydraulic conductivity from a 3D image of porous media has been presented. Image acquisition or thresholding was not considered in this study. This paper aims at analysing structures in 2D or 3D binarized soil images where one colour represents the empty space and the other represents the matrix.

To proceed with the analysis of the image, first the skeleton of the image is computed. Using this skeleton an aperture map, which describes local aperture size for every point of the image, is computed. Then the image is segmented into pores using a growing region algorithm. Each pore is separately analysed to obtain characteristics like aperture, volume, length and connectivity with its neighbours. This information is used to build a pore network model. Using this pore network model, an equivalent hydraulic conductivity and local hydraulic properties can be computed such as local pressure or fluid flow simultaneously. So far the comparison between measured and hydraulic conductivity has been made only on one image. Further work will consist of comparing computed and measured hydraulic conductivities for different soil samples.

More than just computing an equivalent hydraulic conductivity, which can be more easily measured with traditional experimental methods, this method provides a good description of the pore space at the pore scale, which can be used for other purposes such as pore throat detection. In the same way, the computed hydraulic conductivity can be used as a structural indicator, which is directly computed using shape and connectivity information in the pore space and which describes the morphology of the pore space.

References

CONSTANTINIDES, G.N. & PAYATAKES, A.C. 1989. A three-dimensional network model for consolidated porous media basic studies. *Chemical Engineering Communication*, **81**, 55–81.

COUSIN, I., LEVITZ, P. & BRUAND, A. 1996. Three-dimensional analysis of a loamy-clay soil using pore and solid chord distributions. *European Journal of Soil Science*, **47**, 439–452.

DELERUE, J.F., PERRIER, E., TIMMERMAN, A. & RIEU, M. 1999*a*. New computer tools to quantify 3D porous structures in relation with hydraulic properties. *In*: FEYEN, J. & WIYO, K. (eds) *Modelling of Transport Processes in Soils*. Wageningen Press, Wageningen, The Netherlands, 153–163.

DELERUE, J.F., PERRIER, E., YU, Z.Y. & VELDE, B. 1999*b*. New algorithms in 3D image analysis and their application to the measurement of a spatialized pore size distribution in soils. *Physics and Chemistry of the Earth*, **24**, 639–644.

DIXIT, A.B., BUCKLEY, J.S., MCDOUGALL, S.R. & SORBIE, K.S. 2000. Empirical measures of wettability in porous media and the relationship between them derived from pore-scale modelling. *Transport in Porous Media*, **40**, 27–54.

FISHER, U. & CELIA, M.A. 1999. Prediction of relative and absolute permeabilities for gas and water from soil water retention curves using a pore-scale network model. *Water Resources Research*, **35**, 1089–1100.

GAGVANI, N. & SILVER, D. 1997. *Parameter Controlled Skeletonization of Three Dimensional Objects*. Rutgers University Technical Report **CAIPTR216**.

IOANNIDIS, M.A. & CHATZIS, I. 1993. Network modelling of pore structure and transport properties of porous media. *Chemistry Engineering Sciences*, **48**, 951–972.

KLUTE, A. 1965. Laboratory measurement of hydraulic conductivity of saturated soil. *In*: BLACK, C.A. (ed.) *Methods of Soil Analysis. Physical and Mineralogical Properties, including Statistics of Measurement and Sampling. 2nd Edition*. Agronomy Monograph **9**. American Society of Agronomy, Madison, Wisconsin, 210–221.

KNACKSTEDT, M.A., SHEPPARD, A.P. & SAHIMI, M. 2001. Pore Network modelling of two-phase flow in porous rock: the effect of correlated heterogeneity. *Advances in Water Resources*, **24**, 257–277.

LAM, L., LEE, S.W. & SUEN, C.Y. 1992. Thinning methodologies – a comprehensive survey. *IEEE Transactions on Pattern Analysis and Machine Intelligence*, **14**, 869–885.

LEE, T.C., KASHYAP, R.L. & CHU, C.N. 1994. Building skeleton models via 3D medial surface/axis thinning algorithms. *CVGIP, Graphical Models and Image Processing*, **54**, 462–478.

LIN, C.L. & MILLER, J.D. 2000. Network analysis of filter cake pore structured by high resolution X-ray microtomography. *Chemical Engineering Journal*, **77**, 79–86.

LINDQUIST, W.B. & VENKATARANGA, A. 1999. Investigating 3D geometry of porous media from high resolution images. *Physics and Chemistry of the Earth*, **24**, 593–600.

LINDQUIST, W.B., VENKATARANGAN, A., DUNSMUIR, J. & WONG, T. 1999. *Pore and Throat Size Distributions Measured from Synchrotron X-ray Tomographic Images of Fontainebleau Sandstones*. Department of Applied Mathematics and Statistics, SUNY Stony Brook, Technical Report **SUSB-AMS-99-13**.

MAN, H.N. & JING, X.D. 1999. Network modelling of wettability and pore geometry effects on electrical resistivity and capillary pressure. *Journal of Petro leum Science and Engineering*, **24**, 255–267.

MARION-POTY, V. 1994. *3D Thinning Algorithms on Distributed Memory Machines*. Ecole Normale Supérieure de Lyon, Lyon, France, Research Report **94-01**.

MARSILY, G.D. 1986. *Quantitative Hydrology: Groundwater Hydrology for Engineers*. Academic Press, Orlando, Florida.

MILLER, C.T., CHRISTAKOS, G., IMHOFF, P.T., MCBRIDE, J.F., PEDIT, J.A. & TRANGENSTEIN, J.A. 1998. Multiphase flow and transport modelling in heterogeneous porous media: challenges and approaches. *Advances in Water Resources*, **21**, 77–120.

OGNIEWICZ, R.L., SZKELY, G., NAF, M. & KUBLER, O. 1993. Median manifolds hierarchical description of 2D and 3D objects with applications to mri data of human brain. *In*: *Proceedings of the 8th Scandinavian Conference on Image Analysis, Tromsø, Norway*, 875–883.

PEAT, D.M.W., MATTHEWS, G.P., WORSFOLD, P.J. & JARVIS, S.C. 2000. Simulation of water retention and hydraulic conductivity in soil using three dimensional network. *European Journal of Soil Science*, **51**, 65–79.

PERRIER, E., MULLON, C., RIEU, M. & DE MARSILY, G. 1995. Computer construction of fractal soil structures: simulation of their hydraulic and shrinkage properties. *Water Resources Research*, **31**, 2927–2943.

RINGROSE-VOASE, A.J. & BULLOCK, P. 1984. The automatic recognition and measurement of soil pore type by image analysis and computer programs. *Journal of Soil Science*, **35**, 673–684.

ROSENFELD, A. & PFALTZ, J.L. 1966. Sequential operations in digital picture processing. *Journal of the association for computing machinery*, **13**, 471–494.

TIMMERMAN, A., VANDERSTEEN, K., FUSHS, T., CLEYNENBREUGE, J.V. & FEYEN, J. 1999. A flexible and effective pre-correction algorithm for non medical applications with clinical X-ray CT scanners. *In:* FEYEN, J. & WIYO, K. (eds) *Modelling of Transport Processes in Soils*. Wageningen Press, Wageningen, The Netherlands, 121–131.

TSAKIROGLOU, C.D. & PAYATAKES, A.C. 2000. Characterization of the pore structure of reservoir rocks with aid of serial sectioning analysis, mercury porosimetry and network simulation. *Advances in Water Resources*, **20**, 773–789.

VOGEL, H.J. 1997. Morphological determination of pore connectivity as a function of pore size using serial sections. *European Journal of Soil Science*, **48**, 365–377.

VOGEL, H.J. & ROTH, K. 2001. Quantitative morphology and network representation of soil pore structure. *Advances in Water Resources*, **24**, 233–242.

Evaluation of local porosity changes in limestone samples under triaxial stress field by using X-ray computed tomography

C. O. KARACAN, A. S. GRADER & P. M. HALLECK

The Pennsylvania State University, 403 Academic Activities Building, University Park, PA 16802, USA (e-mail: karacan@pnge.psu.edu)

Abstract: The character of reservoir rocks is uncertain and variable at the depths where they are subjected to different tectonic forces and pressure changes due to drilling, well stimulation and production operations. The stress- and time-dependent deformation of the porous structure is expected to change the behaviour of the most important properties of the rock, such as porosity and permeability, which in turn changes the reservoir production predictions.

In this study, we demonstrate the use of X-ray computed tomography (CT) to investigate the porosity and permeability changes of Cordova Limestone samples during deformation in a triaxial cell. The experiments were performed in a specially designed X-ray transparent triaxial test cell, which enables applying stress as well as making flow measurements. Because the presence and value of confining pressure changes the deformation behaviour of the rock, different constant confining stresses were applied to the samples to change the deformation regime. As the axial load was increased, samples were scanned at different locations to determine the stress-dependent local changes in porosity. Absolute permeability during the deformation was also measured.

Results show that the stress condition applied to the porous medium changes the rock and fluid transport properties, compared to measurements taken without stressing the sample. X-ray CT enabled the local porosity changes to be quantified, to locate compaction bands and places where shear location occurs, and to evaluate how the inner structure of the rock changes during different modes of deformation.

Reservoir rocks experience many changes in stress conditions during the production life of the reservoir. During drilling the rock surrounding the wellbore is subjected to high local stresses from the drill bit. After drilling, the surrounding rock must carry the load that was carried by the removed rock. This leads to stress changes in the local *in situ* stress field near the well (commonly called stress concentrations). If the well is drilled in a weak rock, these stress concentrations may collapse the borehole and cause break-outs (Fjaer *et al.* 1992). In cased wells, additional stresses are created by perforating, which causes shock waves that crush the rock grains around the perforation tunnel (Halleck 1997).

During production of the reservoir, pore pressure is depleted because of fluid removal from the pore system. This causes the mean effective stress on the reservoir rocks to increase and deforms the rock by compressing and/or compacting it. Although this provides additional drive energy to produce remaining reserves, compaction can cause production problems (Nagel 2001).

Along with negative impacts like subsidence, all of these stress conditions can alter the permeability of the rock surrounding the wellbore, leading to various types of formation damage. The strains produced by changes in stress may be recoverable if they do not exceed the failure strength of the rock, or permanent if the failure strength is exceeded. Increased mean effective stress can lead to pore compression and ultimately compaction of the rock. Shear stresses or deviatoric stresses result in shear deformation. In the elastic region, there is little volume strain associated with shear deformation, but when approaching shear failure, dilatancy or shear-enhanced compaction can occur. These strains can cause both recoverable and permanent changes in permeability depending greatly on the microscopic failure mechanics of the rock.

It has been generally believed that the failure mechanism is closely related to porosity change. In a study on Solnhofen limestone (3% porosity), brittle-ductile transition was investigated (Baud *et al.* 2000). It was shown that shear-enhanced compaction is appreciable in low porosity rocks too. It was also observed that failure modes show transient behaviour during increasing strain and

From: MEES, F., SWENNEN, R., VAN GEET, M. & JACOBS, P. (eds) 2003. *Applications of X-ray Computed Tomography in the Geosciences*. Geological Society, London, Special Publications, **215**, 177–189. 0305-8719/03/$15.

ultimately lead to shear localization and macroscopic failure.

Deformation experiments in the brittle field showed that low-porosity sandstones experience permeability evolution in triaxial compression that is qualitatively similar to crystalline rocks (Heilend & Raab 2001). In the pre-failure region, permeability is strongly reduced by stress-induced compaction. After the onset of dilatancy, permeability starts to increase. However, when the initial sample volume is regained, permeability is not recovered. Similar results were observed for higher porosity sandstones (15–35%) (Zhu & Wong 1997). In the cataclastic flow regime, a drastic decrease in permeability was triggered by the onset of shear-enhanced compaction caused by grain crushing and pore collapse. It was concluded that before yield stress, permeability and porosity both decrease with increasing effective mean stress, but they are independent of deviatoric stresses. Further increasing the load causes dependency on deviatoric stresses to increase. With the onset of shear-enhanced compaction and development of cataclastic flow, coupling of deviatoric and hydrostatic stresses induces considerable porosity and permeability reduction (Zhu *et al.* 1997). In brittle faulting, both porosity and permeability decrease with effective mean stress. It has been suggested that permeability may actually decrease in the dilation period (Zhu & Wong 1997), as opposed to results presented by Heilend & Raab (2001), and after the peak stress is attained the development of an impermeable shear band causes an accelerated decrease in permeability. Likewise, in rock salt hydrostatic loading gives rise to a marked decrease of permeability (Popp *et al.* 2001). Dilatancy during axial deformation of the compacted salt samples is found to evolve stress dependent in various stages. It has been found that the permeability of the rock salt increases dramatically with progressive dilatancy, followed by a period of constant permeability during strain hardening.

Compaction banding that forms perpendicular to the most compressive principle stress can be another reason for the decrease of permeability within the reservoir. Such bands have been observed in high porosity rocks in the laboratory and in the field (Issen & Rudnicki 2001). It has been shown that axisymmetric compression is the most favourable deviatoric stress state for formation of compaction bands. Shear localization and compaction band formation may occur, depending on the stress path and magnitude of the confining stress (Issen & Rudnicki 2001). However, compaction band formation may be inhibited with higher fractions of feldspar and clay (Klein *et al.* 2001).

Irrespective of the sense for permeability change, the dominant processes are complex and strongly stress path dependent (Ruistuin *et al.* 1996). This observation is of great importance in determining how we operate a well. It means that careful attention to control of well pressures and fluid pressure maintenance in the reservoir may result in improved well productivity over the life of the well.

In summary, the studies cited above have greatly enhanced understanding of the relationships between permeability change and the stress history of the rock. However, existing observations have been based on the external measurements on the samples during deformation or after bringing the samples back to atmospheric pressure conditions. Since deformation localization is known to occur, it is important to observe and investigate these deformations *in-situ*, at different locations in the sample.

Purpose and approach

The aim of this study is to develop methods to map porosity and structural changes in the deforming porous medium under triaxial loading conditions. X-ray computed tomography (CT) allows continuous non-destructive observation of the sample during deformation. With proper calibration the images can be interpreted to produce 3D maps of density and porosity throughout the experiment. Although X-ray CT has proven to be a valuable tool for different fields of geosciences to observe and quantify dynamic processes in the interior of samples, its application to *in-situ* rock deformation studies and rock mechanics has been limited to a small number of studies (Kantzas 1997; Kantzas *et al.* 1992; Gu *et al.* 1991; Vinegar *et al.* 1991). In one of the rare applications of CT to rock mechanics and *in-situ* observations of rock failure under triaxial stress field, Géraud *et al.* (1998) applied this technique to correlate volumetric strains with radiological density measurements and to evaluate the mineralogical effects on the crack location during triaxial loading experiment, as well as measuring porosities non-destructively.

In this study, local changes in the porosity within samples subjected to confining stress and increasing differential stress in a triaxial load cell are investigated using X-ray CT. We also measured the permeability changes and attempted to correlate the effect of structural and porosity changes to the permeability change. The experiments were performed in a specially designed X-ray-transparent triaxial test cell that also allows flow measurements during deformation stages. Since the presence of confining pressure changes the deformation behaviour of the rock, experiments performed with different constant confining stresses were compared to change the deformation regime.

Method

Experiments were conducted on Cordova limestone samples 2.5 cm in diameter and 7 cm in length. Samples were scanned with 2 mm slice thickness and 130 kV energy. The typical resolutions achieved in these experiments were $2 \times 0.5 \times 0.5$ mm.

Figure 1 is a schematic representation of the X-ray-transparent triaxial cell used in the experiments. In this cell, the sample is placed in a heat shrinkage Teflon tubing to pressurize it radially and separate the rock from the confining pressure fluid. After taking a series of scans of the dry core sample, it was vacuum saturated with plain water. After saturating the cores we took the saturated scans to calculate porosity, as described in the next section. Application of confining stress on the sample was achieved when the fluid access ports were open, enabling pore fluid drainage to eliminate pore pressure increase during the hydrostatic compression. Scans were taken at every step of the hydrostatic pressure increase on the sample to see its effect on the sample.

The aluminium body of the cell is capable of withstanding confining pressures up to 40.82 MPa. In these experiments, the pressure was increased up to 3.4 MPa and 20.41 MPa. These hydrostatic stress states on the limestone samples were the initial states before starting the axial loading, at constant confining pressure. The load cell that applies differential stress on the sample has an area ratio of 1/7 compared to piston surface. This ratio enables the load cell to magnify any pressure applied on the piston end by seven and transmits the load to the rock to increase the differential stress. The cores were scanned while increasing the axial load (differential stress) and pressure difference data was also collected between inlet and outlet ports during injection of water at a constant flow rate to calculate the absolute permeability.

Because one of the aims of the experiment was to measure permeability changes during deformation, confining fluid volume change data, due to strains, has not been collected to estimate the overall porosity change. But, even if it was done, this data gives a mean estimate (Géraud *et al.* 1998) and is far from being sufficiently accurate to be used for checking the local porosities calculated from CT measurements.

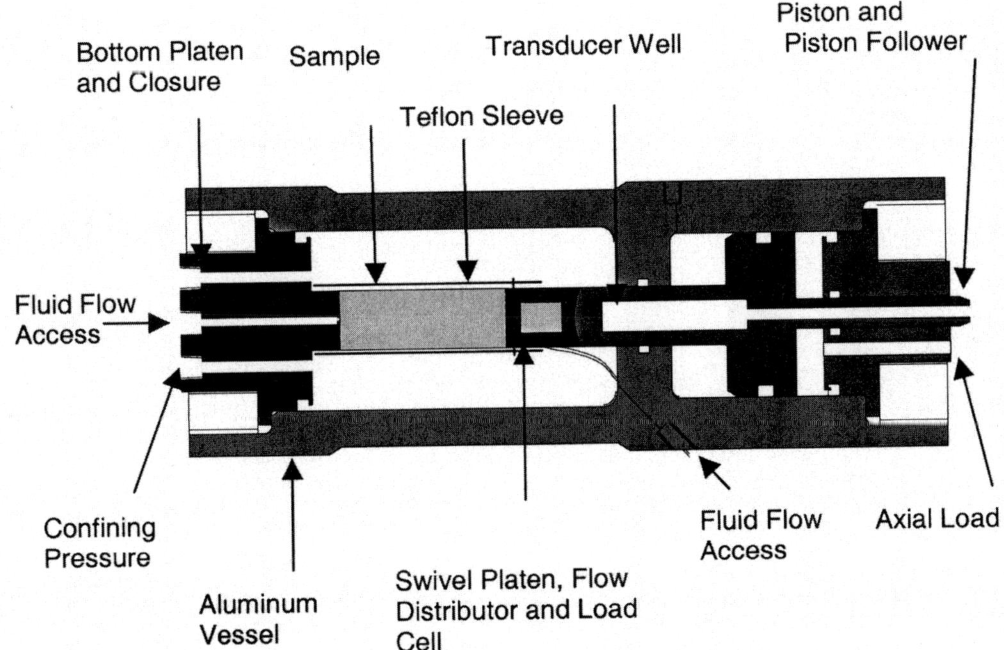

Fig. 1. Schematic representation of the X-ray-transparent triaxial cell.

Calibration of CT scanner and calculation of porosity

The CT images produced in this study contain 512 × 512 matrix of pixels, each representing a 0.5 × 0.5 × 2 mm volume element. Each pixel reading as CT number is an average CT number within that volume and is a function of mean density and effective atomic number. Although the CT number represents a combination of physical parameters, its scale is arbitrary and readings of the CT scanners can be calibrated to get the best and artefact-free response depending on the study, or type of material in question.

For this study a fourth generation medical scanner was used and special attention was paid to the calibration of the scanner. Because an aluminium triaxial cell was being used (which hardens the beam) and a sample whose density is much higher than that of water, a simple or conventional water calibration might not be successful in such an environment and may cause loss of contrast and the appearance of artefacts. Two major calibration routines were carried out to compensate these effects. First, an air calibration was made with the cell, containing everything but the sample, in the scanner. The calibration and reconstruction algorithm of the scanner saves these data as if it was an air scan and subtracts it from all subsequent scans. This helps eliminate the presence of the aluminium cell and other sleeve and fluid material from the images. After this calibration routine a homogeneous material resembling our rock sample (a Teflon rod in this case) was put into the cell, and then scanned to use the data in the beam hardening correction algorithm. This procedure helps to eliminate beam-hardening artefacts due to presence of the cell and the difference in densities between water and limestone. This calibration procedure also changes the scale of CT number readings when the sample is scanned. It is still a linear scale but it will not be a classical CT number reading where water is expected to read zero.

Since this calibration has been used in sample scans during deformation all other calibration scans for the calculation of porosity, for example scans for the CT number readings of the fluids, were taken while the pure fluids were located in the cell. Porosity was calculated by using saturated and dry scans of the sample at the same location, and the pure fluid CT number reading of the saturating fluid:

$$\Phi(\text{fraction}) = \frac{CT_{r,s} - CT_{r,d}}{CT_f - CT_a} \quad (1)$$

In this equation, $CT_{r,s}$ and $CT_{r,d}$ are the saturated and dry CT number readings of the porous medium, CT_f is the CT number of the fluid saturating the porous medium and scanned under the same conditions as the rock; and CT_a is the CT number of air filling the dry rock. This equation can be applied to any voxel in the porous medium or to the average CT number reading of the voxels in a certain region of interest. The latter contains less statistical noise but gives an average value for that region of interest. Application of the formula to the image data depends on the objectives and extent of data mining.

In our calculations of porosity, the above formula has been applied to a circular region of interest to cover almost the whole core. In the sensitivity analysis, which was performed by repeatedly scanning the cell, it was found that average CT number readings within the same region of interest may change by ±1 due to the variations in the X-ray beam. Consequently, there may be variations in reported porosity values, whose maximum and minimum are calculated to be within ±0.0015 fractional porosity. Any value that is greater than ±1 can, therefore, be explained by density (porosity) change in the sample. One of the effective ways of minimizing this variation is to scan the same location several times during the experiment and to perform a voxel averaging to create a single image. Although this reduces the possible noise and averages out X-ray variations, for dynamic processes where there are rapid changes in the samples, this technique may not be practical.

Results and discussion

Effect of confining pressure on sample porosity

Experiments were conducted with 3.4 MPa and 20.41 MPa confining pressures. As discussed in the procedure section, confining pressure was increased stepwise on the saturated sample to determine the changes during increased hydrostatic pressure. The scans and the data obtained at the final stage of confining pressure increase determine the initial stage before increasing the differential stress.

Figure 2 shows the porosity distribution in the sample confined with 20.41 MPa maximum confining stress as a function of confining pressure. The porosities were calculated by using the standard porosity calculation concept from CT images. This figure shows that the initial unconfined porosity distribution of the sample is non-uniform, with a lower porosity zone at one end,

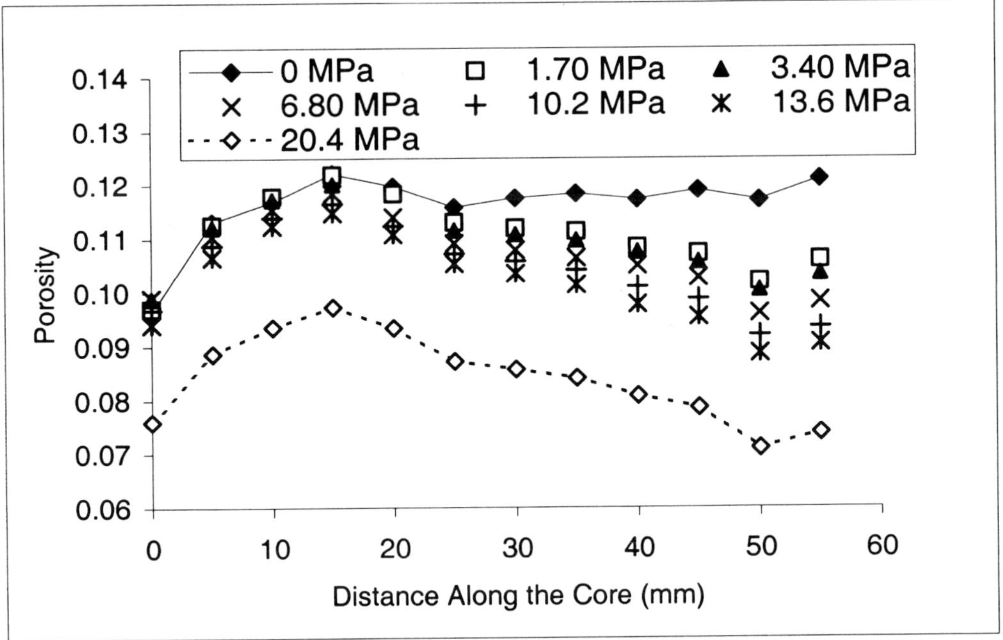

Fig. 2. Porosity distribution along the sample as a function of hydrostatic pressure in the 20.4 MPa confining pressure test sample.

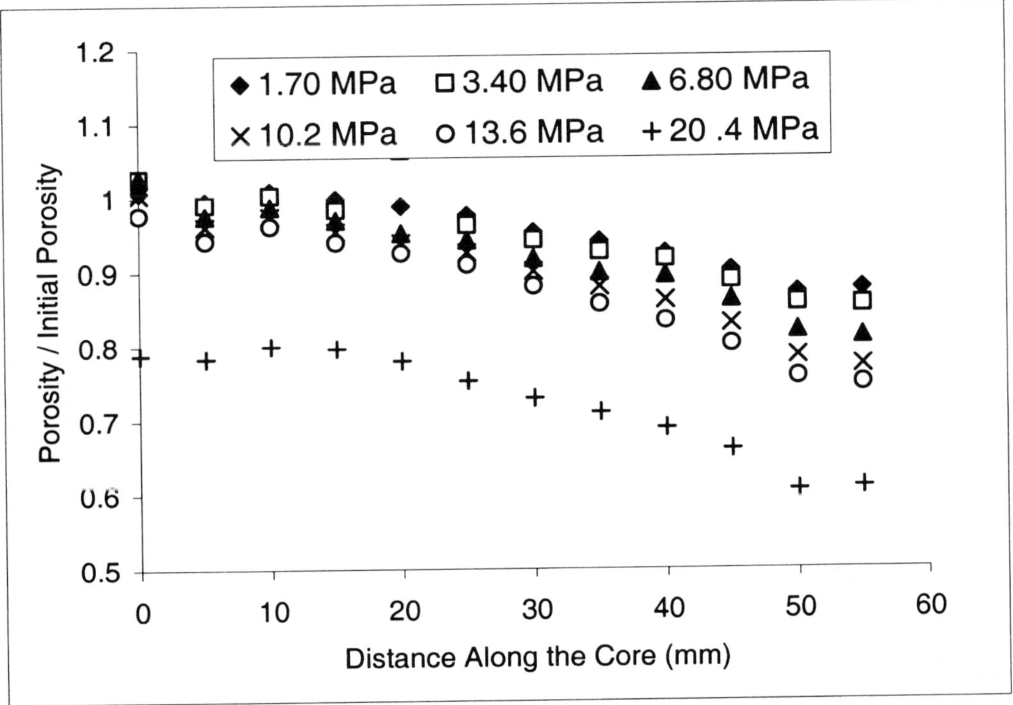

Fig. 3. Normalized porosity change as a function of hydrostatic pressure in the 20.4 MPa confining pressure test sample.

and a comparably higher porosity section on the other end. The tight end of the sample has porosities in the order of 9.5% and increases to about 12% after 15–20 mm. The data presented in this graph show that porosity of the sample decreases significantly with increasing confining pressure. This porosity reduction is not high at the tight end of the sample where the porosity is lower till about 13.6 MPa. However, the high porosity end of the sample is compacted more and the original porosity decreases from 12% to almost 9% even lower than the initially low porosity section of the core. The porosity reduction heterogeneity in the sample may be due to differences in mineralogical composition or the closing of micro-fractures and collapse of larger pores in the high-porosity section.

Increasing the confining stress further to 20.41 MPa affects the porosity decrease significantly and sample is compacted more. At this stress stage, local porosities range from 9.5% to 7%. In order to calculate the percent reduction in porosity with respect to original unconfined porosity, the porosities calculated at each pressure increase were normalized with the original porosity at that location (Fig. 3). This figure better represents the porosity reduction considering the effect of initial porosity on the compaction behaviour under confining stress. This figure shows that the percent decrease in porosity increases within the sample towards the high porosity end of the core. Porosity reduction up to 13.6 MPa was determined as 5% at the tight end of the core and about 20–25% at the more porous or weak end of the core. Also, incremental porosity decrease during stress increase is higher in the porous end compared to the other end of the sample. As we go to 20.41 MPa, additional porosity reduction is very high and the total reduction is about 22% of the original value at one end and about 40% at the other end.

The same type of stepwise increase in the sample confined at 3.4 MPa stress was not performed because this pressure was already quite low to expect an appreciable change in the sample. But we can reasonably estimate from Figure 2 that the decrease should be between 0% and 10% depending on the location within the sample.

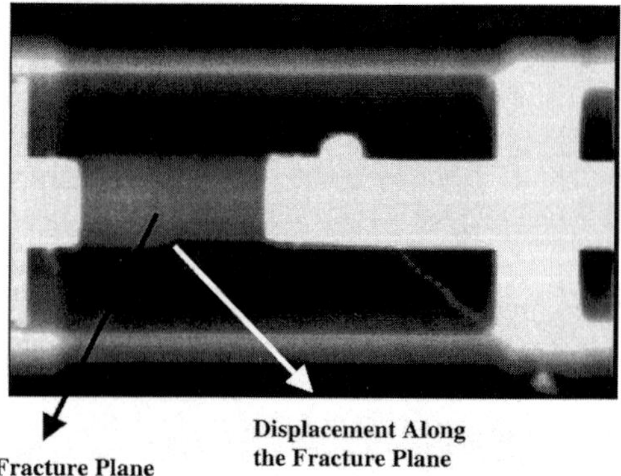

Fig. 4. X-ray radiograph of brittle failure of limestone in triaxial cell at 24.49 MPa.

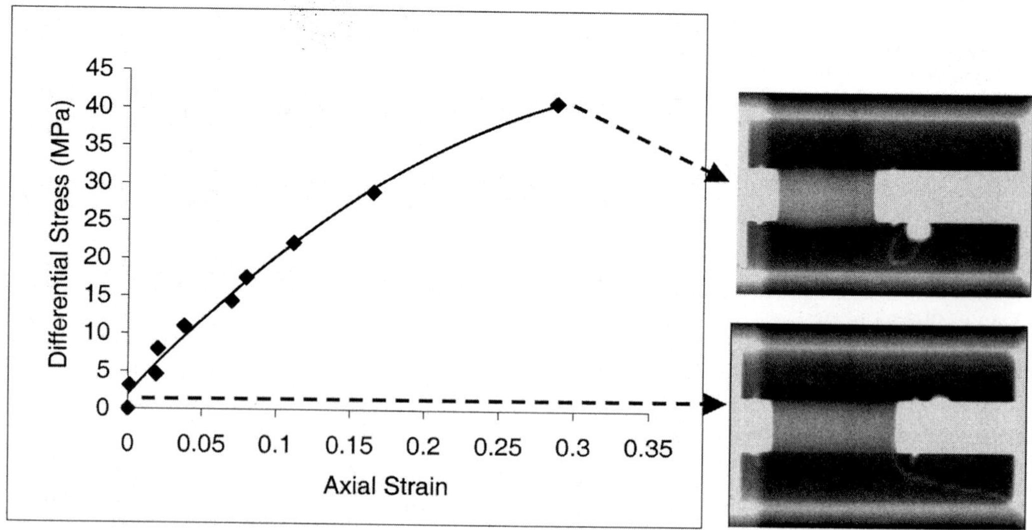

Fig. 5. Stress–strain diagram and X-ray radiographs of ductile failure of limestone, under hydrostatic pressure (top) and under 40.81 MPa differential stress (bottom).

Effect of differential stress increase – porosity and permeability reduction, failure mode

Failure mode. The differential stress on the limestone samples held in hydrostatic stresses of 3.4 MPa and 20.41 MPa was increased by applying fluid pressure to the load piston. The sample was scanned at each stepwise increase in the differential stress. A linear variable displacement transducer (LVDT) was connected to the 20.41 MPa test sample to construct the stress–strain diagram. While increasing the differential stress, the saturating fluid was injected and pressure drop across the sample was continuously measured to calculate the permeability change.

Figures 4 and 5 shows the X-ray radiographs after the samples fail under increased differential stress. Figure 5 also shows the stress–strain diagram for the test performed under 20.41 MPa confining stress. These figures clearly show a great difference in failure mode between the two samples. The limestone under lower confining pressure experiences brittle failure, whereas the sample under high confining pressure exhibits barrelling, which is characteristic of typical ductile failure.

Comparison of local porosities during deformation. Figures 6 and 7 show the change in porosity in the samples during deformation and after failure. Porosities were calculated by using X-ray CT images taken at different positions at every step of differential stress increase.

Figure 6 shows the change in porosity along the sample length at different loading stages. As presented in the previous section, the failure mode in this test was brittle and there is a fracture in the middle of the sample at about 45° to the axis of principle stress. Although the sample was already failed at 20.41 MPa differential stress, the test was continued to 24.49 MPa to create shear on the fracture surface. Figure 6 shows that the sample porosity decreases along the sample, but it is confined locally between 20–30 mm from the end of the sample where the fracture is created. In this region, porosities are about 70% lower than the initial values measured at hydrostatic stress. Further shearing the fracture surface does not change the porosity too much. Porosity recovers to some extent at 24.49 MPa differential pressure at two sides of the fracture region. This is probably due to the stress redistribution and release in these regions after failure. This confirms the non-homogeneous nature of compaction and dilatation within the sample as a result of non-homogeneous distribution of stress and strain fields. Another interesting observation is that the local porosity starts to decrease at a particular position (25 mm, point B) at 13.6 MPa differential stress and minimum porosity locates itself at 30 mm (point C) even before failure occurs. This region is probably the compaction band where shear localization and pre-fracture compaction occur. The images in Figure 6 were recorded at locations marked as A, B and C in the graph to show the change in structure and porosity distribution in the shear localization and fracture region. From these images we can see how the porosity evolves at different differential stresses in the shear localization area and at the two flanks of the fracture zone after brittle failure occurs.

Figure 7 shows all CT images at different positions along the core before and after deformation. Comparison of the images shows very clearly that the local porosities within the sample change also in the sections outside the shear area. This change is not as great as in the shear zone, but the effect of compaction can still be observed.

In the ductile failure mode observed in the 20.41 MPa confined test, the porosity change is totally different from what was observed for brittle failure in Figure 6. In this sample, porosity starts to decrease at all locations along the sample, as seen in Figure 8. In this test the sample is compressed initially and undergoes barrelling with increasing differential stress, as shown in the X-ray radiograph given in Figure 5. During the compaction stage, porosity decreases to almost 20–30% of the initial hydrostatic porosity in the middle section of the core (Fig. 8) until 28.56 MPa differential stress. The decrease in porosity until this level is due to the compaction of the pore space and probably grain breakage. The extent of porosity reduction at the ends of the sample, where it is highly compacted, is even greater. Part of the reduction at the end plates may be due to the end effects. After a critical point, where the sample is plastically deformed, is passed (between 28.56 and 40.82 MPa) porosity increase is observed, especially in the middle section of the sample; this is probably due to a great amount of barrelling and creation of micro/macro-fractures in the sample. In this section, porosity is recovered back by as much as 3% (which corresponds to 50% recovery) at some locations. Figure 8 also shows that the strain is non-homogeneous along the core axis – it is higher in the central part of the sample and decreases towards both loading surfaces, as also observed by Géraud *et al.* (1998).

The images given for 3.13 MPa, 17.48 MPa, 22.11 MPa, 28.91 MPa and 40.81 MPa differential stress levels at 20 mm from the end of the sample show that the initial structure of the sample contains some high porosity regions (point A).

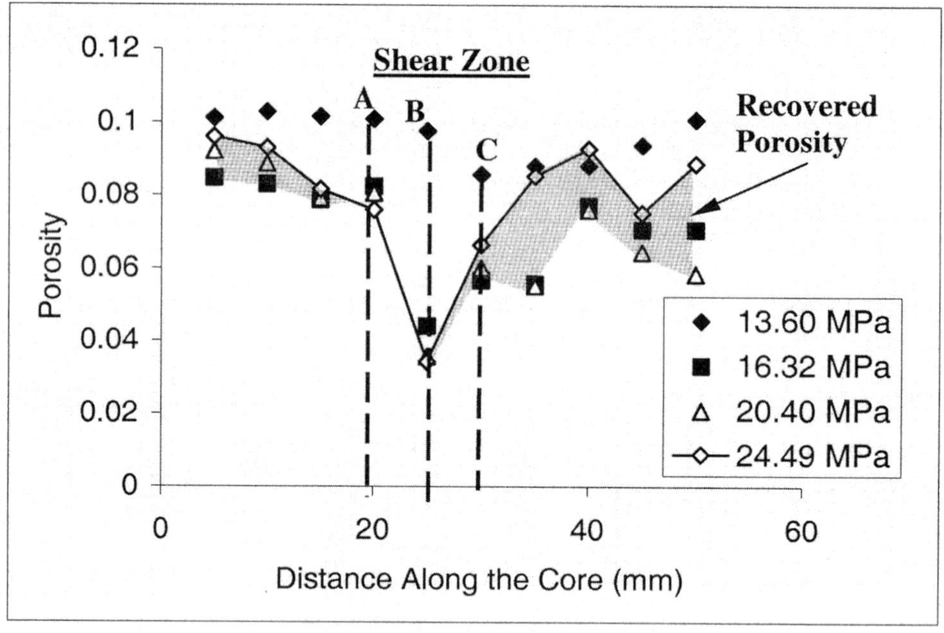

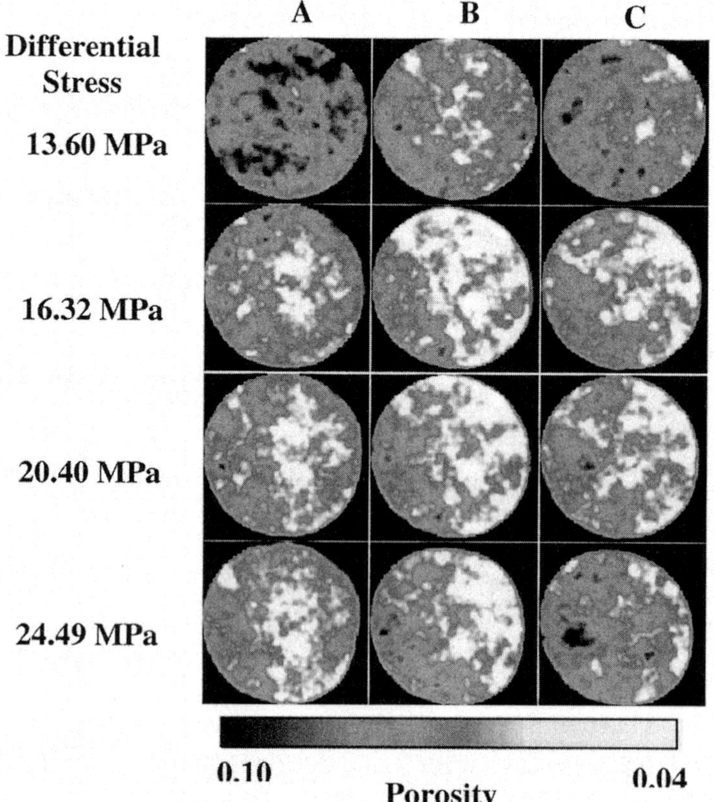

Fig. 6. Porosity evolution as a function of differential stress and images showing the structure changes in the shear location and eventually failure zone in the limestone sample that experienced brittle failure.

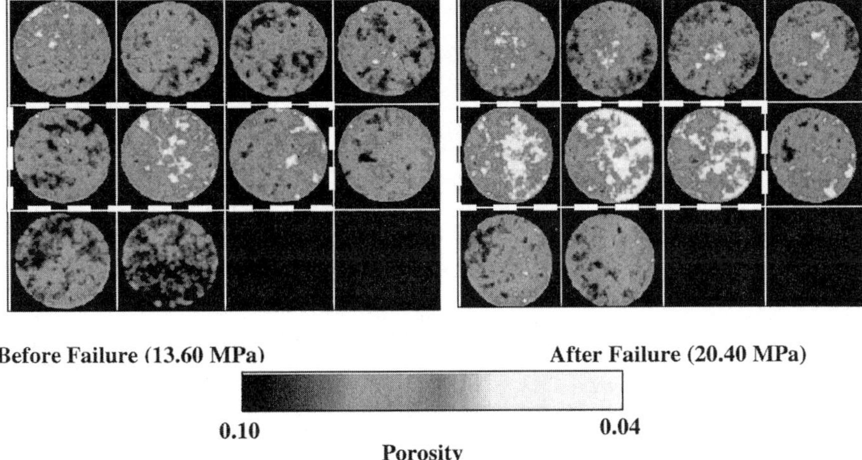

Fig. 7. CT images showing different locations before and after brittle failure (dashed box shows locations A, B and C).

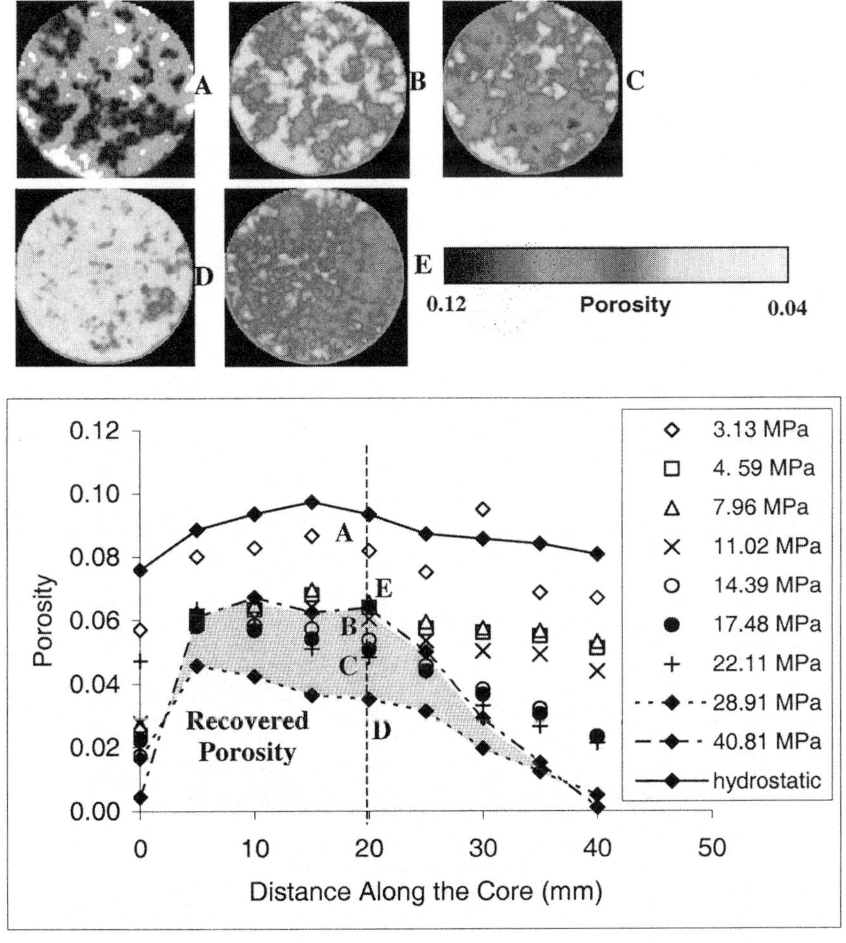

Fig. 8. Porosity evolution as a function of differential stress in the limestone sample that experienced ductile failure.

With the increase of differential stress compactions within the sample starts, as observed in the images until 28.91 MPa (point D). After this stress level, the sample dilates (point E) and porosity is recovered to some extent. The internal structure of the sample completely changes as a result of plastic deformation, as illustrated by images A–E in Figure 8.

Figure 9 shows the porosity evolution only at a 20 mm position in the sample for some sample differential stress conditions. The graph and images show that the porosity decreases and then increases again after failure. A close examination of point E at 40.81 MPa differential stress, confirms that the inner structure of the sample has changed and it has lost the general structure observed in A. It looks like the tight or low porosity structures observed in A were crushed and fractured. Also, initially porous structures probably collapsed and compacted. This also confirms that the deformation at this level is plastic.

Comparison of permeability change. Permeabilities of the samples were calculated by measuring the pressure drop in the sample while injecting the saturating fluid during deformation. Figures 10 and 11 show the permeability values as a function of increased differential stress on the samples. First of all, it should be noticed that the initial permeabilities of the samples at hydrostatic pressure conditions are different from each other. This is the effect of the confining pressure on the Cordova limestone samples. Under 3.4 MPa confining stress, it has a permeability of about 21-md, but permeability decreases to about 6-md at 20.41 MPa confining stress.

Permeability change behaviours are also different from each other. In 3.4 MPa confining pressure test permeability reduction is sudden, with the onset of compaction banding shear fracturing, and further increase in differential stress does not decrease permeability significantly (as in the case of porosity). The change in

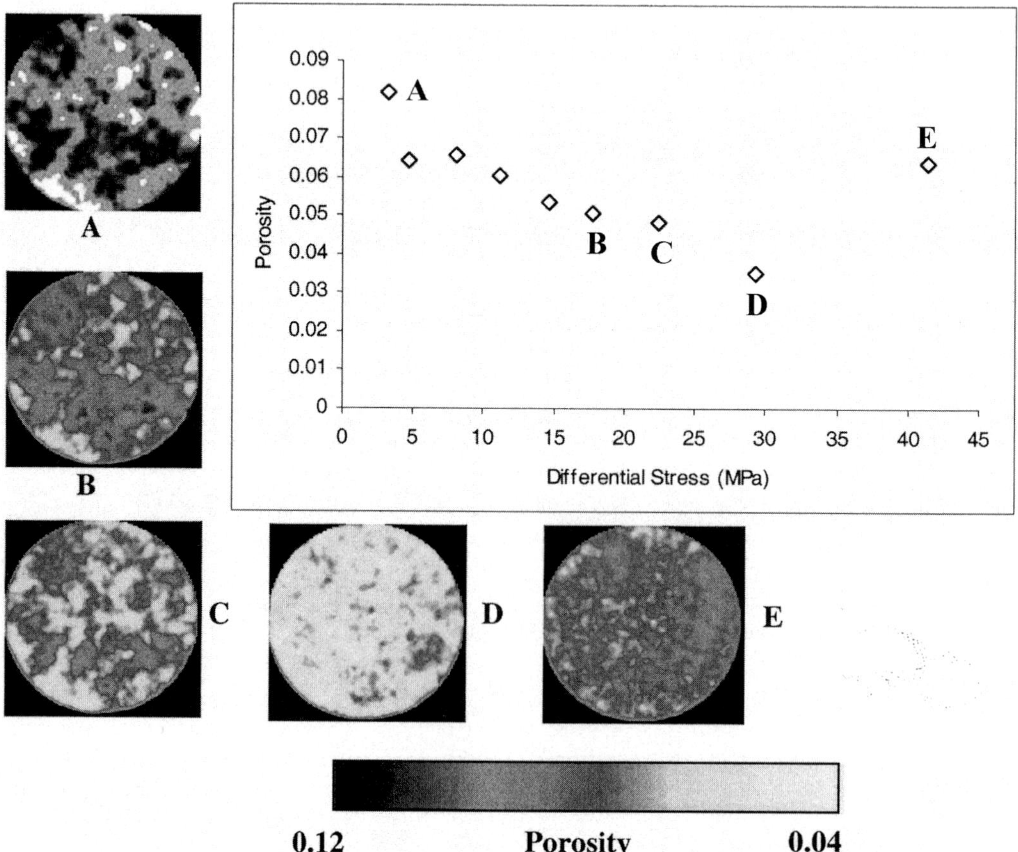

Fig. 9. Porosity change as a function of differential stress and CT images showing structural change, at 20 mm location in the sample experienced ductile failure.

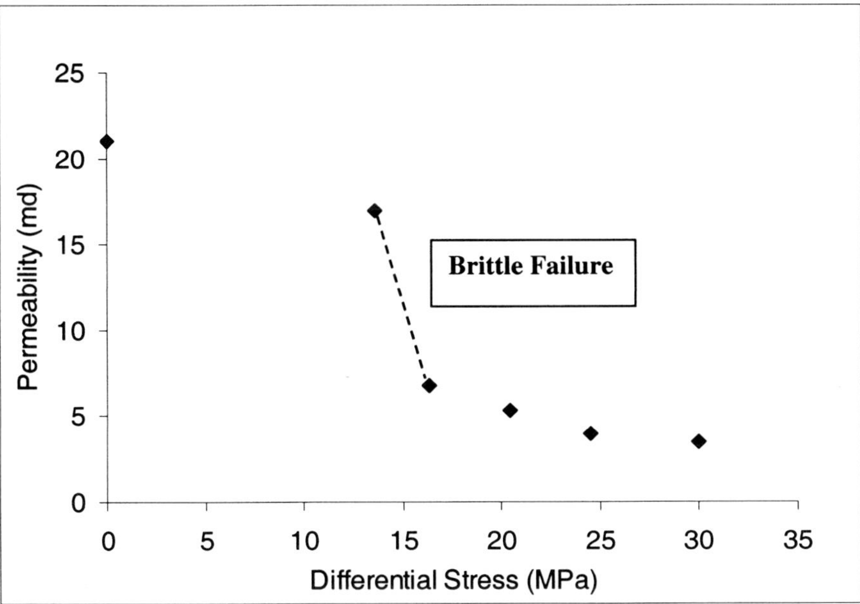

Fig. 10. Permeability change during brittle failure.

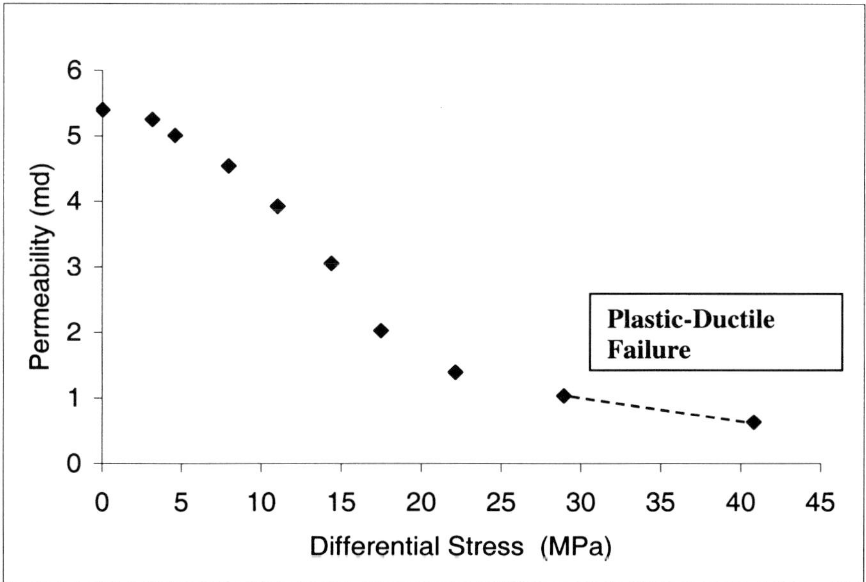

Fig. 11. Permeability change during ductile failure.

permeability is about 75% of the original permeability (from 20-md to 5-md). In the ductile failure test, permeability decreases smoothly as the sample deforms and reaches a minimum value at 40.82 MPa differential stress. The permeability decreases from 5-md to below 1-md. Unlike porosity, permeability cannot be recovered after deformation. This shows that in ductile-plastic deformation, the structure of the sample changes by crushing, microfracture formation, etc., to produce a rise in porosity in the middle of the sample. However, because the ends of the sample are highly deformed to very low porosity values (Fig. 8), permeability cannot be recovered. Part of the reason why porosity is so low close to end plates may be due to end effects. Unfortunately, this may have an additional effect on the measured permeabilities and it may be the reason why we cannot see permeability enhancement.

Conclusions

The following conclusions can be drawn from the results of this study:

(1) In brittle failure, shear is localized mostly to a specific plane where compressive stresses are maximum and where compaction band develops. Sample fails from this region.
(2) In brittle failure, porosity reduction is particularly pronounced in the shear region. Although porosity in the rest of the sample is decreased also due to compaction, this decrease is not as high as it is in the local shear region.
(3) In ductile failure, porosity is decreased at all locations along the sample length during compression. It increases again after a certain stress limit where the sample barrels with ductile-plastic deformation. At this instant, previous low porosity regions are fractured to yield a higher porosity, especially in the middle of the sample. Porosity at the ends of the sample cannot be recovered because of high compaction.
(4) Permeability in a brittle failure regime drops suddenly with the onset of compression banding and shear localisation. It reaches almost its minimum with the failure.
(5) In ductile failure, the decrease in permeability is more gradual. It reduces almost following the reduction in porosity, due to compression within the sample. It cannot be recovered back, although porosity increases in the middle of the sample after a threshold because of high compaction at the ends of the sample.
(6) Percent reduction in permeability is greater in ductile deformation.
(7) Confining pressure affects both the failure regime and the initial porosity and permeability of the sample before starting axial loading.
(8) X-ray CT is a valuable tool to study the local changes in the fluid transport related properties of deforming porous media. Data quality and resolution could be enhanced by use of microfocus CT devices in similar type of studies.

References

BAUD, P., SCHUBNEL, A. & WONG, T.F. 2000. Dilatancy, compaction and failure mode in Solnhofen limestone. *Journal of Geophysical Research*, **105**, 19 289–19 303.

FJAER, E., HOLT, R.M., HORSRUD, P., RAAEN, A.M. & RISNES, R. 1992. *Petroleum Related Rock Mechanics*. Elsevier, Amsterdam.

GÉRAUD, Y., MAZEROLLE, F., RAYNAUD, S. & LEBON, P. 1998. Crack location in granitic samples submitted to heating, low confining pressure and axial loading. *Geophysical Journal International*, **133**, 553–567.

GU, R.J., HOVANESIAN, J.D. & HUNG, Y.Y. 1991. Calculations of strains and internal displacement fields using computerized tomography. *Journal of Applied Mechanics*, **58**, 24–27.

HALLECK, P.M. 1997. Recent advances in understanding perforator penetration and flow performance. *Society of Petroleum Engineers, Drilling and Completions*, 19–25.

HEILEND, J. & RAAB, S. 2001. Experimental investigation of the influence of differential stress on permeability of a Lower Permian (Rotliegend) sandstone deformed in the brittle deformation field. *Physics and Chemistry of the Earth Part A*, **26**, 33–38.

ISSEN, K.A. & RUDNICKI, J.W. 2001. Theory of compaction bands in porous rock. *Physics and Chemistry of the Earth Part A*, **26**, 95–1000.

KANTZAS, A. 1997. Stress strain characterization of sand packs under uniform loads as determined from computer assisted tomography. *Journal of Canadian Petroleum Technology*, **36**, 48–52.

KANTZAS, A., ROTHENBURG, L. & BARRETT, S.V.L. 1992. Determination of stress strain characteristics of sand packs under uniform loads by the use of computer assisted tomography and finite element modeling. Society of Petroleum Engineers, Paper **23791**.

KLEIN, E., BAUD, P., REUSCHLE, T. & WONG, T.F. 2001. Mechanical behavior and failure mode of Bentheim Sandstone under triaxial compression, *Physics and Chemistry of the Earth Part A*, **26**, 21–25.

NAGEL, N.B. 2001. Compaction and subsidence issues within the petroleum industry: From Wilmington to Ekofisk and beyond. *Physics and Chemistry of the Earth Part A*, **26**, 3–14.

POPP, T., KERN, H. & SCHULZE, O. 2001. Evolution of dilatancy and permeability in rock salt during hydrostatic compaction and triaxial deformation. *Journal of Geophysical Research*, **106**, 4061–4078.

RUISTUIN, H., TEUFEL, L.W. & RHETT, D. 1996. Influence of reservoir stress path on deformation and permeability of weakly cemented sandstone reservoirs. Society of Petroleum Engineers, Paper **36535**.

VINEGAR, H.J., DE WAAL, J.A. & WELLINGTON, S.L. 1991. CT studies of brittle failure in Castlegate Sandstone. *International Journal of Rock Mechanics and Mining Sciences & Geomechanical Abstacts*, **28**, 441–448.

ZHU, W.L., & WONG, T.F. 1997. The transition from brittle faulting to cataclastic flow: permeability evolution. *Journal of Geophysical Research*, **102**, 3027–3041.

ZHU, W.L., MONTESI, L.G.J. & WONG, T.F. 1997. Shear-enhanced compaction and permeability reduction: triaxial extension tests on porous sandstone. *Mechanics of Materials*, **25** 199–214.

Monitoring void ratio redistribution during continuous undrained triaxial compression by X-ray computed tomography

P. R. THOMSON & R. C. K. WONG

Department of Civil Engineering, University of Calgary, 2500 University Drive NW, Calgary, Alberta T2N 1N4, Canada (e-mail: rckwong@ucalgary.ca)

Abstract: It is usually implied that void ratio of a soil specimen is homogeneous and unchanging during an undrained triaxial test. In this study, X-ray computed tomography was used to measure void ratio redistribution during undrained triaxial compression of a cohesionless soil sample. A specially designed triaxial apparatus was used to scan the same sample at different axial strain levels, while the axial loads and confining pressures were maintained. Significant variation in sample void ratio was observed at the end of consolidation and sample uniformity was found to increase with increasing axial strain. The process of void ratio redistribution can be effectively illustrated in void ratio-effective stress space.

The undrained behaviour of cohesionless soils, such as potentially liquefiable sand deposits, is usually studied within the framework of steady state soil mechanics. By definition, steady state is reached when: (i) a soil mass is continuously deformed at a constant volume, normal effective stress, shear stress, and velocity; (ii) all particle orientations are statistically constant; and (iii) particle breakage, if any, has stopped (Poulos 1981).

Accurate void ratio measurement is necessary in steady state soil mechanics because the slope of the steady state line in void ratio-mean effective stress space is very weak. Small measurement errors or changes in void ratio have a large effect on predicted soil response (McRoberts & Sladen 1992). Use of the state parameter Ψ, a measure of the proximity of given soil conditions to those existing at steady state, requires accurate determination of the steady state line in void ratio-mean effective stress space (Been *et al.* 1991; Vaid & Thomas 1995). In addition, a scatter of $\pm 50\%$ of the average stress at steady state for a given void ratio has been observed in laboratory testing (Konrad 1990). It has been suggested that a technique for examining the structure of sands as they collapse and flow is needed since the effects of fabric may affect triaxial test results in an unknown manner (Been *et al.* 1991).

An investigation of the fundamental behaviour of cohesionless soil samples during undrained triaxial testing is required. This paper presents an investigation of one of the traditional assumptions regarding undrained testing, namely the notion that sample void ratio remains unchanged during axial deformation.

Void ratio redistribution

Destructive methods

Several methods have been used to measure local void ratio and to estimate void ratio redistribution in cohesionless soil samples. Destructive methods involve stabilization of the soil fabric using some physical technique, followed by dissection of the sample into smaller elements for void ratio determination. Resin impregnation has been used to stabilize samples followed by element void ratio determination with chord intercept analysis (e.g. Hird & Hassona 1990; Muhunthan *et al.* 1996), or with digital image analysis of thin sections (e.g. Ruzyla 1986; Frost & Jang 2000). Element void ratio has also been determined using gravimetric methods with samples that were preserved with uniaxial freezing (e.g. Gilbert 1984; Ayoubian & Robertson 1998) or replacement of the pore fluid with gelatine (e.g. Emery *et al.* 1973; Kuerbis & Vaid 1988).

One of the inherent limitations of destructive methods is the need for sufficiently large elements to minimize errors during weighing, chord intercept analysis, or digital image analysis. The most severe limitation of destructive methods is that the void ratio distribution of a single sample cannot be assessed at different axial strain levels. Different samples must be taken to progressively

From: MEES, F., SWENNEN, R., VAN GEET, M. & JACOBS, P. (eds) 2003. *Applications of X-ray Computed Tomography in the Geosciences*. Geological Society, London, Special Publications, **215**, 191–198. 0305-8719/03/$15. © The Geological Society of London.

higher strain levels and each sample is analysed separately to estimate redistribution, assuming that all of the samples were similar at the end of consolidation.

X-ray computed tomography

The basic principles of X-ray computed tomography (CT) are widely available in existing publications (e.g. Zatz 1981; Bossi *et al.* 1990) and are not be presented here. CT has been used to estimate localized bulk density values within soil samples and it has been shown that the measured linear attenuation coefficients are linearly correlated with bulk density for X-ray energies between 0.1 MV and 1.0 MV (Wellington & Vinegar 1987; Desrues *et al.* 1996; Otani *et al.* 2000; Pralle *et al.* 2001).

Void ratio redistribution in cohesionless soil samples subjected to drained triaxial compression has been reported (Colliat-Dangus *et al.* 1988; Desrues *et al.* 1996). In these studies the effective confining stresses were applied without triaxial cells using negative pore pressures; axial loads were removed before scanning, and negative pore pressures were used during the scanning procedure to maintain an effective confining stress. It has been suggested that a triaxial system that can be placed within the scanning field should be developed in order to avoid stress relaxation (Otani *et al.* 2000).

To the authors' knowledge, the investigation presented in this paper is the first attempt in using CT to quantify void ratio redistribution during undrained triaxial compression of cohesionless soil. A specially designed triaxial cell has been constructed that is placed directly inside the CT scanner during undrained triaxial testing.

Experimental methods

Triaxial apparatus

Detailed descriptions of conventional triaxial equipment are widely available in literature (e.g. Bishop & Henkel 1962; Craig 1997, pp. 108–113) and are not repeated here. Several features of conventional triaxial equipment are unsuitable for CT scanning.

Firstly, the steel tie bars that are used to connect the upper and lower caps of a conventional triaxial cell would result in measurement disparities between adjacent detectors. The reconstruction algorithm is sensitive to abrupt changes in measured attenuation from one X-ray source position to the next, changes that will occur if one X-ray beam is transmitted through a steel tie bar and the adjacent beam is not. The resulting CT image has streaks of false high density, which originate near the steel tie bars. These streaks are a type of geometric artefact and the resulting CT image is unsuitable for quantitative analysis (Joseph 1981).

Secondly, an external reaction frame is required when using a conventional triaxial cell. This frame is connected to an actuator, which is used to displace the cell piston and to load the soil sample. If used with a medical CT scanner, an external reaction frame would be required to span the width of the scanner itself in order for no part of the frame to be inside the scanning field. Although feasible if a dedicated CT scanner was available for research, the use of an external reaction frame was undesirable in this investigation since the CT scanner to be used was dedicated to patient examinations during regular hours.

Thirdly, conventional triaxial systems, including the system pressure sources and data acquisition systems, are not portable. Instead, they are mounted on the floor or a workbench. It was necessary that the entire triaxial system, especially the triaxial cell, be portable and compact since this research would occur outside a conventional geotechnical laboratory.

The triaxial system constructed for this research has been described in detail elsewhere (Thomson & Wong 1999). The system is designed for displacement-controlled, monotonic compression and extension of undrained samples. An aluminium pipe with blind flanges at each end is used as a cell, so tie bars and an external reaction frame are not required (Fig. 1). The aluminium pipe, i.e. the cell wall, is also used to filter low-energy X-ray quanta and minimize beam hardening artefacts (Joseph 1981; Kantzas 1990). The pressure sources include compressed nitrogen bottles and a constant-volume pump, which are mounted in wheeled cabinets with the data acquisition system. The instrumentation is located outside the scanning field and consists of a load cell and transducers for confining pressure, pore water pressure and displacement. Axial load is applied to the sample using a hydraulic actuator and a piston, which enters the cell via the lower flange.

Sample preparation

The soil used in this experiment was Ottawa sand (designation C778). It is a uniform sand with a mean grain size (d_{50}) of 0.35 mm and gradation indices of $C_u = 1.7$ and $C_c = 1.1$. The maximum and minimum void ratios, as determined by methods ASTM D4253 and D4254, were 0.732

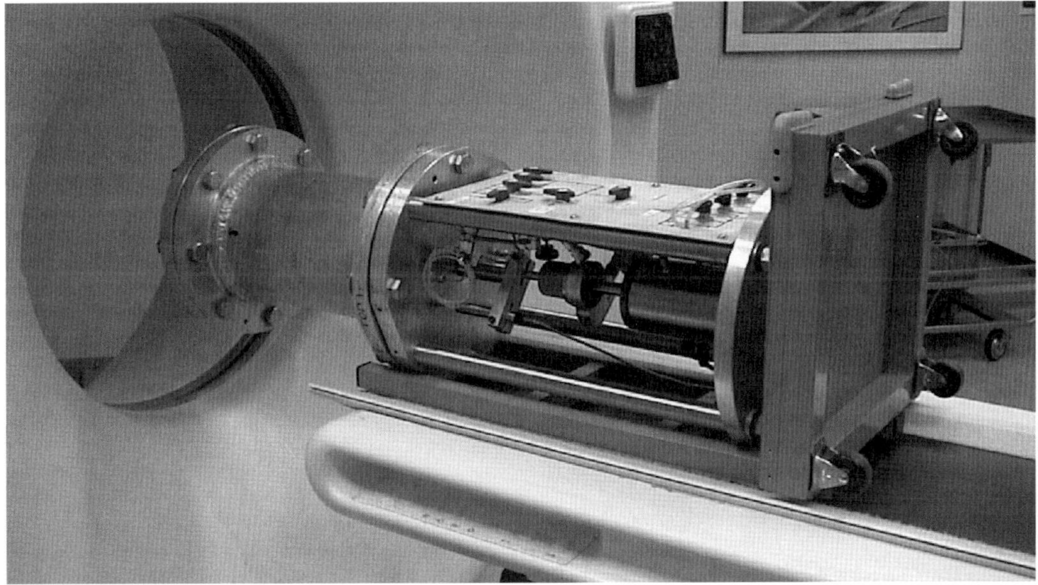

Fig. 1. Triaxial apparatus on CT scanner bed.

and 0.556, respectively. It is often convenient to state void ratio measurements in terms of relative density, given by:

$$D_r = \frac{e_{max} - e}{e_{max} - e_{min}} \times 100(\%) \quad (1)$$

where D_r is relative density, e is the measured void ratio and e_{max} and e_{min} are the maximum and minimum void ratios determined above.

A moist tamping undercompaction method was used to construct the soil sample in six layers of equal dry mass and a moisture content of 5%. The sample forming mold had an interior diameter of 76 mm and an internal height of 92 mm. Lubricated ends, which consisted of two layers of latex membrane with vacuum grease between them, were used. Oversize platens with 19 mm diameter, centralized porous stones were also employed. The upper and lower platens are constructed of mirror-finished aluminium and acrylic, respectively.

An effective confining stress of 20 kPa was applied to the sample while the forming mold was removed. After assembling the triaxial cell around the sample, the cell was filled with distilled, deaired water. A minimum effective confining stress of 50 kPa was maintained while the sample was flushed with carbon dioxide; flushed with deaired, distilled water; and subjected to a back-pressure of 650 kPa to saturate the sample. The measured B-value, an indicator of sample saturation, was 0.992.

The sample was consolidated to an effective stress of 75 kPa and data acquisition was initialized. The pressure control panel and triaxial apparatus were rolled into the adjoining room and, while under pressure, the triaxial apparatus was tilted onto a ramp and was raised onto the CT scanner bed. The triaxial cell was positioned in the centre of the scanning field with the sample centreline lying in the horizontal plane (Fig. 1). Finally, the sample was allowed to consolidate to an effective stress of 100 kPa and the external pore fluid drainage line was closed. Sample volume change during consolidation was measured using the computer-controlled pump.

Triaxial compression and CT scanning

The pore fluid drainage line was closed and the sample was scanned from bottom to top to obtain the first slice set. The CT scanner settings were: 120 kV, 150 mA, 0.8 s exposure and a 25.0 cm scanner field of view. Consecutive 1 mm thick slices at a spacing of 1 mm were recorded. All subsequent slice sets were obtained with the same scanner settings.

Before initiating the compression test it was observed that the measured pore pressure had increased slightly. The drainage line was opened and sample volume change (consolidation) was recorded using the computer-controlled pump. The drainage line was closed and the measured volume change was used to back-calculate sample void ratio before axial compression began. For simplicity, the CT data obtained above will be referred to as having being obtained at the end of consolidation.

Monotonic undrained axial compression was performed in displacement-control at a rate of 0.5 mm/minute. Slice sets were obtained at axial strains of 0.7, 2.2, 5.9, 11.4 and 21.2%. To obtain slice sets during the compression test, the hydraulic actuator was paused for approximately 5 minutes in order to halt axial displacement. The total confining stress was maintained at a constant value throughout the test. After the sample was scanned at 21.2% axial strain the pore pressure valves outside the cell were closed. The triaxial cell was dismantled and valves on the pore pressure lines inside the cell were closed to further isolate the sample.

The sample, platens and internal pore pressure lines and valves were disconnected from the triaxial apparatus. The sample was wrapped with flexible plastic tubing and was then covered with insulating material. Coolant was circulated through the plastic tubing for 12 hours at a temperature of −25°C in order to freeze the entire sample. After freezing was complete, the platens and membrane were removed and the overall sample void ratio was determined gravimetrically.

Data manipulation

CT slices that were seen to intercept the platens at either end of the sample were removed. The grey-scale pixel values in the original data files were converted to computed tomography numbers, CTn, using the rescale intercept and slope values that were available in the file headers. Although beam hardening was not observed, the CTn values were found to increase from that of water, the confining fluid, to that of the sample over an interval of 5 voxels at the sample boundary. The data files were cropped using a minimum CTn cutoff of 950 HU.

Results

The overall relative density at the end of the compression test was determined gravimetrically to be 11.4%. The overall relative density before axial compression began, i.e. for the first slice set, was back-calculated using pore fluid volume measurements and found to be 10.8%. In light of the results presented below, this discrepancy can be considered insignificant.

It was assumed that the sample deformed as a right cylinder during compression. Sample area and axial stresses were calculated accordingly.

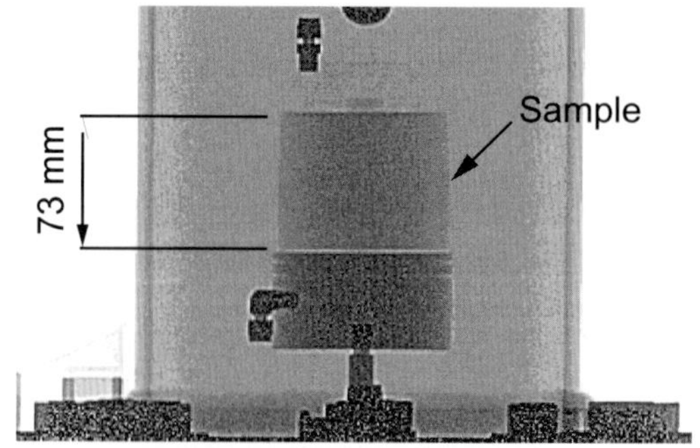

Fig. 2. X-ray radiograph at the end of the compression test.

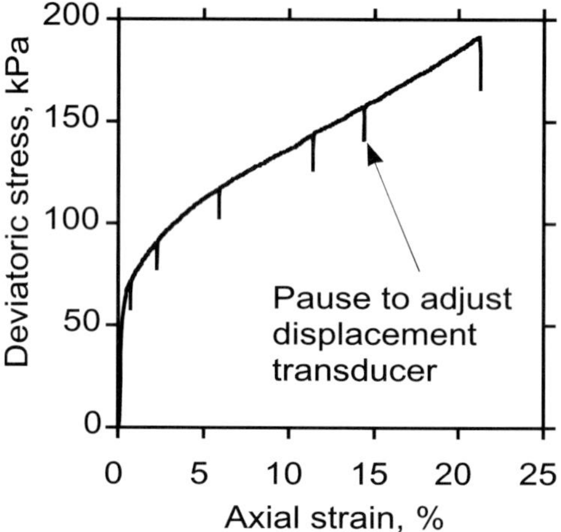

Fig. 3. Load-deformation response.

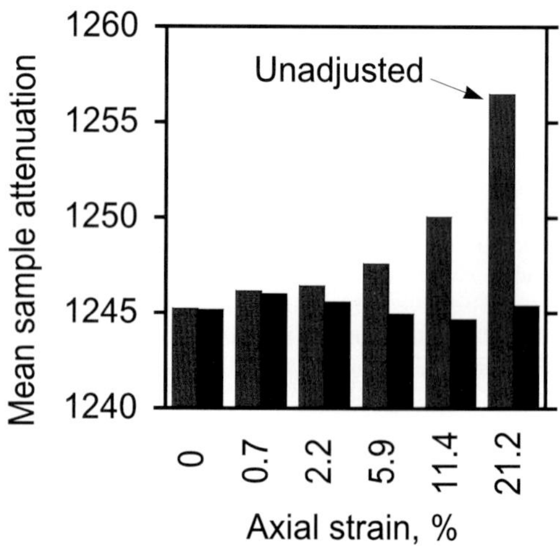

Fig. 4. Relationship between attenuation and sample diameter.

An X-ray radiograph that supports this assumption was recorded using the CT scanner after the last slice set was obtained (Fig. 2).

The load-deformation response of this sample is illustrated in Figure 3. Strain-hardening behaviour is apparent, despite the low overall relative density of the sample (11.4%). Slight sample relaxation occurred when axial displacement was paused to obtain CT data.

The mean sample attenuation value increased with increasing axial strain, i.e. increasing sample diameter (Fig. 4). In order to correlate attenuation values with bulk density, the measured attenuation values in each slice were adjusted for slice diameter using the relation:

$$\mathrm{CTn_a} = \mathrm{CTn_m} \times \left[1.0 + 6.91 \times 10^{-2}\left(1.0 - \frac{\phi_\mathrm{m}}{\phi_\mathrm{r}}\right)\right] \quad (2)$$

where $\mathrm{CTn_a}$ is the adjusted voxel attenuation value, $\mathrm{CTn_m}$ is the measured voxel attenuation value, ϕ_m is the calculated slice diameter in voxels and ϕ_r is the mean calculated sample diameter at the reference state, taken as 0.7% axial strain, in voxels. The mean attenuation values of the adjusted slice sets are shown in Figure 4 for each strain level.

As mentioned previously, it has been shown that attenuation values are linearly correlated with bulk density for X-ray energies between 0.1 MV and 1 MV (Wellington & Vinegar 1987; Desrues *et al.* 1996; Otani *et al.* 2000; Pralle *et al.* 2001). In the present study the measured bulk density and mean attenuation value of the confining fluid, water, were used as the first reference point. The mean adjusted sample attenuation at 0.7% axial strain and measured sample bulk density were used as the second reference point. The resulting relationship between bulk density and attenuation was:

$$\rho_\mathrm{b} = 1.001 + 7.73 \times 10^{-4}\mathrm{CTn} \quad (4)$$

where ρ_b is bulk density, g/cm^3 and CTn is the attenuation value. The calculated bulk density values for each slice were converted into void

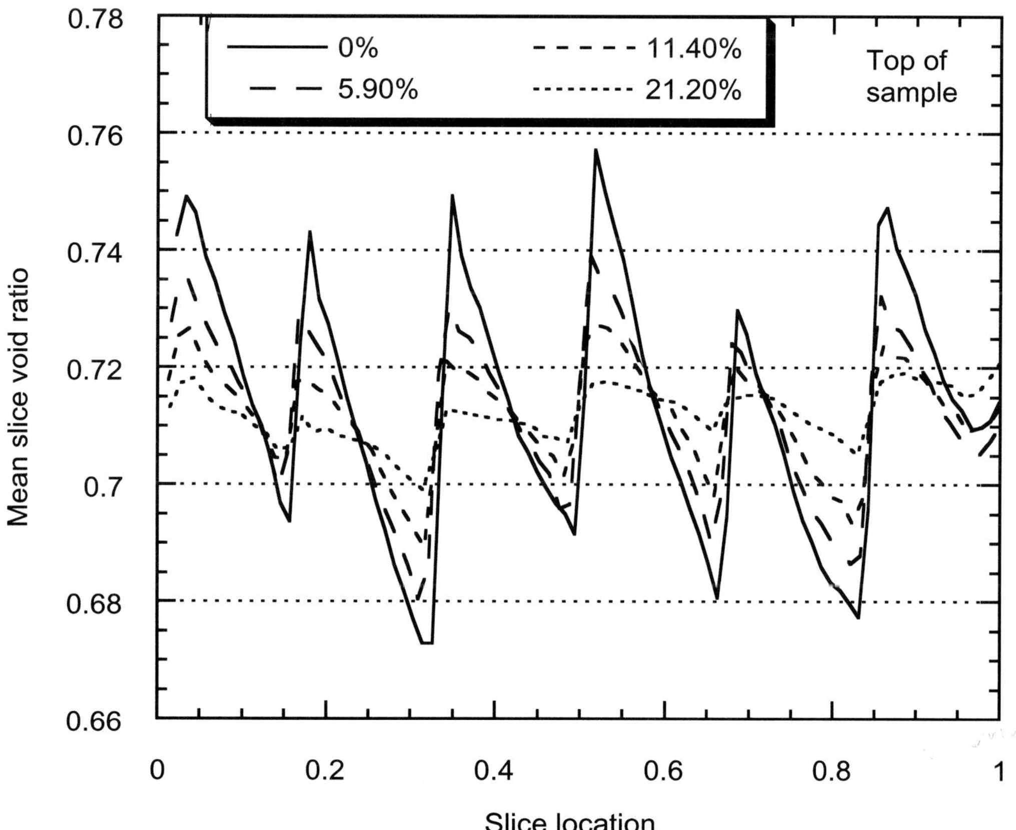

Fig. 5. Variation in slice void ratio with axial strain level.

ratio values assuming complete sample saturation, based on the measured B-value of 0.992.

Estimated mean slice void ratios along the sample length are presented for different axial strain levels in Figure 5. The slice location is presented as a fraction of the overall sample height at that axial strain level. For clarity, slice void ratio data are not presented for axial strain levels of 0.7% and 2.2%, as these values were very close to those at 0%.

Discussion and conclusions

Our experiments show that mean slice void ratios varied considerably at the end of consolidation (see Fig. 5). Before triaxial compression began, the maximum and minimum slice void ratios were 0.757 and 0.673, respectively. The overage sample void ratio during these scans was 0.713. In terms of relative density, although the overall relative density was 10.8%, at the end of consolidation the relative densities of individual slices was found to vary between −14.3% and 33.5%. These findings are in agreement with the suggestions of Yamamuro & Lade (1997), who reported that moist tamping might result in overconsolidation of regions within the sample. The demarcation between individual lifts is clearly visible on Figure 5.

As illustrated by Figure 5, significant void ratio redistribution occurred during triaxial compression. Sample uniformity increased with increasing deformation. This observation may be more clearly illustrated using the maximum and minimum slice void ratios which were determined for each slice set. These upper and lower bounds for slice void ratio were fitted with continuous functions in order to interpolate over all strain levels. In effective stress space, the measured sample response is given by:

$$q = \sigma'_1 - \sigma'_3 \tag{4}$$

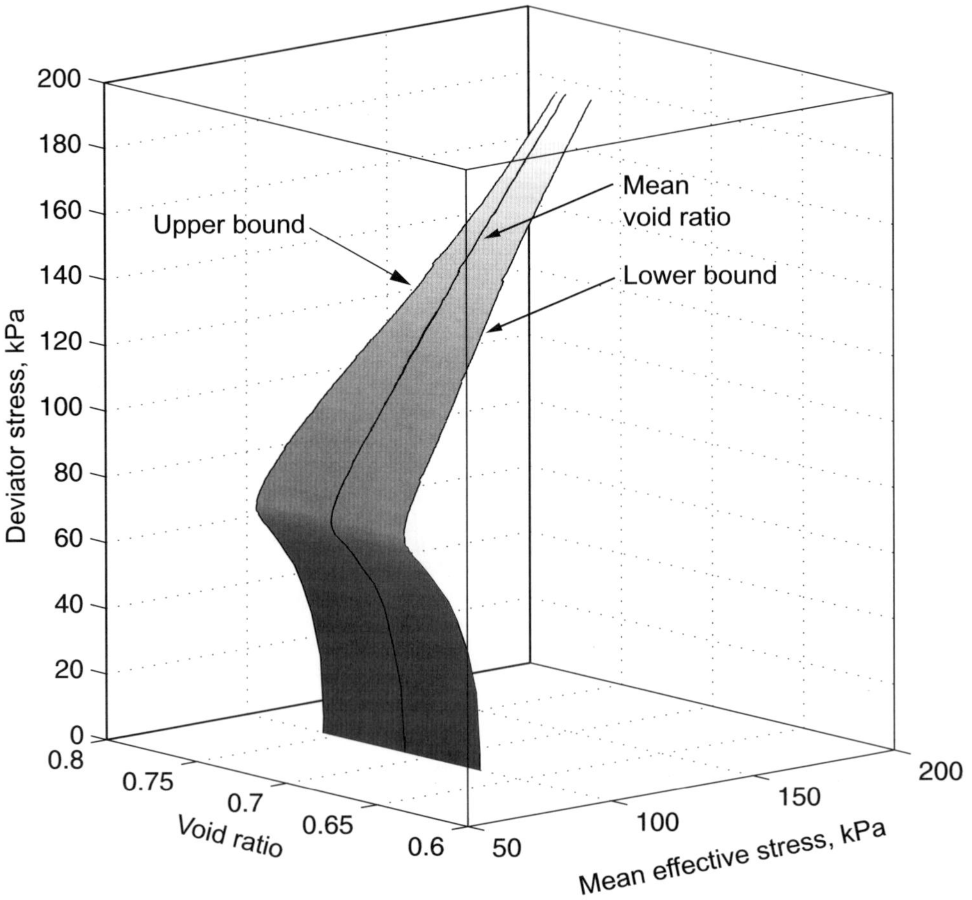

Fig. 6. Void ratio redistribution along the stress path.

and

$$p' = \frac{(\sigma'_1 + 2\sigma'_3)}{3} \quad (5)$$

where q is deviator stress, p' is the mean effective stress and σ'_1 and σ'_3 are the major and minor principal effective stresses, respectively. The interpolated upper and lower bounds for slice void ratio were incorporated with measured sample response to illustrate how sample uniformity varied in void ratio-effective stress space (Fig. 6). Slight sample relaxation occurred when axial displacement was paused to obtain CT data and the resulting discontinuities in the stress path have been omitted from Figure 6 for the purpose of clarity.

The results presented in Figure 6 raise several issues. Although the overall relative density of the sample was 11.4% at an axial strain of 2.2%, a strain near or above that at which peak shear stress is mobilized for strain softening samples, the relative density of individual slices was found to vary between approximately −8.8% and +33.6%. The void ratio distribution did not change abruptly when the sample was taken past this point on the stress path. Instead, void ratio redistribution was an ongoing process (see Fig. 6).

The soil sample presented here was shown to exhibit strain-hardening behaviour. It is not known if void ratio redistribution occurs in the same continuous, gradual manner for soil samples that exhibit a peak strength. Also unknown is the degree to which peak stress is affected by an initial void ratio distribution, i.e. the sample reconstitution procedure.

Finally, the relationship between external, visible deformation patterns – such as necking in axial extension or barrelling in axial compression – and internal void ratio redistribution has yet to be clarified since this sample was observed to deform as a right cylinder. Further investigation into the effects of sample preparation method and loading direction on the undrained response of cohesionless soils is required.

This research would not have been possible without the support of the Natural Sciences and Engineering Research Council of Canada (NSERC). The assistance of L. Curtis of Foothills Medical Center, and D.S. Phillips of University of Calgary Computing Services, is gratefully acknowledged. The efforts of machinists at the University of Calgary is also appreciated.

References

AYOUBIAN, A. & ROBERTSON, P.K. 1998. Void ratio redistribution in undrained triaxial extension tests on Ottawa sand. *Canadian Geotechnical Journal*, **35**, 351–359.

BEEN, K., JEFFERIES, M.G. & HACHEY, J. 1991. The critical state of sands. *Géotechnique*, **41**, 365–381.

BISHOP, A.W. & HENKEL, D.J. 1962. *The Measurement of Soil Properties in the Triaxial Test, 2nd Edition.* Edward Arnold Ltd., London.

BOSSI, R.H., FRIDDELL, K.D. & LOWREY, A.R. 1990. Computed tomography. *In*: SUMMERSCALES, J. (ed.) *Non-destructive Testing of Fibre-reinforced Plastic Composites. Volume 2.* Elsevier, London, 201–252.

COLLIAT-DANGUS, J.L., DESRUES, J. & FORAY, P. 1988. Triaxial testing of granular soil under elevated cell pressure. *In*: DONAGHE, R.T., CHANEY, R.C. & SILVER, M.L. (eds) *Advanced Triaxial Testing of Soil and Rock.* American Society for Testing and Materials, Special Technical Publication, **977**, 290–310.

CRAIG, R.F. 1997. *Soil Mechanics, 6th Edition.* E & FN Spon, London.

DESRUES, J., CHAMBON, R., MOKNI, M. & MAZEROLLE, F. 1996. Void ratio evolution inside shear bands in triaxial sand specimens studied by computed tomography. *Géotechnique*, **46**: 529–546.

EMERY, J.J., FINN, W.D.L. & LEE, K.W. 1973. Uniformity of saturated sand specimens. *In*: SELIG, E.T. & LADD, R.S. (eds) *Evaluation of Relative Density and Its Role in Geotechnical Projects Involving Cohesionless Soils.* American Society for Testing and Materials, Special Technical Publication, **523**, 182–194.

FROST, J.D. & JANG, D.J. 2000. Evolution of sand microstructure during shear. *Journal of Geotechnical and Geoenvironmental Engineering*, **126**, 116–130.

GILBERT, P.A. 1984. *Investigation of Density Variation in Triaxial Test Specimens of Cohesionless Soil Subjected to Cyclic and Monotonic Loading.* Department of the Army, Waterways Experiment Station, Corps of Engineers, Vicksburg, Mississippi. Technical Report **GL-84-10**.

HIRD, C.C. & HASSONA, F.A.K. 1990. Some factors affecting the liquefaction and flow of saturated sands in laboratory tests. *Engineering Geology*, **28**, 149–170.

JOSEPH, P.M. 1981. Artifacts in computed tomography. *In*: NEWTON, T.H. & POTTS, D.G. (eds) *Radiology of the Skull and Brain: Technical Aspects of Computed Tomography. Volume 5.* The C.V. Mosby Company, St Louis, Missouri, 3956–3992.

KANTZAS, A. 1990. Investigation of physical properties of porous rocks and fluid flow phenomena in porous media using computer assisted tomography. *In Situ*, **14**, 77–132.

KONRAD, J.M. 1990. Minimum undrained strength of two sands. *Journal of Geotechnical Engineering*, **116**, 932–947.

KUERBIS, R. & VAID, Y.P. 1988. Sand sample preparation – the slurry deposition method. *Soils and Foundations*, **28**, 107–118.

MCROBERTS, E.C. & SLADEN, J.A. 1992. Observations on static and cyclic sand-liquefaction analysis. *Canadian Geotechnical Journal*, **29**, 650–665.

MUHUNTHAN, B., CHAMEAU, J.L. & MASAD, E. 1996. Fabrics effects of the yield behaviour of soils. *Soils and Foundations*, **36**, 85–97.

OTANI, J., TOSHIFUMI, M. & OBARA, Y. 2000. Application of X-ray CT method for characterization of failure in soils. *Soils and Foundations*, **40**, 111–118.

POULOS, S.J. 1981. The steady state of deformation. *Journal of Geotechnical Engineering*, **107**, 553–562.

PRALLE, N., BAHNER, M.L. & BENKLER, J. 2001. Computer tomographic analysis of undisturbed samples of loose sands. *Canadian Geotechnical Journal*, **38**, 770–781.

RUZYLA, K. 1986. Characterization of pore space by quantitative image analysis. *Society of Petroleum Engineers Formation Evaluation*, **1**, 389–398.

THOMSON, P.R. & WONG, R.C.K. 1999. Design of a triaxial cell for computer tomography scan. *Proceedings of the 52nd Annual Canadian Geotechnical Conference, Canadian Geotechnical Society, Regina, October 24–27*, 525–528.

VAID, Y.P. & THOMAS, J. 1995. Liquefaction and postliquefaction behaviour of sand. *Journal of Geotechnical Engineering*, **121**, 163–173.

WELLINGTON, S.L. & VINEGAR, H.J. 1987. X-ray computerized tomography. *Journal of Petroleum Technology*, **39**, 885–898.

YAMAMURO, J.A. & LADE, P.V. 1997. Static liquefaction of very loose sands. *Canadian Geotechnical Journal*, **34**, 905–917.

ZATZ, L.M. 1981. Basic principles of computed tomography scanning. *In*: NEWTON, T.H. & POTTS, D.G. (eds) *Radiology of the Skull and Brain: Technical Aspects of Computed Tomography. Volume 5*. The C.V. Mosby Company, St Louis, Missouri, 3853–3876.

The use of X-ray computed tomography in the investigation of the settlement behaviour of compacted mudrock

M. A. O'NEILL[1], A. K. GOODWIN[1] & W. F. ANDERSON[2]

[1] *School of Environment & Development, Sheffield Hallam University, Sheffield S1 1WB, UK (e-mail: m.oneill@shu.ac.uk)*

[2] *Department of Civil & Structural Engineering, University of Sheffield, Sheffield S1 3JD, UK*

Abstract: The material used in the restoration of opencast coal mines is made up predominantly of mudstone particles ranging in size from less than 60 μm to more than 60 mm nominal diameter. Post-compaction settlements are mainly sub-divided into short-term 'collapse' and long term 'creep' components, both of which can be very large and often significantly reduce the development potential of restored opencast sites. As the mechanics of these movements are poorly understood at present, X-ray computed tomography (CT) was used to investigate the fundamental nature of particulate interactions within specimens of backfill undergoing long-term creep settlement.

The preliminary results from a series of CT analyses of a number of large scale specimens are presented. Different CT procedures were used and an indication for best practice for scanning of large samples has been found. Qualitative interpretation of the images obtained to date is presented in terms of fill structure, particulate changes during settlement and effects of arching. Indications are that particle breakage is a less significant mechanism than previously thought, and that local collapse of voids, particle sliding and particle rotation are the dominant mechanisms. Quantitative data to be extracted from the images may lead to a probabilistic approach to the prediction of settlement. This possibility is noted, as is the potential for the wider use of CT in geotechnical engineering.

Opencast mining involves excavating the material overlying the coal to enable its extraction and the subsequent replacement of the overburden into the worked void. Opencast sites in the UK may encompass an area in excess of 300 hectares (Scott Wilson Kirkpatrick 1995) and may be more than 100 m deep and accomplished with extraction ratios of overburden to coal greater than 25 to 1 (Goodwin *et al.* 2001). The demand for re-development of these large sites after backfilling means that post-compaction settlements must be controlled and settlements predicted. The latter is generally accomplished using empirical correlations and relationships, many of which were devised for use in situations away from the opencast mining industry. This empirical approach to settlement prediction is a symptom of the lack of understanding of the fundamental behaviour of particulate soils.

In principle, particulate soils are thought to carry load through the particle contacts, as shown indicatively in Figure 1. In equilibrium under an applied load, the soil particles remain in their relative positions. Further loading and/or particulate changes due to weathering after compaction will cause the soil particles to re-order themselves into another state of equilibrium. This will result in the reduction of void space and an increase in bulk density of the particulate mass, which will be manifested as surface settlements that may adversely affect the stability and serviceability of any post-restoration developments.

At a more detailed level, the fundamental mechanisms of soil deformation at a particulate level are not well understood. Mechanisms that

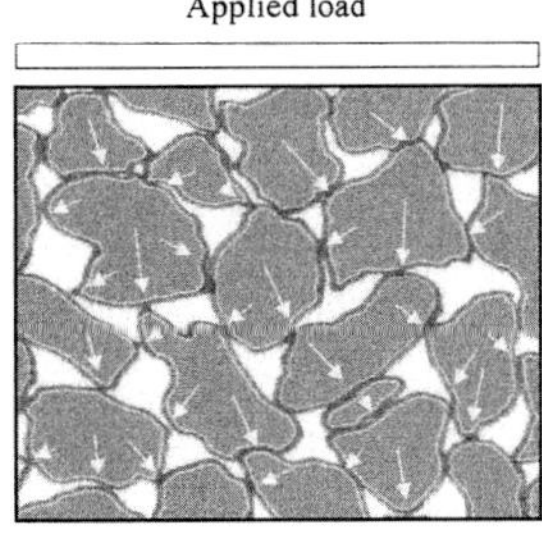

Fig. 1. Indicative two-dimensional particulate soil model.

From: MEES, F., SWENNEN, R., VAN GEET, M. & JACOBS, P. (eds) 2003. *Applications of X-ray Computed Tomography in the Geosciences*. Geological Society, London, Special Publications, **215**, 199–204. 0305-8719/03/$15.

could be at work in a particulate medium include: (i) crushing of particle contacts (Sowers *et al.* 1965); (ii) variations of particle stiffness and strength with time due to weathering, softening, slaking and expansion (Bally 1988); (iii) particle splitting and breakage (Marsal 1973); (iv) re-arrangement and/or rotation of particles (Hills & Denby 1996); (v) asperity indentation at contacts and inter-particle ploughing (Scholz & Engelder 1976); and (vi) variation of inter-particle friction due to local moisture content variations (Pigeon 1969). Other mechanisms, thought to be less important in opencast fills, include inter-particle lubrication by wetted clay- and silt-sized particles, and viscous yielding of particle contacts.

Considering opencast backfill in particular, two additional complexities arise, namely scale and material variability. The scale of the opencasting works results in the overburden being broken down from its intact state during excavation, transportation and subsequent backfilling into a wide range of particle sizes. Particles range in size from large boulders greater than 2 m in nominal diameter, to silt- and clay-sized particles of less than 60 μm nominal diameter. The backfill also varies in mineralogical composition. Although mudstones dominate in the UK, the relative proportions of siltstone and sandstone may be significant. Furthermore, the mud-stones in particular are susceptible to rapid weathering (Blanchfield & Anderson 2000), which may introduce a time-dependent element to the particulate interactions that cause post-restoration movements.

Hills & Denby (1996) probably represent the consensus view of industry when they state that the settlement of opencast fills results primarily from the gradual re-arrangement of the material fragments due to the crushing of highly stressed contact points. However, there is little direct evidence for this view and the actual mechanics in the variable fill commonly produced on an opencast coal site may be far more complex than this. The use of X-ray computed tomography (CT) presents an opportunity to assess directly the relative importance of the various mechanisms and to recognize any changes in importance with time.

Material and methods

The experimental cells are large modified Rowe cells, as shown in Figure 2. Two cell sizes have been manufactured: 0.60 m diameter cells for large scale creep testing, and 0.23 m diameter cells for relatively short-term compressibility tests in combination with CT scanning. The cells are designed to transmit the applied hydraulic load to the soil specimen via the loading platen and soil compression is measured via the loading piston shaft. The applied vertical pressure of up to 800 kPa was achieved in 100 kPa increments over an eight hour period and monitored through ports in the top plate. The cell walls are formed from plastic pipe typically used in water transportation and rated to carry liquid at pressures of up to 1600 kPa. These walls are stiff enough to provide sufficient restraint against volume

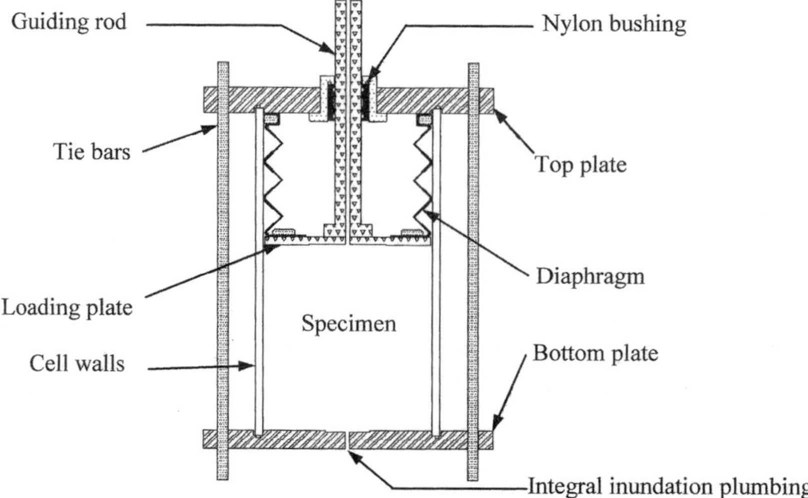

Fig. 2. Modified Rowe cell (not to scale).

change in the minor principal axis of the sample to ensure that the specimens are effectively subjected to one-dimensional compression only.

The specimens are made up of material collected locally from virgin mudstone strata exposed during opencast coal mining operations. Particle size distribution, natural moisture content and compaction behaviour of the mechanically excavated material were determined using the appropriate BS 1377: 1990 test procedures. The particle size distribution was then modified (or 'scalped') by removing particles greater than 75 mm nominal diameter. The scalped grading was used to form three equal layers of soil in the cells, each layer being subjected to a compaction effort of 25 blows with a 4.5 kg hammer (see BS 1377: 1990 for details of the hammer used; British Standards Institution 1990). This level of compactive effort produced a specimen with a bulk density of 1.9 Mg/m^3 (± 0.10 Mg/m^3), which is comparable to densities achieved in site restorations (Scott Wilson Kirkpatrick, 1995). In some series of tests, particles of siltstone were added to the middle layers of the test specimens to provide materials with greater strength and stiffness against which the softer mudstone could interact. The samples were otherwise treated in the same manner.

After loading, the soil was transported to the CT scanning facility (Royal Hallamshire Hospital, Sheffield), where cross-sectional scans of the samples were completed using a GE Medical Systems B75202EZ HiSpeed FX/I scanner. Scanning was completed shortly after compaction of the specimens and again one week to three months later. As such, any changes seen reflect particulate and structural changes due to creep compression.

The best scanning protocol was determined on the basis of image quality and time. Initially, helical scanning was undertaken but, as discussed below, it was determined that better quality images could be obtained using an axial technique. Axial scanning gave 3 mm slices taken every 2 mm along the sample length, from which 2 mm slices were restored. Each slice was set to take 3 seconds at 120 kV and 100 mA. This re-quired the CT scanner to be halted halfway through the scan to allow the X-ray tube to cool. Longitudinal-sections were also restored from the cross-sectional images and analysed.

Results and discussion

Examples of CT images are presented in Figures 3 to 5. Figure 3 represents transverse images of two sectional planes before and after compression, achieved using a helical scanning protocol. Figure 4 presents similar data for one plane, as obtained with an axial scanning protocol. Figure 5 shows longitudinal sections through two different specimens based on the reconstruction of axial scanning data. All images show the larger particles of the fill, down to approximately 2 mm nominal diameter. Finer particles appear as a grey mass that is not possible to resolve due to the 1 mm^3 resolution of the scanning equipment used.

Qualitative observations and interpretation of these images are made below in respect of four general aspects: comparison of axial and helical scanning protocols, observations on fill structure, particulate changes during settlement and effects of arching.

Helical scanning protocols were developed in order to achieve an image of adequate quality for medical diagnosis whilst minimizing the radiation dose given to the patient. This protocol also had the added benefit of reducing the heat load on the X-ray tube of the scanner. The radiation exposure used in the helical protocol, coupled with the large specimen size and the high density of the fill, meant that the images obtained for the studied samples were grainy and lacked good contrast (compare Figs 3 and 4).

As expected, the images obtained using the axial technique are of a higher quality, with a higher degree of contrast. However, the heat load on the equipment increased, compared with the helical protocol and at times scanning times almost doubled. Furthermore, the longitudinal images presented in Figure 5 still show a change in contrast along a central core, which is believed to be due to a combination of beam hardening, the greater concentration of fine particles and a lack of digital image optimization.

The observed distribution of larger particles in Figures 3 to 5 appears random, with greater concentrations in some areas than others. Similarly, the voids (black areas in the images) appear random in distribution. These general observations are significant, as they provide direct evidence to support the widely held belief that stress is transferred non-uniformly through a particulate medium dependent on its structure. Furthermore, the observations support the view that a probabilistic approach to the prediction of the compacted structure and the subsequent settlement behaviour of such materials may present a better avenue for theoretical modelling than discrete physical models of particulate interactions.

Interpretation of particulate interactions using two-dimensional transverse CT images of the specimens (such as Figs 3, 4 and 5) was attempted. Similar sections from scans completed

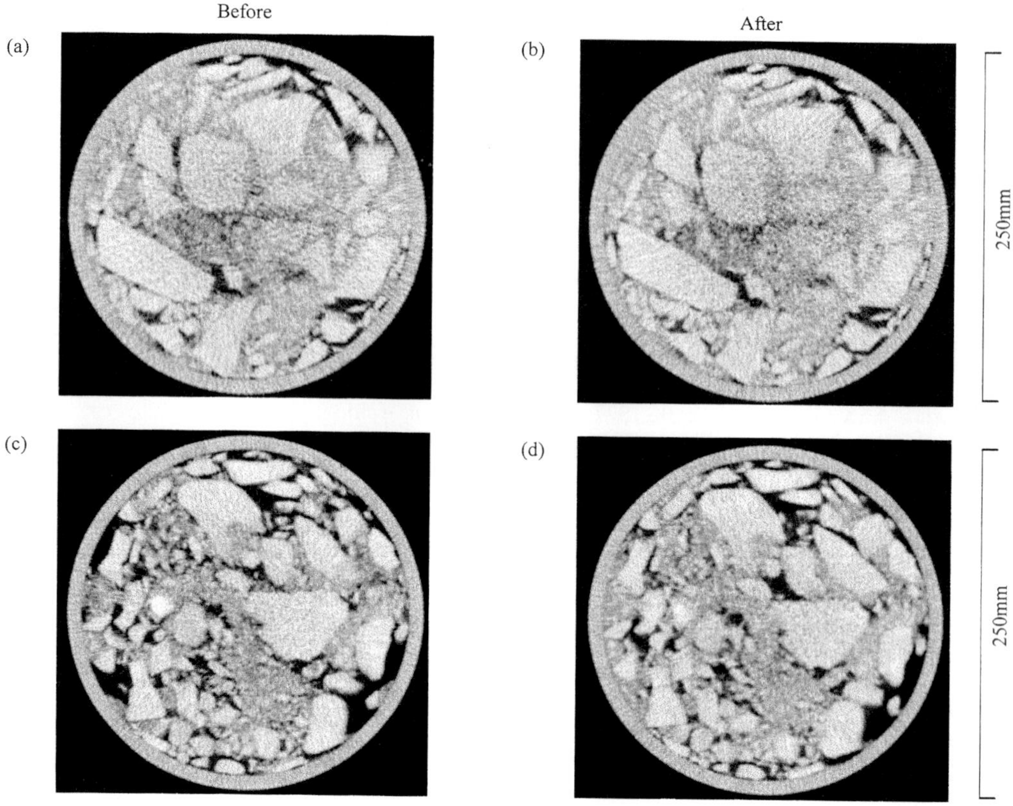

Before

After

(a) (b)

250mm

(c) (d)

250mm

Fig. 3. CT images of a specimen before and after compression, obtained using helical scanning (after Goodwin *et al.* 2001).

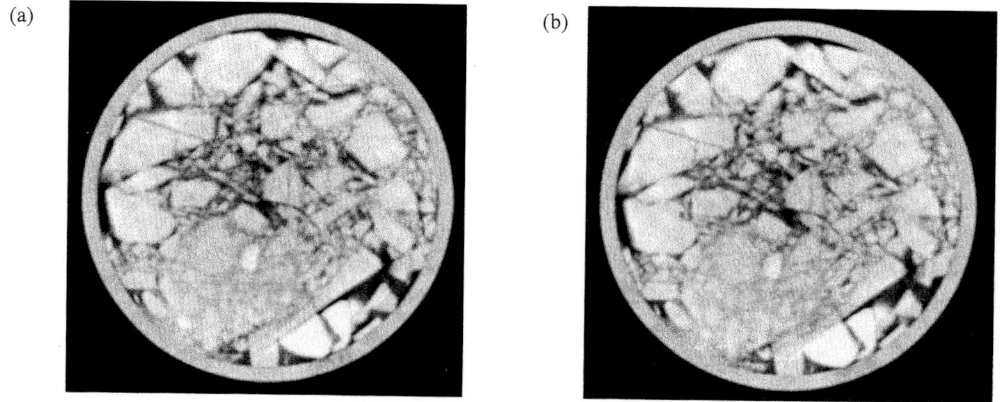

Fig. 4. Cross-sectional image of approximately the same section (**a**) before and (**b**) after compression, obtained using axial scanning.

before and after compression loading were first identified based on their relative distance from the base of the cell, which was included in all scans as a reference. Individual particles were then found based on size, shape, orientation and proximity to the cell walls. The images obtained using the helical technique (Fig. 3) indicate that the dominant mechanisms at work are local 'collapse' of the fill structure into voids left during compaction, relative inter-particle sliding

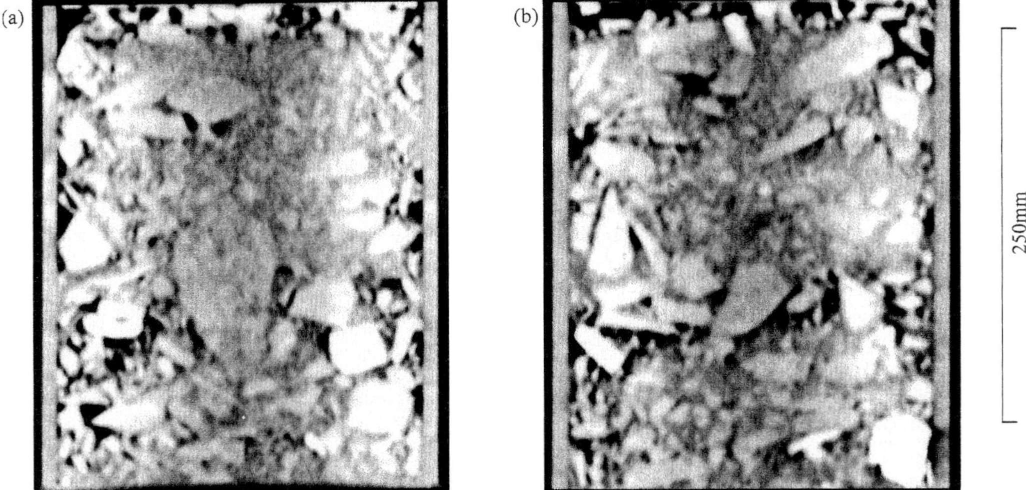

Fig. 5. Longitudinal sectional images reconstructed from axial scans.

and rotation of particles. These mechanisms appear to have been facilitated in part by the presence of some large local voids, but the mere presence of a large void does not mean movements that will occur will be local to that void. For example, comparison of Figures 3a and 3b shows that whilst some voids have changed in shape and size (e.g. voids below centre and below left of centre) others have not (e.g. above centre by cell wall). Similarly, the lower left quadrant indicates that particles have rearranged and moved further apart, whilst in the lower right quadrant the fill structure is virtually unchanged. Overall, the findings support the view that stress transmission and settlements occur non-uniformly within a particulate fill depending on the local structure and its relation to the global structure.

Splitting of mudstone particles is discernible on some of the transverse sections, for example for the large particle in the lower left quadrant in Figures 3c and 3d. Further indication of the prevalence of breakage is provided by cross-sectional images obtained using the axial scanning protocol, for example in Figure 4. The images in this figure represent the same cross-sectional plane, within achievable limits, before and after one week of sustained constant load. Some particles can again be seen to have suffered breakage, but this phenomenon appears to be less prevalent than expected before the tests. Other images not reproduced here confirm this and suggest that particle crushing is a less significant settlement mechanism than previously thought, at least for particles greater than the voxel size achieved in these tests. This is despite the expected stress concentrations associated with non-uniform stress transmission and despite the general view that splitting occurs predominantly in larger particles.

The longitudinal sections derived from the axial scanning data (Fig. 5) are from two different samples. In these images, a band of voids can be seen approximately one third of the way up one of the images (see Fig. 5b). This layer corresponds with the likely position of an interface between two of the three compaction layers. This band is sheltered by an arch of large particles that spans the cell diameter in this plane, an effect that may be expected to occur in places given the large size of the particles.

Aside from confirming the occurrence of arching despite high levels of compaction, this visualization of arching structures with voids below is potentially significant in terms of long-term settlement. Weathering of the predominantly mudstone fill may be conjectured to lead to a progressive collapse of the arch, with associated settlements.

Conclusions

There has been some discussion of possible creep mechanisms in the past, but it has not been possible before to directly determine their relative importance. The report made has provided good evidence that an X-ray CT is a usable non-destructive tool capable of providing data that can used to give support to fundamental soil behaviour theories. The axial scanning technique has been found more suitable than helical scanning.

Findings from work in progress have been qualitatively reported and assessed in the context of the settlement of opencast coal mine backfills, though it is noted that further investigations are needed before the general applicability of the findings can be assessed. The work thus far has highlighted the importance of visualization of the fill mass and has shed some light on fill structure. The indications to date contradict to some degree the consensus of opinion in geotechnical engineering that particle crushing is a major mechanism of creep settlement. Rather, local collapse into small voids left by compaction and relative sliding and rotation of particles, seem to be the dominant factors affecting compression and creep of opencast fills. A potential for settlement due to collapse of arching structures has also been identified.

Further work is needed to confirm the indications to date, including quantitative analysis of the data. It is likely that probabilistic approaches to settlement prediction offer much potential and such work could be based on the quantitative interpretation of observed particulate arrangements. These and other benefits are clearly offered to geotechnical engineering by CT scanning, which is considered by the researchers to be a potentially major growth area in ground related research over the next decade.

The assistance of S. Linskill in the geotechnics laboratory at Sheffield Hallam University has been greatly appreciated, as has the assistance and advice freely offered by staff at the Royal Hallamshire Hospital, Sheffield. Thanks go also to S. Haylock, a former undergraduate student at Sheffield Hallam University, who undertook some of the particle characterization tests.

References

BALLY, R.J. 1988. Some specific problems of wetted loessial soils in civil engineering. *Engineering Geology*, **25**, 303–324.

BLANCHFIELD, R. & ANDERSON, W.F. 2000. Wetting collapse in opencast coalmine backfill. *Proceedings of the Institution of Civil Engineers, Geotechnical Engineering*, **147**, 139–149.

BRITISH STANDARDS INSTITUTION 1990. *BS 1377: 1990, Methods of Testing Soils for Civil Engineering Purposes*. British Standard Institution, UK.

GOODWIN, A.K., O'NEILL, M.A. & ANDERSON, W.F. 2001. Investigation of the fundamental mechanisms affecting creep settlement of restored opencast coal mine sites. *In*: *Proceedings of the 3rd British Geotechnical Association Geoenvironmental Engineering Conference*, 204–209.

HILLS, C.W.W. & DENBY, D. 1996. The prediction of opencast backfill settlement. *In*: *Proceedings of the Institution of Civil Engineers, Geotechnical Engineering*, **119**, 167–176.

MARSAL, R.J. 1973. Mechanical properties of rockfill. *In*: HIRSCHFELD, R.C. & POULOS, S.J. (eds) *Embankment Dam Engineering, Casagrande Volume*. John Wiley & Sons, New York, 109–200.

PIGEON, Y. 1969. *The compressibility of rockfill*. PhD thesis, Imperial College, London.

SCHOLZ, C.H. & ENGELDER, J.T. 1976. The role of asperity indentation and ploughing in rock friction. *International Journal of Rock Mechanics & Mining Sciences*, **13**, 149–154.

SCOTT WILSON KIRKPATRICK 1995. *State of the Art Review of the Compaction of Opencast Backfill*. Report for British Coal Opencast.

SOWERS, G.F., WILLIAMS, R.C. & WALLACE, T.S. 1965. Compressibility of broken rock and the settlement of rockfills. *In*: *Proceedings of the 6th International Conference on Soil Mechanics & Foundation Engineering*, **2**, 561–565.

Industrial X-ray computed tomography studies of lake sediment drill cores

A. FLISCH[1] & A. BECKER[2]

[1] *Swiss Federal Laboratories for Materials Testing and Research (EMPA), CH-8600 Dübendorf, Switzerland (e-mail: alexander.flisch@empa.ch)*
[2] *Swiss Federal Institute of Technology (ETH), Institute for Geophysics, CH-8093 Zurich, Switzerland (e-mail: becker@seismo.ifg.ethz.ch)*

Abstract: An industrial X-ray computed tomography (CT) system was used for the study of sedimentary and deformation structures in weakly consolidated late Pleistocene and Holocene lake sediments. CT analysis revealed details of structures that could not be detected by X-ray radiography or visual core logging. Examples include sand dykes, dropstones and plant remains, which are potentially important in palaeoseismological and palaeoenvironmental research. The CT images also help to discriminate between drill core artefacts and natural structures. X-ray tomography was also used for the determination of physical properties, particularly for bulk density measurements based on calibrated grey-scale values.

X-ray computed tomography (CT) is a technique that is increasingly used for geological investigations (Denison *et al.* 1997; Ketcham & Carlson 2001). A general outline of its principles and application is given by Moore (1990), Wells *et al.* (1994) and Kak & Slaney (1998). CT allows the observation and analysis of internal structures without the need for manual serial-sectioning of samples, which is time-consuming, destructive and not appropriate for valuable specimens. Industrial CT systems are the preferred instruments for analysing large geological objects with high X-ray attenuation coefficients, because high-energy radiation can be used and a high spatial and contrast resolution is offered. In this paper, results are presented for the application of industrial CT for the study of lake sediment drill cores from northern Switzerland.

Method and scanner specification

The CT system of the Swiss Federal Laboratories for Materials Testing and Research (EMPA) is a medium-sized device that is particularly suitable for investigating geological samples. The 450 kV X-ray tube has the capability to penetrate up to 20 cm of sedimentary rock. The tube is highly stable and has a small optically effective focal plane. The achievable spatial resolution of the CT scanner is in the range of 0.5‰ to 1‰ of the object diameter. To minimize radiation scattering within the object and to achieve a high signal-to-noise ratio, the X-ray beam is collimated to a 2 mm thick fan.

The line detector consists of 125 cadmium–tungsten scintillators. A detector collimator made of tungsten reduces diffuse scattering reaching the scintillators and improves image quality. The collimator aperture can be set vertically and horizontally. A small horizontal aperture provides a high spatial resolution. The vertical opening determines the thickness of the investigated slice.

The object manipulator can handle samples that are up to 40 cm in diameter, 60 cm in height and 25 kg in weight. During the measurement, the object is rotated several times over 360° (fan beam mode). The higher the required spatial resolution the more rotations are needed to fill the gaps between the detector channels. If the object diameter is larger than the line detector, which is 25 cm long, the sample has to be moved back and forth several times in order to obtain the required projections (parallel beam mode). For image reconstruction, fast Fourier transformation (FFT) algorithms are used (filtered back projection).

The selection of acquisition parameter was different for the scans that are reported in the different parts of this paper: Lake Seewen (dropstone study) – 450 kV, 2 mA, no filter, 0.20×0.50 mm detector aperture, 600×800 pixels, 0.15×0.15 mm pixel size; Lake Seewen (sand dyke study) – 400 kV, 2.25 mA, 1.5 mm brass

From: MEES, F., SWENNEN, R., VAN GEET, M. & JACOBS, P. (eds) 2003. *Applications of X-ray Computed Tomography in the Geosciences*. Geological Society, London, Special Publications, **215**, 205–212. 0305-8719/03/$15.

filter, 0.35 × 0.50 mm detector aperture, 800 × 500 pixels, 0.15 × 0.15 mm pixel size; Lake Bergsee (density study) – 150 kV, 6 mA, no filter, 0.35 × 0.75 mm detector aperture, 360 × 240 pixels, 0.25 × 0.25 mm pixel size.

Investigation of lake sediment drill cores

Two lakes in the Basle region, Lake Seewen in northern Switzerland and Lake Bergsee in southern Germany, were sampled. The drill core samples, with a diameter between 5 and 10 cm, were taken as part of a palaeoseismological research project searching for deformation features (seismites) related to strong pre-historical earthquakes, occurring in Holocene and late Pleistocene lake deposits. Earthquakes may create characteristic soft-sediment deformation structures, sand dykes and fractures or changes in physical properties of the sediment such as bulk density.

In addition to visual logging of the core samples, X-ray CT was used to identify features that are not visible to the naked eye. The core samples were first split parallel to the long core axis for visual logging and then scanned by X-ray radiography, followed by X-ray CT analysis. X-ray radiography was used as a first pass technique because it quickly and inexpensively provides information about the internal structure of the core samples and also helps to define intervals that are most interesting for CT investigations. All CT images presented in this paper are slices of half-cores scanned perpendicular to the long core axis.

Grey-scale value graphs along selected lines in CT images were used to clarify the nature of recognized features (Fig. 1). This is possible because grey-scale values are closely related to density, especially for sediments with a nearly uniform composition. Because EMPA's industrial CT is not calibrated to Hounsfield Units, the (arbitrary) absolute grey-scale values are not indicated on Figure 1. The qualitative grey-scale curves help to clarify the nature of objects and their origin. The short lines A–B, C–D, E–F and G–H in Figure 1 mark the positions along which the grey scale values were determined, which are presented in the curves A′–B′, C′–D′, E′–F′ and G′–H′. The dark grey object along line A–B in Figure 1 showing grey-scale values in curve A′–B′, which are lower than those of the surrounding silty sands but higher than those of air, is a piece of wood. In contrast, the white feature along line C–D, marked by higher grey-scale values in curve C′–D′ and thus by a higher density than the surrounding silty sand, is a rock

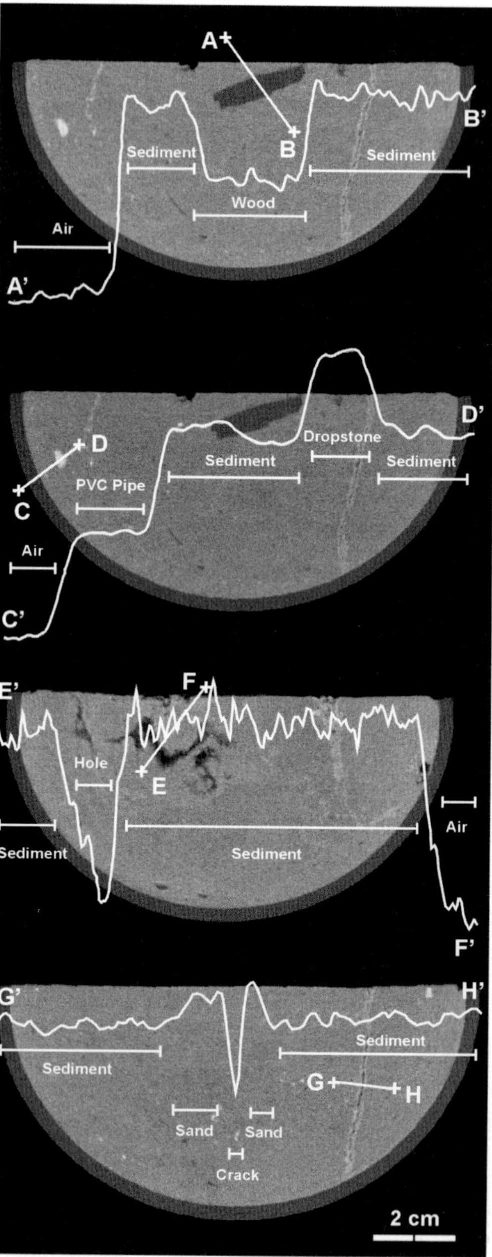

Fig. 1. Examples of the use of grey-scale values for qualitative investigations of natural features and artefacts in CT images. Transects are indicated by lines A–B, C–D, E–F, G–H, and the corresponding grey-scale value graphs by lines A′–B′, C′–D′, E′–F′, G′–H′. The interpretations (sediment, wood, ...) refer to the grey-scale graphs.

fragment embedded in fine-grained lacustrine deposits. The dark grey inclusion along line E–F could be an enrichment of plant chaff in a sandy clay silt layer, but the grey-scale curve E′–F′ identifies it as a hole, with a grey-scale value close to that of air. It is most likely an artefact, because layers with low cohesion are preferred sites of failure during core recovery or handling. Line G–H transects a sand dyke, which again was the site of a core failure creating a crack that follows the sand fill of the dyke, characterized by a high density and low cohesion.

Example 1: dropstones

Detailed investigation of the Lake Seewen sediments revealed several examples of mainly angular rock fragments with a size of up to 3 cm, embedded in the fine-grained lake deposits (Becker *et al.* 2000). Figure 2 shows three CT images with examples of angular, (a), or weakly rounded, (b), rock fragments, seen as light patches with a diameter of up to 13 mm, in a matrix of silty clay. The smaller, slightly rounded components indicate transportation of the fragments over short distances before deposition in the lake. In addition to the dropstones, large plant remains can be seen as dark patches in slice (b) and especially in slice (c).

The presence of large rock fragments in fine-grained lake deposits is not possible under normal lacustrine hydraulic conditions. However, transport of fragments of all sizes is possible by floating ice or wood. The occurrence of large rock fragments can therefore be related to: (i) the melting of icebergs derived from glaciers in an ice-dammed lake; (ii) the melting of the ice of a seasonally frozen periglacial lake covered with rock debris by spring floods or slope failures; or (iii) the decay of tree trunks and root bales that float on the lake surface.

Most of the dropstones in the Lake Seewen cores occur in late Pleistocene deposits and only a few (small) stones occur in the Holocene sediments. Because the Lake Seewen area was not glaciated during the Würm ice age, the most plausible origin of the dropstones is related to deposition of rock debris on the ice-covered surface of the seasonally frozen periglacial lake. However, transportation in root bales cannot be excluded, as indicated by the frequent co-occurrence of plant remains and dropstones.

Example 2: sand dyke

Figure 3 shows the most prominent sand dyke in the Lake Seewen cores. Visual logging of the core sample reveals a 52 cm long, vertical feature, which widens upward from about 1 mm to 4 mm width (Fig. 3, left hand column). It has an associated yellowish-brown alteration zone, which is most prominent in the upper half of the

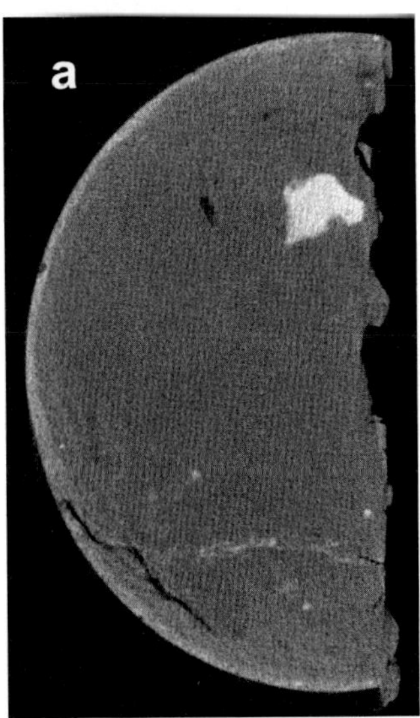

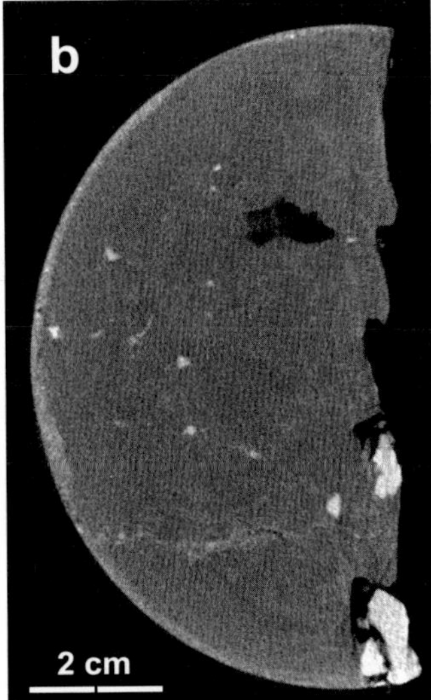

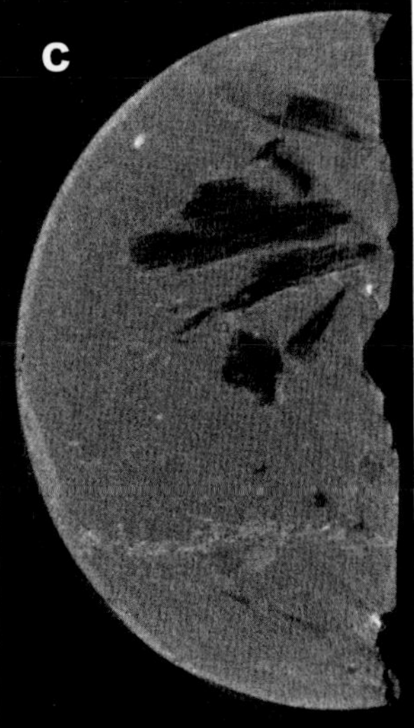

Fig. 2. Three CT images, up to 5 cm apart, showing angular to sub-angular dropstones (white) (**a**, **b**), large plant remains (dark grey) (**b**, **c**) and a sand dyke in the lower half of the pictures (**a**, **b**, **c**). The pale grey rim of the image is not caused by beam hardening but a drilling artifact, due to the drag of a sand-rich layer along the tube margin.

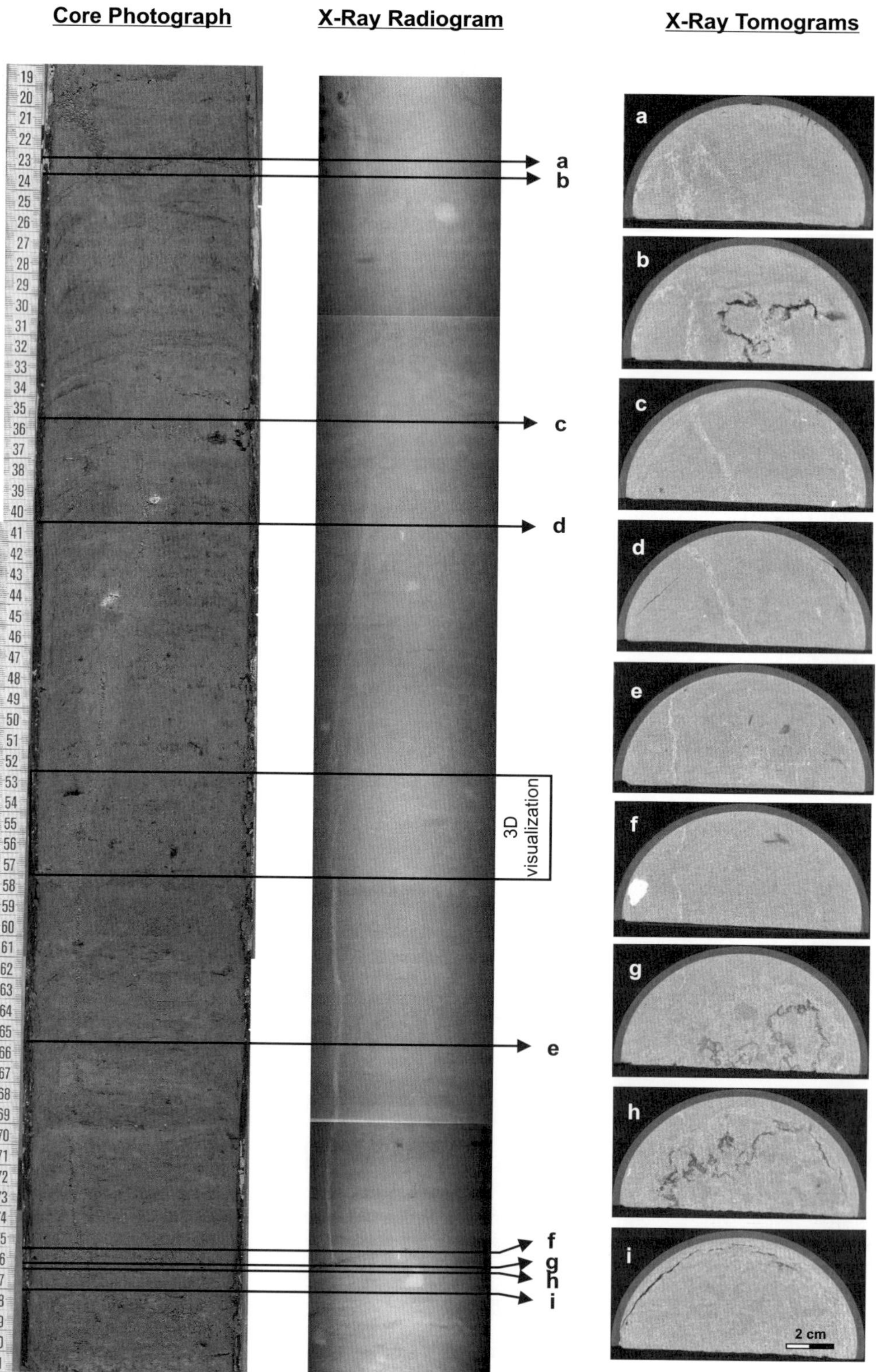

Fig. 3. Photograph (left), X-ray radiograph (centre) and representative CT images (right) of a drill core split along the long core axis, showing a 52 cm long sand dyke in lake deposits.

dyke where it is up to 30 mm wide, becoming narrower and less pronounced at depth. The sand dyke consists of pure calcareous coarse silt to fine sand. It rises from a 5 cm thick layer of silty sand to silty clay sand and passes upward through silty sands and sandy clay silts, which contain some dropstones and plant remains. The weakly developed layering in the surrounding lake deposits shows no offset across the sand dyke. The dyke changes in character when it reaches a thin silty sand layer at 23.5 cm, where it shows a lateral offset of about 2 cm reaching a maximum thickness of 8 mm about 4 cm above this level, at the highest point of its occurrence in the core. Below the continuation of the dyke, the silty sand layer shows a reduced thickness (Fig. 3).

In the X-ray radiogram (Fig. 3 central column) the onset of the dyke is clearly visible as a white line starting abruptly at 75.6 cm. In the middle of the core section, the dyke appears to split into branches, whereas in the upper section it can be hardly recognized except at the top left side of the core. X-ray radiogram reveals the layering of the surrounding lacustrine deposits, with also some dropstones and plant fragments.

Additional information about the structure of the sand dyke was obtained by CT scanning. 147 slices were scanned perpendicular to the long core axis, nine of which are shown in Figure 3 (right hand column, slices (a) to (i)). The sand dyke first clearly appears as a short, thin white line at the top left margin in slice (g), although a light grey inclusions in slice (h) could correspond to a lower occurrence. The sand dyke becomes wider from slice (f) to (c) and turns counterclockwise by about 25° in slices (d) and (c). In slices (a) and (b) the dyke is wider, less well confined and split into two branches. Slices (h), (g) and (b) show dark grey structures, corresponding to slightly irregular sand layers in cross-sections that are nearly parallel to the bedding.

The section of the dyke between 52 and 57 cm was scanned in some detail for 3D visualization (Fig. 4). 101 slices with a spacing of 0.5 mm were recorded, using a scan time of 11 minutes. The reconstruction shows dropstones and twigs, as well as layering disturbed by minor core drag along the tube walls. The sand dyke is a prominent feature, traversing the whole width of the core sample before it is cut abruptly by the core casing.

The CT slices and the 3D visualization show that the sand dyke is a planar structure and not a cylindrical de-watering or de-gassing tube (Guhman & Pederson 1992; Neumann-Mahlkau 1976), or a root canal. The core casing cuts

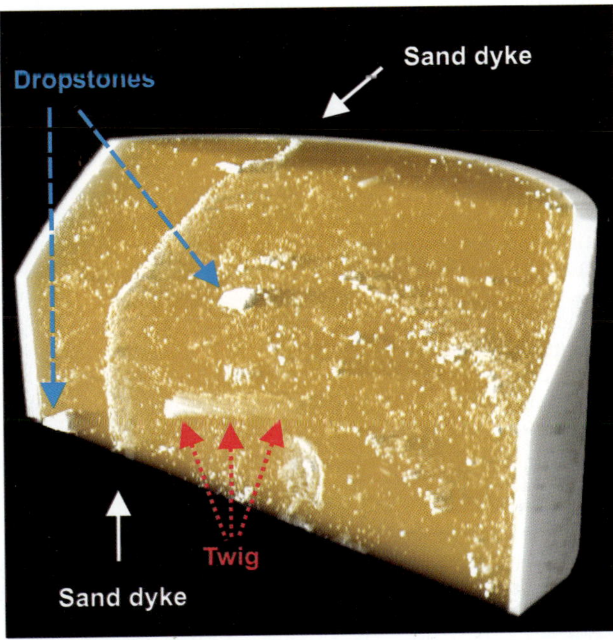

Fig. 4. 3D visualization of a 5 cm high section of the core sample shown in Figure 3. The image has been created using the volume rendering technique.

the dyke abruptly and there is no relationship with the slight disturbance of the sediments near the edge of the core. We therefore conclude that the dyke is a natural feature rather than an artefact. The orientation of the dyke shows only minor variations, which is also observed for other sand dykes in the Lake Seewen cores (e.g. Fig. 2). Such a consistent orientation is another argument in favour of a geological origin (Rodríguez-Pascua *et al.* 2000). The dyke is rooted in a sand-rich layer, which shows peculiar structures that are similar to those described by Anketell *et al.* (1970) for unstably-bedded sediments. As indicated by radiocarbon dates and pollen frequencies (Becker *et al.* 2000, 2002), the sand dyke was formed in an environment with low to normal sedimentation rates. It formed in a very shallow lake environment, as demonstrated by the occurrence of tube-like carbonate concretions (Magny 1992) and by the gastropod fauna. In addition, there are no indications of sub-aquatic slumps in the layers topping the sand dyke or a sudden increase in water level during emplacement of the sand dyke.

It is believed that the sand dyke has not formed from top to base as a Neptunian dyke: (i) the very 'clean' dyke fill containing only sand and coarse silt may indicate that clay and fine silt particles were washed out; (ii) the dyke is very narrow and shows no grading of the sediment fill; (iii) a sufficiently large source of sand is lacking at the top of the sand dyke; and (iv) the base of the sand dyke is connected to a sand-rich layer. Therefore, it is believed that the sand dyke was formed by the upward injection of sand facilitated by liquefaction and fluidization processes. A sudden increase of pore-water pressure in the water-saturated lake deposits, to a point which equals the confining pressure may have caused liquefaction of the sand layer at a depth of 76 to 81 cm. This caused instabilities in the sand layer, recorded by the irregular sediment structures around 76–76.5 cm and resulting in the injection of a sediment suspension into the overlying sediments. The increase of the width of the dyke from base to top may be explained by erosion or by a decrease in density of the confining sediments closer to the former water-sediment interface. However, it is believed that the dyke did not reach the former lake bottom. As soon as the density of the surrounding lake deposits reached the density of the water-sand suspension in the dyke, the injected silty sand started to intrude laterally, creating sill-like sand intrusions or sand pockets only a few centimetres below the water/sediment interface. Evidence for a sill-like intrusion is found in the increase in thickness of the thin sandy layer where the sand dyke shows a 2 cm lateral offset close to the top (Fig. 3, left column). Furthermore, patchy sediment structures can be seen where sand intruded laterally (Fig. 3, slice (b)), similar to those in the sand layer at the base of the dyke (Fig. 3, slices (g) and (h)). Alternatively, it is possible that the sand dyke could not pass through sediments along the former lake bottom, which were completely root ridden. However, there is no evidence in the sediments that would support this idea.

We believe that the formation of the sand dyke can be most easily achieved by strong earthquake shocks, which increases pore pressure in the slightly cohesive sand deposits at the base of the sand dyke to such a level that liquefaction can take place and, in addition, facilitate the opening of cracks for sand injection due to shearing. Contemporaneous structures such as mushroom structures, disrupted layers and fractures in the Lake Seewen deposits, together with the sand dyke described here, further confirm the occurrence of a strong prehistorical earthquake of a magnitude between $M \geq 5.5$ and $M \leq 6.5$ (Becker *et al.* 2002).

Example 3: density determination

Following Orsi *et al.* (1994) and Amos *et al.* (1996), CT data were used to determine the density of almost purely organic sediments (gyttja) to establish a high resolution density profile for the deposits of Lake Bergsee, southern Germany (Fig. 5). 3905 slices were recorded, with a slice distance between 2 and 5 mm and a 3 min scan time. The quality of the slices can be seen in Figure 5. The resolution is low and not really suitable for the investigation of structures, but it is sufficient for density determinations. For every slice, the average of the grey-scale values of 16 counting squares was determined for the undisturbed central part of the slice. For every core section the density of the sediments were determined conventionally by the Geotechnical Institute at ETH using a 5 cm thick sample. These density values were used to calibrate the grey-scale values for each core section.

The diagram in Figure 5 shows an example of the density determination for an approximately 9 cm long profile along a typical core section. The average density for all counting squares is indicated, as well as the maximum and minimum density. Slice 1 is a typical example of an almost homogeneous core section. Slices 2 to 4 include some single mineral grains and holes,

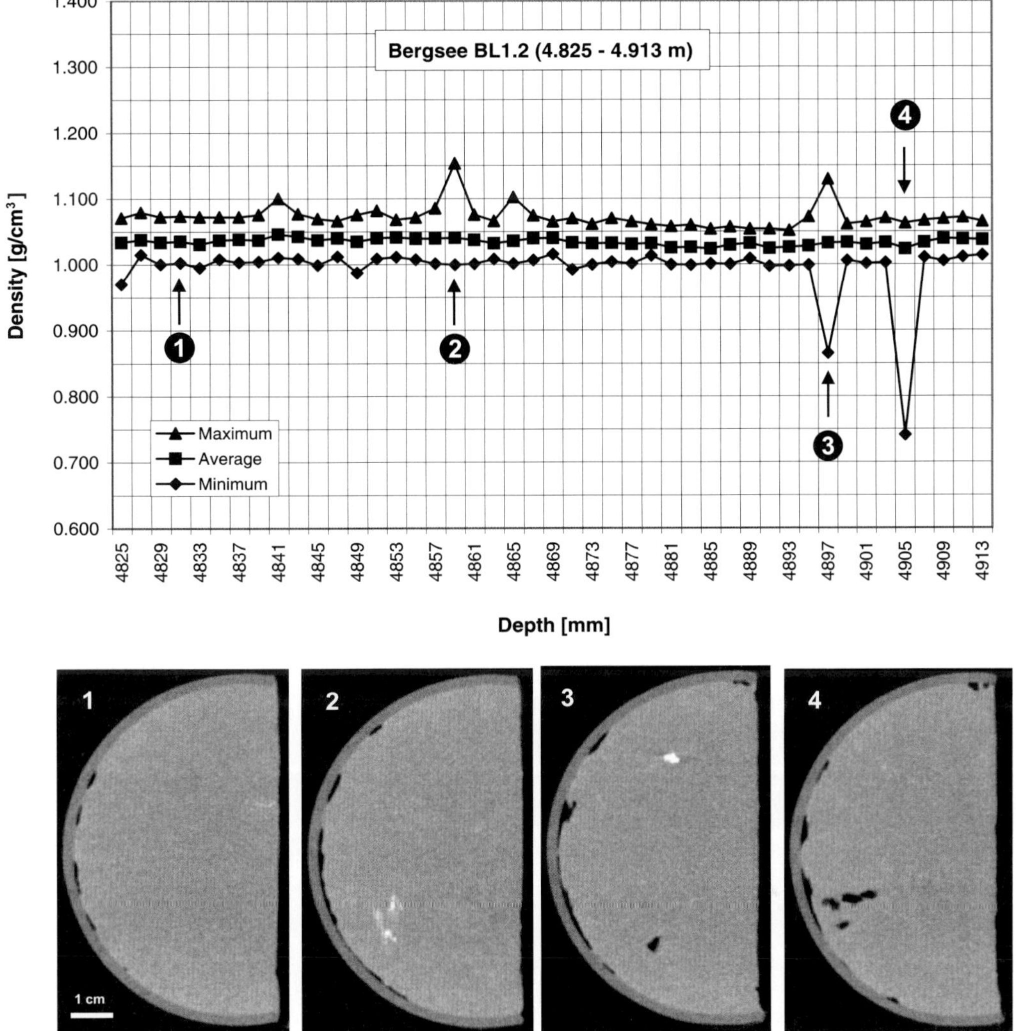

Fig. 5. Density profile of an approximately 9 cm long core section using calibrated grey-scale values determined by CT analysis (top), and four representative CT images (bottom), showing the effects of single mineral grains (white) and holes (black) on the density curves of a largely homogeneous organic lake deposit (grey).

recognized as light and dark patches, which influence the reported maximum or minimum densities. Because only few of these features occur, they do not strongly affect the average density values. No significant variations or trends in average density of the gyttja are recognized.

These measurements were carried out because gyttja, with its weak jelly-like consistency and very high water content, may have a structure that is susceptible to collapse following earthquake shocks. However, a general increase of the density which might have indicated such changes could not be found in the Lake Bergsee deposits.

Conclusions

Industrial CT has been applied to the investigation of sedimentary and deformation structures in lake drill cores. CT analysis, which is entirely non-destructive, is especially suited for the study of soft-sediment structures of this type.

It preserves (expensive) drill core samples for additional investigations and it is also a rapid method compared with classical methods based on thin section preparation.

High resolution CT images reveal various types of sedimentary structures in drill cores, which can be used for a palaeoenvironmental and palaeoseismological analysis of lake deposits. They also help to discriminate between drill core artefacts and natural structures, which is of special interest in palaeoseismological investigations using drill core samples. For the determination of physical properties of sediment samples, CT images with a lower resolution can be used. Calibrated grey-scale values supply continuous information about density variations in extremely thin core sections, which is a type of information that cannot be obtained by conventional geotechnical methods. In addition, the CT images supply direct information about the origin of the density changes, which is an advantage compared with classical methods of density determination for core samples using gamma ray radiography.

This research was funded by the SNF (Swiss National Science Foundation) and the ETH (Swiss Federal Institute of Technology). We gratefully acknowledge the measurements of soil mechanical properties of gyttja deposits from Bergsee by the Geotechnical Institute of ETH Zürich, F. Bucher and many inspiring discussions with C. Davenport, University of East Anglia. This is Institut für Geophysik, ETH Zürich, contribution no. 1236.

References

AMOS, C.L., SUTHERLAND, T.F., RADZIJEWSKI, B. & DOUCETTE, M. 1996. A rapid technique to determine bulk density of fine-grained sediments by x-ray computed tomography. *Journal of Sedimentary Research*, **66**, 1023–1024.

ANKETELL, J.M., CEGŁA, J. & DŻUŁYŃSKI, S. 1970. On the deformational structures in systems with reversed density gradients. *Annales de la Société Géologique de Pologne*, **XL**, 3–30.

BECKER, A., DAVENPORT, C.A., HAEBERLI, W., BURGA, C., PERRET, R., FLISCH, A. & KELLER, W.A. 2000. The Fulnau Landslide and former Lake Seewen in the northern Swiss Jura Mountains. *Eclogae geologicae Helvetiae*, **93**, 291–305.

BECKER, A., DAVENPORT, C.A. & GIARDINI, D. 2002. Palaeoseismicity studies on end-Pleistocene and Holocene lake deposits around Basle, Switzerland. *Geophysical Journal International*, **149**, 659–678.

DENISON, C., CARLSON, W.D. & KETCHAM, R.A. 1997. Three-dimensional quantitative textural analysis of metamorphic rocks using high-resolution computed X-ray tomography: part I. Methods and techniques. *Journal of Metamorphic Geology*, **15**, 29–44.

GUHMAN, A.I. & PEDERSON, D.T. 1992. Boiling sand-filled artesian springs, Dismal River, Nebraska: agents for geomorphic change and formation of vertical cylindrical structures. *Geology*, **20**, 8–10.

KAK, A.C. & SLANEY, M. 1998. *Principles of Computerized Tomographic Imaging*. IEEE Press, New York. World Wide Web Address: http://www.slaney.org/pct/

KETCHAM, R.A. & CARLSON, W.D. 2001. Acquisition, optimization and interpretation of X-ray computed tomographic imagery: applications to the geosciences. *Computers & Geosciences*, **27**, 381–400.

MAGNY, M. 1992. Holocene lake-level fluctuations in Jura and the northern subalpine ranges, France: regional pattern and climatic implications. *Boreas*, **21**, 319–334.

MOORE, J.F. 1990. Evolution of computed tomography. *Materials Evaluation*, **48**, 630–640.

NEUMANN-MAHLKAU, P. 1976. Recent sand volcanoes in the sand of a dike under construction. *Sedimentology*, **23**, 421–425.

ORSI, T.H., EDWARDS, C.M. & ANDERSON, A.L. 1994. X-ray computed tomography: a nondestructive method for quantitative analysis of sediment cores. *Journal of Sedimentary Research*, Section A: *Sedimentary Petrology and Processes*, **64**, 690–693.

RODRÍGUEZ-PASCUA, M.A., CALVO, J.P., DE VICENTE, G. & GÓMEZ-GRAS, D. 2000. Soft-sediment deformation structures interpreted as seismites in lacustrine sediments of the Prebetic Zone, SE Spain, and their potential use as indicators of eaarthquake magnitudes during the Late Miocene. *Sedimentary Geology*, **135**, 117–135.

WELLS, P., DAVIS, J. & MORGAN, M. 1994. Computed Tomography. *Materials Forum*, **18**, 111–113.

Analysis of analogue models by helical X-ray computed tomography

G. SCHREURS[1], R. HÄNNI[1], M. PANIEN[1] & P. VOCK[2]

[1] *Institute of Geological Sciences, University of Bern, Baltzerstrasse 1 CH-3012 Bern, Switzerland (e-mail: schreurs@geo.unibe.ch)*
[2] *Institute of Diagnostic Radiology, Inselspital, CH-3010 Bern, Switzerland*

Abstract: The aim of analogue model experiments in geology is to simulate structures in nature under specific imposed boundary conditions using materials whose rheological properties are similar to those of rocks in nature. In the late 1980s, X-ray computed tomography (CT) was first applied to the analysis of such models. In early studies only a limited number of cross-sectional slices could be recorded because of the time involved in CT data acquisition, the long cooling periods for the X-ray source and computational capacity. Technological improvements presently allow an almost unlimited number of closely spaced serial cross-sections to be acquired and calculated. Computer visualization software allows a full 3D analysis of every recorded stage. Such analyses are especially valuable when trying to understand complex geological structures, commonly with lateral changes in 3D geometry. Periodic acquisition of volumetric data sets in the course of the experiment makes it possible to carry out a 4D analysis of the model, i.e. 3D analysis through time. Examples are shown of 4D analysis of analogue models that tested the influence of lateral rheological changes on the structures obtained in contractional and extensional settings.

Analogue models have been used since the 19th century to simulate various kinds of geological structures in nature. The main objective of analogue models is to create a geometric analogue of a natural structure. Early analogue model experiments include those of Hall (1815), Favre (1878), Daubrée (1879) and Cadell (1890). Experiments on horizontal shortening were conducted by Cadell (1890), who described folding and thrusting of a layered model consisting of sand and plaster of Paris (Fig. 1) and suggested similarities with structures in the Highlands of Scotland and the Alps.

Analogue modelling offers the opportunity to determine the relation between imposed boundary conditions and resulting structures. Specific controlling parameters can be varied in order to assess their importance for the structure being considered. In analogue models, materials are used whose rheological properties are assumed to be similar to those of rocks in nature, e.g. sand, clay, paraffin, wax or silicone. One major advantage of analogue model experiments is the ability to analyse progressive deformation, from the initial undeformed state to the final deformed state with a spatial and temporal resolution that is not possible in natural field examples. The experiments are especially meant to stimulate the conception of testable hypotheses regarding the development of natural deformation structures. Results of analogue modelling have been applied to both small and large-scale natural structures.

Analogue models have provided important new insights into the geometry and progressive evolution of geological structures. In some cases, experimental studies proved to be fundamental for understanding natural deformation structures. For example, using clay models, Riedel (1929), Cloos (1928, 1930) and later Tchalenko (1970) and Wilcox *et al.* (1973) proposed an explanation for fractures and faults associated with strike-slip faulting. Experiments on folding by Biot *et al.* (1961) and Ramberg (1964) revealed that fold wavelength is a function of relative viscosity and bed thickness. Experiments by McClay & Ellis (1987), Ellis & McClay (1988) and McClay (1990), documented in detail fundamental fault geometries and roll-over anticlines within the hanging walls of listric extensional fault systems. More recently, analogue modelling by Chemenda *et al.* (1995) revealed possible mechanisms for syn-collisional exhumation.

Until the late 1980s, analogue models were analysed either by recording the surface evolution, by monitoring the structures through transparent side panels (where undesirable boundary effects are largest), and/or by physically cutting a series of sections through the consolidated model. Such methods of analysis did not allow a full 3D reconstruction of the model at the end of

From: MEES, F., SWENNEN, R., VAN GEET, M. & JACOBS, P. (eds) 2003. *Applications of X-ray Computed Tomography in the Geosciences*. Geological Society, London, Special Publications, **215**, 213–223. 0305-8719/03/$15.

Fig. 1. Experimental modelling of mountain-building processes by Cadell (1890), who shortened an initially horizontally layered sequence consisting of plaster of Paris and sand to investigate thrusting and folding. Reproduced by permission of the Royal Society of Edinburgh.

the experiment, nor was it possible to study the internal geometry of the model in the course of the experiment. This was especially a disadvantage when analysing the geometry and evolution of non-cylindrical structures. X-ray computed tomography (CT) overcomes this disadvantage because this technique makes it possible to visualize the interior of an analogue model without destroying it.

X-ray computed tomography

X-ray CT generates cross-sectional slices through an analogue model and thus allows a detailed analysis of its internal geometry and kinematics. X-ray attenuation is mainly a function of material density, atomic number and thickness. Layering in the analogue model is produced by using analogue materials with different attenuation values. The number of cross-sectional slices that can be obtained for a particular stage in the evolution of the model depends mainly on the X-ray dose intensity needed for proper visualization and the performance of the X-ray source. The time between consecutive runs is dictated by the required cooling of this source below a certain threshold value. Because of the time involved in CT data acquisition and the long cooling duration of the X-ray source, early medical scanners could only record a limited number of cross-sectional slices. Another limiting factor was the computational capacity needed to calculate images from projectional raw data profiles. Periodic acquisition of sections at similar positions during deformation made it nevertheless possible to follow the 2D evolution of structures in time. However, the time-consuming recording of closely spaced sequential cross-sectional slices, necessary for a full 3D analysis of the model, was generally not carried out until the end of the experiment (e.g. Mandl 1988; Colletta *et al.* 1991; Philippe 1995; Schreurs & Colletta 1998).

X-ray CT has in the past been used to study structures resulting from shortening and oblique shortening (Colletta *et al.* 1991; Wilkerson *et al.* 1992; Philippe 1994; Guillier *et al.* 1995; Letouzey *et al.* 1995; Philippe 1995; Mugnier *et al.* 1997; Schreurs 1997; Philippe *et al.* 1998; Schreurs & Colletta 1998; Nieuwland *et al.* 2000; Schreurs *et al.* 2001), extension and oblique extension (Mandl 1988; Naylor *et al.* 1994; Schreurs & Colletta 1998), strike-slip (Richard 1990; Schreurs 1994, 2003; Richard *et al.* 1995; Ueta *et al.* 2000), and inversion (Roure *et al.* 1992; Vially *et al.* 1994; Letouzey *et al.* 1995).

Recent technological improvements have resulted in more powerful X-ray CT scanners. The acquisition system of helical or spiral scanners (Fig. 2) revolves around the object as the object moves in the longitudinal direction, orthogonal to the scanning plane. In this manner, 3D volume raw data on analogue models are easily acquired. This technique makes it possible to calculate an almost unlimited number of closely spaced serial cross-sections retrospectively from the raw data. Data acquisition time for such a 3D data set depends on the X-ray dose necessary to adequately penetrate the analogue material (i.e. material composition and thickness), detector quality and the performance of the entire acquisition system (e.g. rotation time, table speed,

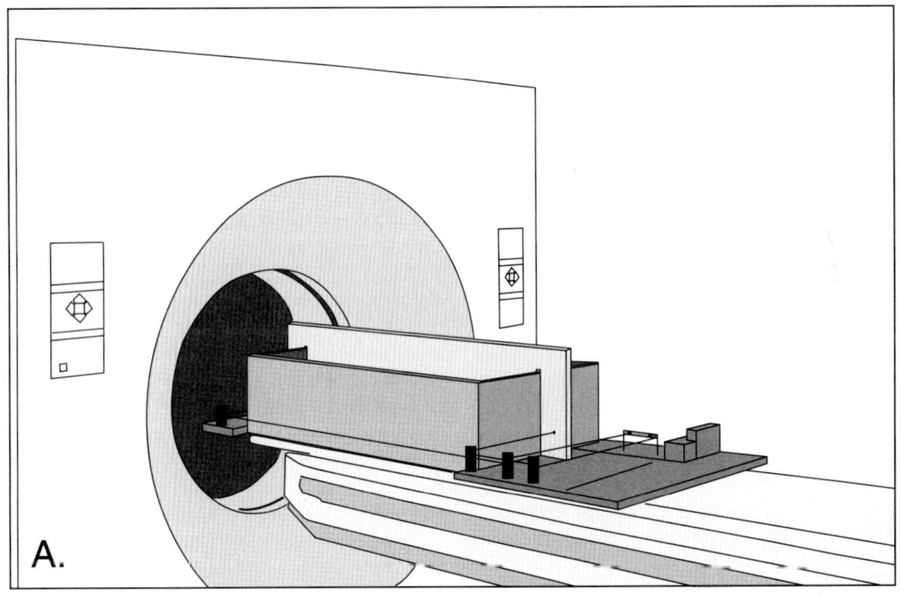

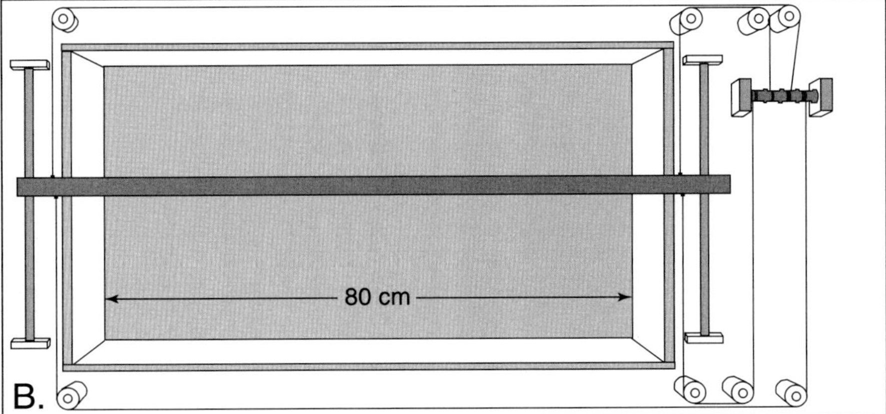

Fig. 2. (**a**) Helical X-ray CT scanner and experimental apparatus. During rotation of the acquisition system the experimental apparatus is translated horizontally through the field of investigation making it possible to obtain three-dimensional volume raw data. (**b**) Top view of experimental apparatus used to test the influence of the shape and position of viscous layers during shortening or extension of a brittle-viscous analogue model.

data transfer, computation speed). Slice spacing will only affect the post-processing time. Periodic acquisition of such volumetric data sets makes it possible to follow the 3D evolution of models from the initial undeformed stage to the final deformed stage. This opens new and exciting perspectives for a complete 4D analysis (3D through time) of analogue models (Schreurs *et al.* 2001). Using computer visualization software, the digital data also allow the reconstruction of any desired section, be it horizontal, vertical or oblique.

Experimental procedure

In normal gravity experiments, dry granular materials with low cohesion are commonly used as analogue materials for brittle rocks in the upper crust because they obey the Mohr-Coulomb criterion of failure. Viscous materials are generally used to simulate viscous flow of evaporites in the upper crust or rocks in the lower crust. X-ray attenuation values of different analogue materials were first determined by Colletta *et al.* (1991), who selected sand, glass powder and glass microbeads with an average grain size of about 100 μm for brittle modelling experiments. Attenuation values of granular materials depend mainly on mineral composition of the grains, on grain size and on compaction (Colletta *et al.* 1991).

X-ray attenuation values of several granular materials with grain sizes between 80 and 200 μm were tested for potential use in analogue modelling experiments (Table 1). Quartz sand (80–200 μm) and corundum sand ('Normalkorund', 80–125 μm) were selected as brittle analogue materials. The angles of internal friction of quartz and corundum sand are about 35° and 37°, respectively. These materials can be used to simulate upper crustal rocks that have comparable angles of internal friction at low normal stresses (Byerlee 1978). As viscous analogue material, polydimethyl-siloxane (PDMS) was used, which is a non-toxic, transparent material with a linear viscosity of 5×10^4 Pa · s at room temperature and at strain rates below $3 \times 10^{-3}\,s^{-1}$ (Weijermars 1986). The different attenuation values for PDMS (70 Hounsfield Units, HU), quartz sand (870 HU) and corundum sand (1240 HU) make it possible to construct layered brittle and brittle-viscous models. Attenuation values of adjacent analogue materials should differ by at least 200 HU for optimal visualization of the boundaries between the two materials. During deformation of the model, faults develop in granular materials. Faults correspond to dilatant zones along which the grains have been perturbed. Dilatancy along faults results in a lower density and thus in a reduction of the attenuation compared to the surrounding unfaulted material. This contrast in attenuation makes faults clearly visible in CT images.

The experimental set-up includes a rectangular wooden box with a length of 80 cm, a width of 43 cm and a height of 17 cm (Fig. 2). A vertical mobile wall driven by a motor produces either extension or shortening. In contrast to wood, metal objects strongly attenuate X-rays and cause important artefacts in image computing.

Table 1. *Material parameters and distributors of analogue materials*

Material	Grain size (μm)	Density (g/cm^3)	Attenuation value (HU)	ϕ_{peak} (°)	Distributor
Air	–	0.0013	−1000	–	–
Water	–	1.0	0	–	–
Quartz sand	80–200	1.56*	870	35	(1)
Quartz sand (ref. BDF Gallardon 28)	~140	~1.45*	740	–	(2)
Corundum sand ('Edelkorund')	105–149	1.89*	1070	36	(1)
Corundum sand ('Normalkorund')	88–125	1.90*	1240	37	(1)
Glasspowder ('G2', ref. 585-10)	88–125	1.09*	360	36	(3)
Glasspowder (recycled glass, white)	100–210	1.24*	690	42	(1)
Glass microbeads	70–110	1.48*	950	22	(4)
Polydimethylsiloxane (GUM SGM 36)	–	0.96	70	–	(5)

Asterisk indicates that the material was sifted into a container before density measurement. Average attenuation values are in Hounsfield units (HU), measured with a Toshiba Asteion CT scanner, at 120 kV X-ray tube energy. ϕ_{peak} is the angle of internal peak friction, measured for all granular materials (except quartz sand, ref. BDF Gallardon 28) using a ring-shear tester. (1) www.carloag.ch; (2) Ets. Bervialle, 11 rue Teheran, F-75008 Paris; (3) V.R.C., rue des Tilleuls, F-95870 Bezons; (4) www.worf.de; (5) S.T.I./Dow Corning, 19 rue d'Arsonval, F-69680 Chassieu.

For this reason, the motor was placed adjacent to the wooden box. Analogue models were analysed by single- or multi-slice helical X-ray CT scanners. The Siemens Somatom Plus 4 single-slice helical scanner used a 120 kV voltage, a 130 mA current and 2 mm X-ray beam width to differentiate analogue material attenuation. Rotation time of source and detector was 1 s and table feed per rotation was 4.2 mm. During one rotation, the X-ray fan examines a slice of the object that is 2 mm thick. The diameter of the circular opening of the X-ray scanner was 55 cm, but the area of investigation is a circle 45 cm in diameter. In contrast to the single-slice scanner, the Toshiba Asteion multi-slice helical CT scanner investigates four slices through the model simultaneously during one rotation. Each slice has a selectable thickness between 0.5 and 8 mm. We used a 120 kV voltage, 100 mA current and 8 mm beam width. During one rotation, four 2 mm slices were examined. Rotation time was 1 s and table feed per rotation was 11 mm. The multi-slice scanner has a circular opening 50 cm in diameter, whereas the field of investigation is a circle 40 cm in diameter. There is no noticeable difference in image quality or resolution between the two instruments used. However, the advantage of the multi-slice CT is that this scanner allows the acquisition of a larger 3D volume raw data set during one scan.

Deformation of the analogue model takes place in the investigation field of the helical X-ray CT scanner. 3D volumetric data sets were acquired at regular time intervals in order to follow the 3D evolution of models from the initial undeformed stage to the final deformed stage. From the 3D volume raw data, cross-sectional slices with 2 mm spacing were retrospectively computed. Attenuation values were computed for a 512×512 matrix of volume elements (voxels). The computed values produce a 512×512 pixel grey-scale image. Digital image processing and computer visualization software allow the reconstruction of sections in any direction through a 3D data set and the production of movies that show, for example, successive serial cross-sections through one particular deformation stage or the temporal evolution of model deformation in 2D or 3D.

Using brittle-viscous analogue models, we investigated the influence of the shape and position of a viscous layer on the resulting structures during shortening and extension. In the shortening experiment, initial width of the model was 27 cm and its initial thickness was 3 cm. In the extension experiment, initial width and thickness were 22.5 cm and 3.5 cm, respectively. Very high width/thickness ratios should be avoided, as these result in poorly defined images, because image computation is optimized for objects with round cross-sections.

Modelling results

Influence of basal rheological changes during shortening

The experimental set-up for the shortening experiment is shown in Figure 3. A rectangular layer of viscous PDMS (simulating an evaporite layer), 40 cm long, 17 cm wide and 5 mm thick was placed on top of the wooden base next to the mobile wall. Alternating layers of quartz and corundum sand (simulating a sequence of competent sedimentary rocks) were poured on top.

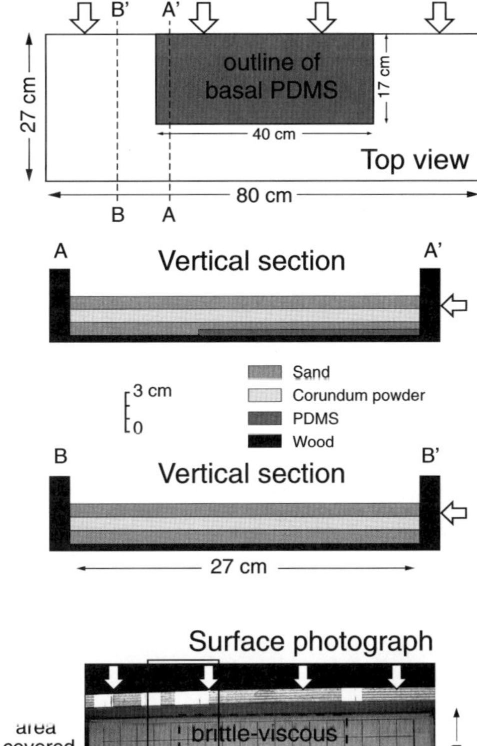

Fig. 3. Initial experimental set-up for testing the influence of basal rheological changes during shortening, illustrated by a top view, two vertical sections and surface photograph.

Purely for descriptive purposes, we refer to the domain containing the quartz-corundum-quartz sand layers as the brittle domain and to the domain with both quartz-corundum-quartz sand layers and a basal viscous PDMS layer as the brittle-viscous domain. Initially, the undeformed analogue model was 27 cm wide. By displacing the vertical mobile wall the model was shortened by 9 cm. CT 3D raw data acquisition was performed at the initial undeformed stage and at every full cm of progressive shortening.

There was a marked difference in structural evolution between the brittle and brittle-viscous domains (Fig. 4; Movies A-1 and A-2). In the brittle domain, thrust faults were closely spaced and a narrow and high fold-and-thrust belt formed. The sequence of thrusting propagated forward and the belt had a dominant vergence of thrusts and associated folds. In the brittle-viscous domain, however, the thrust belt was wider and lower, and there was no consistent vergence of thrusts and folding. A transfer zone formed in the transition zone between the brittle and brittle-viscous domains (Fig. 5; Movies A-3 and A-4). It linked thrusts that formed at different locations and stages in the evolution of the model. The transfer zone rooted in the viscous layer and its surface strike was parallel to the orientation of the boundary between viscous and brittle material at depth, and parallel to the shortening direction. The dip of the transfer zone was shallow (<30°) and the angle of dip changed along strike (Fig. 6; Movie A-5). Forward thrusting in the brittle domain propagated along strike into the brittle-viscous domain and resulted in out-of-sequence thrusts in the latter domain. Back thrusts forming above the basal PDMS interfere with forward thrusts that developed in the purely brittle domain. These laterally propagating forward and backward thrusts contributed to the complexity of the fault pattern (Movie A-6). The experiments suggest that location and orientation of transfer zones in nature may be controlled by basal rheological changes.

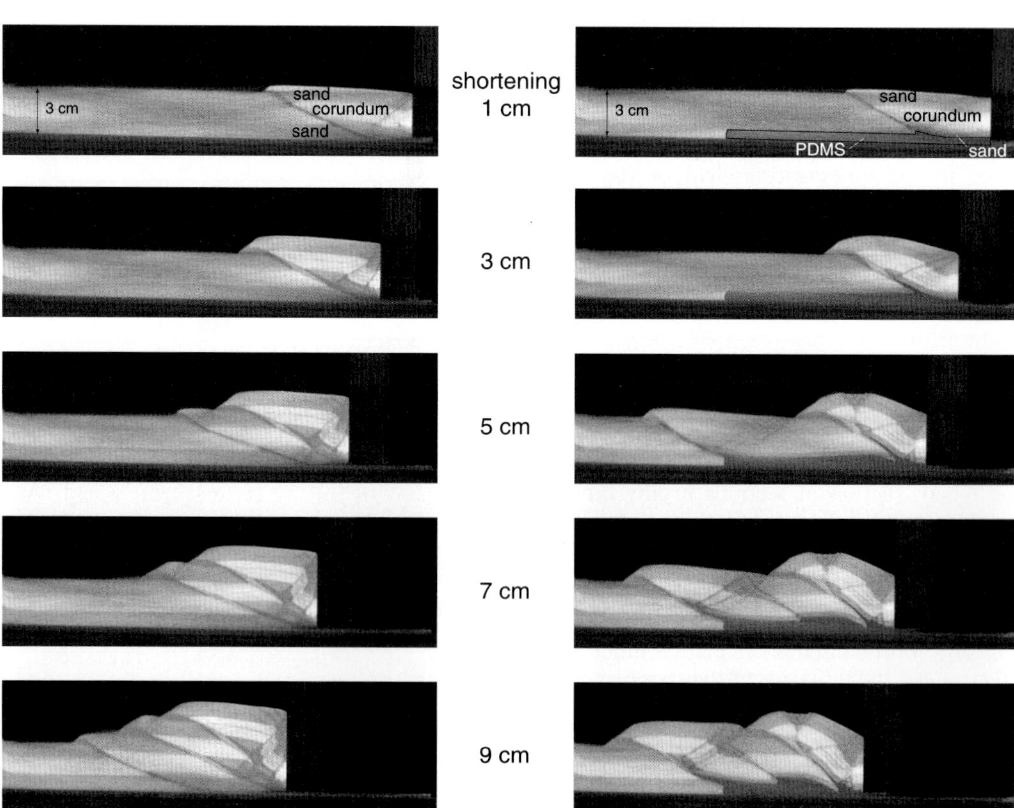

Fig. 4. Comparison of structural evolution between brittle and brittle-viscous domains for the shortening experiment by successive transverse computed X-ray CT images of the progressively deformed model.

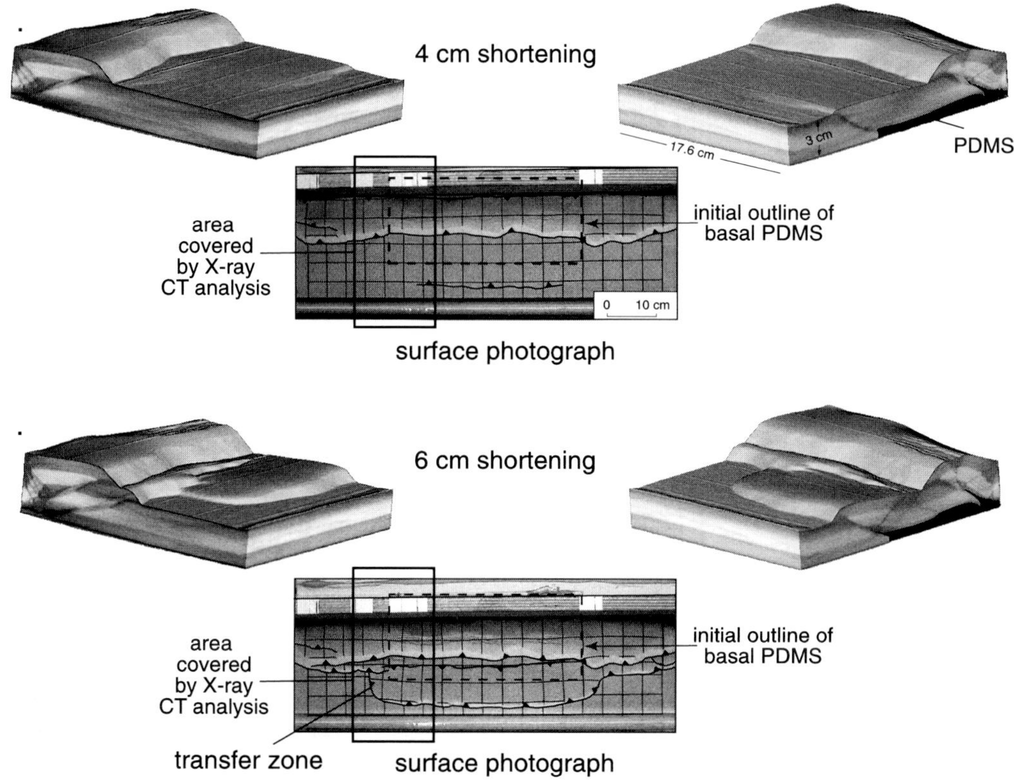

Fig. 5. Surface photograph with superposed interpretation and two oblique 3D views, shown for two consecutive stages of the shortening experiment, illustrating the development of a transfer zone, at 4 cm shortening (**a**) and 6 cm shortening (**b**). The rectangle indicates the area of three-dimensional analysis of one particular stage. The dashed rectangle indicates the initial outline of the basal viscous silicone.

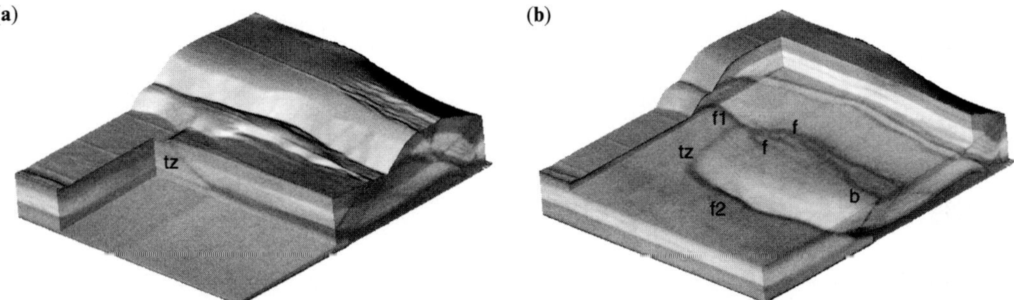

Fig. 6. Two three-dimensional views of an analogue model at 6 cm shortening. (**a**) Longitudinal section (perpendicular to shortening direction) showing a shallow dip angle of the transfer zone (lateral ramp). (**b**) Horizontal section showing the position of the transfer zone, which connects the frontal ramp in the brittle domain, (f1), with the frontal ramp in the brittle-viscous domain, (f2); back thrusts in brittle-viscous domain disappear laterally or interfere with forward thrusts that originated in the brittle domain and propagated sideways into the brittle-viscous domain; tz = transfer zone (lateral ramp); f = forward directed thrust; b = back thrust.

Influence of lateral rheological changes during extension

Analogue modelling on the influence of lateral rheological changes during extension was done with the experimental set-up shown in Figure 7. The base of the rectangular wooden box is overlain by an alternation of nine plexiglass bars (each bar is 5 mm wide, 5 cm high and 80 cm long) and eight foam bars (each bar is 3 cm wide, 5 cm high

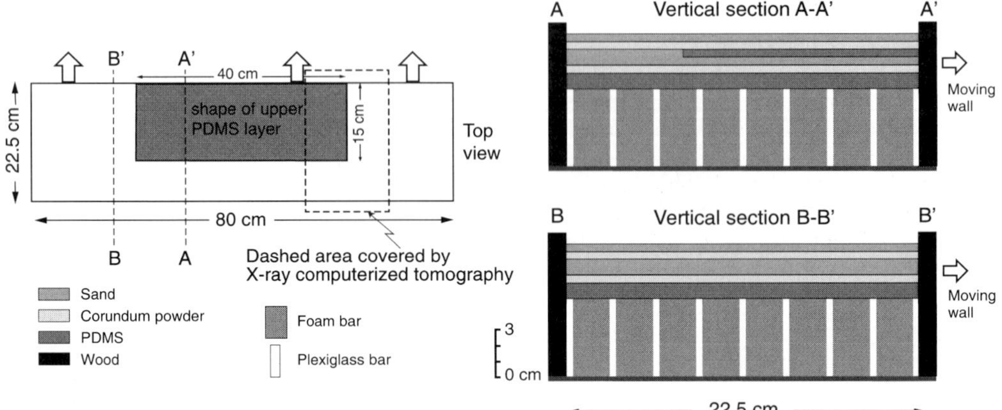

Fig. 7. Initial experimental set-up for testing the influence of a viscous layer embedded in brittle strata during extension of a brittle-viscous multilayer model.

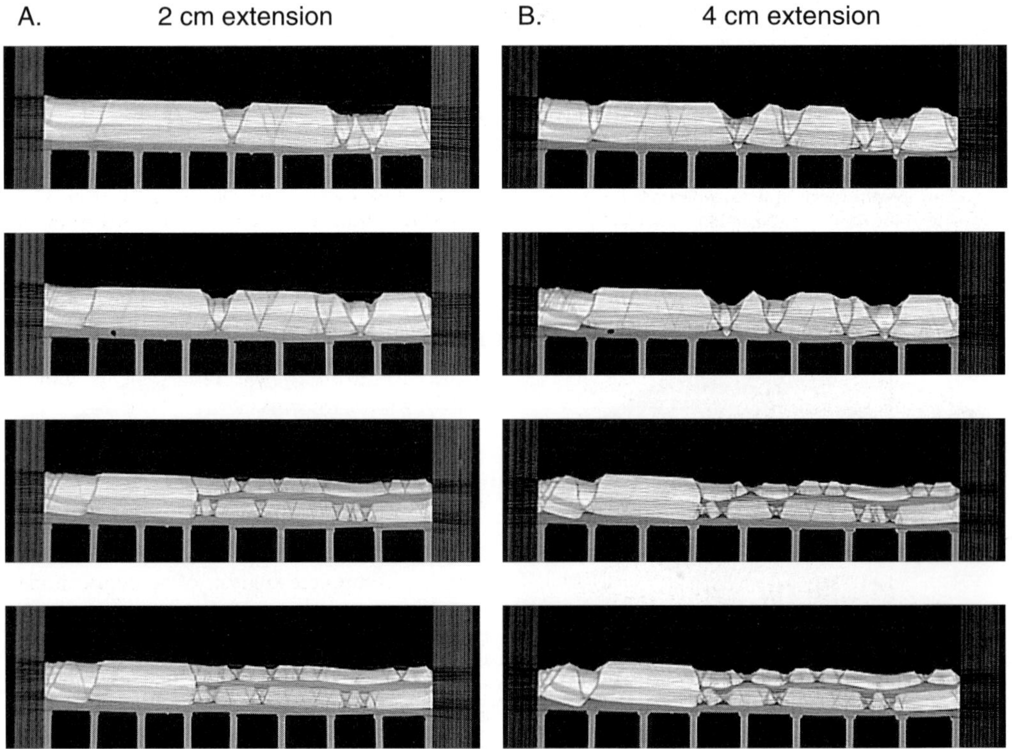

Fig. 8. Vertical sections through the brittle-viscous multilayer model at 2 cm extension, (**A**), and 4 cm extension, (**B**). The upper two sections are taken from the domain with only a basal viscous layer; the lower two sections are taken from the domain with two viscous layers. Note that the presence of a second viscous layer caused the development of decoupled conjugate normal fault systems in the upper and lower brittle compartments.

and 80 cm long). Before constructing the stratified brittle-viscous analogue model, the assemblage of plexiglass and foam bars was shortened by 6 cm. In the shortened state, the width of each foam bar was reduced to circa 2.25 cm, whereas the width of the plexiglass bars remained unchanged. The brittle-viscous multilayer model was then constructed on top of the shortened assemblage and consisted of a 1 cm thick basal viscous layer and an upper, smaller viscous layer embedded in brittle strata made up of corundum and quartz sand (Fig. 7). The upper viscous layer was placed near the extending mobile wall and covered only part of the underlying brittle layers. The presence of foam bars alternating with plexiglass bars in combination with the overlying basal viscous layer prevented localization of deformation near the extending mobile wall. The initial width of the model was 22.5 cm and total extension was 5 cm. During extension, the foam bars 'decompressed' and deformation

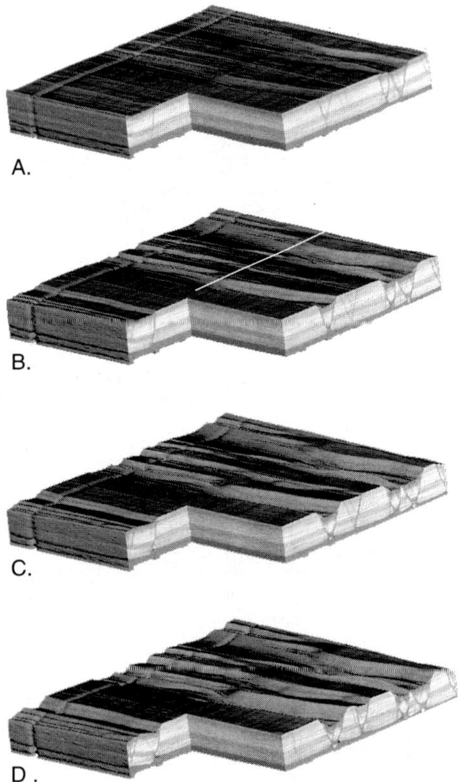

Fig. 9. Progressive evolution of structures at 1 cm (**a**), 2 cm (**b**), 3 cm (**c**) and 4 cm (**d**) extension, illustrated by 3D views. The white line in (**b**) indicates the position of the extensional transfer zone. The discontinuity near the left-hand side of each block results from missing cross-sections.

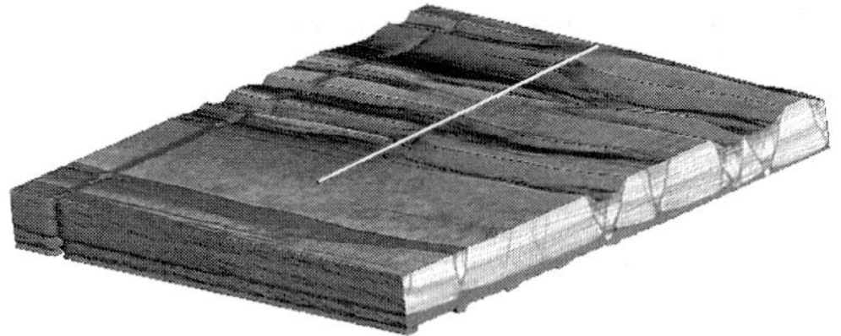

Fig. 10. Horizontal section through a 3D view of the analogue model at 4 cm extension. The location of the extensional transfer zone (indicated by a white line) is directly linked to the lateral termination of the interbedded viscous layer.

in the model was distributed over the width of the model. CT 3D raw data were acquired of the initial undeformed stage and at every 5 mm of progressive extension.

The shape and position of the upper viscous layer had a profound influence on the structural style that developed in the course of the experiment. The upper viscous layer embedded within brittle strata initially caused the development of two independent, decoupled conjugate normal fault systems in upper and lower brittle compartments (Fig. 8; Movies B-1 and B-2). The width between conjugate normal faults reflects the depth to detachment (viscous layer). In the part of the model where there was only one basal viscous layer, the width between conjugate faults was large, whereas in the part of the model with a second viscous layer embedded in brittle layers, this width was small (Fig. 9; Movie B-3). The location and orientation of extensional transfer zones, visible in horizontal sections (Fig. 10; Movie B-4) and in surface view, was directly linked to the geometry of the interbedded viscous layer. These transfer zones connect widely spaced conjugate normal fault systems with less widely spaced systems.

Concluding remarks

This paper shows the great potential of CT scanning for a complete analysis of analogue models. The 3D volumetric raw data allow the computation of contiguous cross-sectional slices parallel to the transport direction. Computer visualization software makes it possible to reconstruct the analogue model in three dimensions and to reconstruct sections in any direction. On the basis of 3D reconstructions, animations can be created in order to study the spatial evolution of structures at a specific stage, or to study the 4D evolution of the model.

CT volume scans are especially important when modelling complex structural regimes in which lateral changes in 3D geometry are common. 3D imaging of models can provide constraints for interpretations of complex zones based on 2D and 3D seismic data, which are often fragmentary or difficult to interpret. Understanding the 3D evolution through time can help geologists and seismic interpreters in developing kinematic models.

The experiments suggest that the position and orientation of transfer zones in nature might be directly linked to initial lateral rheological changes within a layered sequence. Basal rheological changes in the shortening experiment resulted in shallowly dipping transfer zones (lateral ramps) that root directly at the lateral contact between viscous and brittle layer. In the extensional experiment, vertical transfer zones formed directly above the lateral boundary between upper viscous layer and adjacent brittle layers.

Movies

Ten short movies on the accompanying CD-ROM (back cover of this Special Publication) illustrate the structural evolution of the shortening and extension experiment. For more details see text file on CD-ROM.

Funding by Hochschulstiftung Bern and Swiss National Science Foundation Grant 2000-0554.11 98/1 is gratefully acknowledged. The manuscript benefited from constructive reviews by M. Sintubin and A. Vervoort. H.-P. Bärtschi is thanked for technical assistance, A. Schneider and C. Seiler for assistance in CT data acquisition, and J. Lohrmann, N. Kukowski and O. Oncken for allowing M. Panien to use the ringshear tester at the GeoForschungsZentrum in Potsdam to measure mechanical parameters of our granular analogue materials.

References

BIOT, M.A., ODÉ, H. & ROEVER, W.L. 1961. Experimental verification of the theory of folding of stratified viscoelastic media. *Geological Society of America Bulletin*, **72**, 1621–1632.

BYERLEE, J.D. 1978. Friction of rocks. *Pure Applied Geophysics*, **116**, 97–115.

CADELL, H.M. 1890. Experimental researches in mountain building. *Transactions of the Royal Society of Edinburgh*, **35**, 337–357.

CHEMENDA, A.I., MATTAUER, M., MALAVIEILLE, J. & BOKUN, A.N. 1995. A mechanism for syncollisional rock exhumation and associated normal faulting: Results from physical modelling. *Earth and Planetary Science Letters*, **132**, 225–232.

CLOOS, H. 1928. Experimente zur inneren Tektonik. *Zentralblatt für Mineralogie, Geologie und Paläontologie*, **12**, 609–621.

CLOOS, H. 1930. Zur experimentellen Tektonik V. Vergleichende Analyse dreier Verschiebungen. *Geologische Rundschau*, **21**, 353–367.

COLLETTA, B., BALE, P., BALLARD, J.F., LETOUZEY, J. & PINEDO, R. 1991. Computerized X-ray tomography analysis of sandbox models: examples of thin-skinned thrust systems, *Geology*, **19**, 1063–1067.

DAUBRÉE, A. 1879. *Etude synthétique de géologie expérimentale*. Dunod, Paris.

ELLIS, P.G. & MCCLAY, K.R. 1988. Listric extensional fault systems. Results of analogue model experiments. *Basin Research*, **1**, 55–70.

FAVRE, A. 1878. The formation of mountains. *Nature*, **19**, 103–106.

GUILLIER, B., BABY, P., COLLETTA, B., MENDEZ, E., LIMACHI, R., LETOUZEY, J. & SPECHT, M. 1995. Analyse géometrique et cinématique d'un 'duplex' issu d'un modèle analogique visualisé en 3D par tomographie aux rayons X. *Comptes Rendus de l'Academie des Sciences, Serie II*, **321**, 901–908.

HALL, J. 1815. On the vertical position and convolution of certain strata and their relation with granite. *Transactions of the Royal Society of Edinburgh*, **7**, 79–108.

LETOUZEY, J., COLLETTA, B., VIALLY, R. & CHERMETTE, J.C. 1995. Evolution of salt-related structures in compressional settings. *In*: JACKSON, M.P.A., ROBERTS, D.G. & SNELSON, S. (eds) *Salt Tectonics: A Global Perspective*. American Association Petroleum Geologists Memoir, **65**, 41–60.

MANDL, G. 1988. *Mechanics of Tectonic Faulting: Models and Basic Concepts*. Elsevier, Amsterdam.

MCCLAY, K.R. 1990. Extensional fault systems in sedimentary basins: a review of analogue model studies. *Marine and Petroleum Geology*, **7**, 206–233.

MCCLAY, K.R. & ELLIS, P.G. 1987. Geometries of extensional fault systems developed in model experiments. *Geology*, **15**, 341–344.

MUGNIER, J.L., BABY, P., COLLETTA, B., VINOUR, B., BALE, P. & LETURMY, P. 1997. Thrust geometry controlled by erosion and sedimentation: a view from analogue models, *Geology*, **25**, 427–430.

NAYLOR, M.A., LARROQUE, J.M. & GAUTHIER, B.D.M. 1994. Understanding extensional tectonics: insights from sandbox models. *In*: ROURE, F. (ed.) *Geodynamic Evolution of Sedimentary Basins*. Editions Technip, Paris, 69–83.

NIEUWLAND, D.A., LEUTSCHER, J.H. & GAST, J. 2000. Wedge equilibrium in fold-and-thrust belts: prediction of out-of-sequence thrusting based on sandbox experiments and natural examples. *Geologie en Mijnbouw, Netherlands Journal of Geosciences*, **79**, 81–91.

PHILIPPE, Y. 1994. Transfer zone in the southern Jura thrust belt (eastern France): geometry, development and comparison with analogue modelling experiments. *In*: MASCLE, A. (ed.) *Hydrocarbon and Petroleum Geology of France*. European Association of Petroleum Geologists, **4**, 327–346.

PHILIPPE, Y. 1995. *Rampes Latérales et Zones de Transfert dans les Chaînes Plissées: Géometrie, Conditions de Formations et Pièges Structuraux Associés*. PhD thesis, Université de Savoie.

PHILIPPE, Y., DEVILLE, E. & MASCLE, A. 1998. Thin-skinned inversion tectonics at oblique basin margins: example of the western Vercors and Chartreuse Subalpine massifs (SE France). *In*: MASCLE, A., PUIGDEFÀBREGAS, C., LUTERBACHER, H.P. & FERNANDEZ, M. (eds) *Cenozoic Foreland Basins of Western Europe*. Geological Society, London, Special Publications, **134**, 239–262.

RAMBERG, H. 1964. Selective buckling of composite layers with contrasted rheological properties: a theory of simultaneous formation of several orders of folds. *Tectonophysics*, **1**, 307–341.

RICHARD, P. 1990. Champs de failles au dessus d'un décrochement de socle: modélisation expérimentale. *Mémoires et Documents du Centre Armoricain d'Etude Structurale des Socles*, **34**.

RICHARD, P., NAYLOR, M.A. & KOOPMAN, A. 1995. Experimental models of strike-slip tectonics, *Petroleum Geoscience*, **1**, 71–80.

RIEDEL, W. 1929. Zur Mechanik geologischer Brucherscheinungen. *Zentralblatt für Mineralogie, Abteilung B*, 354–368.

ROURE, F., BRUN, J.P., COLLETTA, B. & VAN DEN DRIESSCHE, J. 1992. Geometry and kinematics of extensional structures in the Alpine Foreland Basin of southeastern France. *Journal of Structural Geology*, **14**, 503–519.

SCHREURS, G. 1994. Experiments on strike-slip faulting and block rotation. *Geology*, **22**, 567–570.

SCHREURS, G. 1997. Experiments on faulting in zones of oblique shortening. *The Leading Edge*, 1159–1163.

SCHREURS, G. 2003. Fault development and interaction in distributed strike-slip shear zones: an experimental approach. *In*: STORTI, F., HOLDSWORTH, R.E. & SALVINI, F. (eds) *Intraplate Strike-slip Deformation Belts*. Geological Society, London, Special Publications, **210**, 35–52.

SCHREURS, G. & COLLETTA, B. 1998. Analogue modelling of faulting in zones of continental transpression and transtension. *In*: HOLDSWORTH, R.E., STRACHAN, R.A. & DEWEY, J.F. (eds) *Continental Transpressional and Transtensional Tectonics*. Geological Society, London, Special Publications, **135**, 59–79.

SCHREURS, G., HÄNNI, R. & VOCK, P. 2001. 4-D Analysis of analog models: experiments on transfer zones in fold-and-thrust belts. *In*: KOYI, H.A. & MANCKTELOW, N.S. (eds) *Tectonic Modeling: A Volume in Honor of Hans Ramberg*. Geological Society of America Memoir, **193**, 179–190.

TCHALENKO, J.S. 1970. Similarities between shear zones of different magnitudes. *Geological Society of America Bulletin*, **81**, 1625–1640.

UETA, K., TANI, K. & KATO, T. 2000. Computerized X-ray tomography analysis of three-dimensional fault geometries in basement-induced wrench faulting. *Engineering Geology*, **56**, 197–210.

VIALLY, R., LETOUZEY, J., BÉNARD, F., HADDADI, N., DESFORGES, G., ASKRI, H. & BOUDJEMA, A. 1994. Basin inversion along the North African margin of the Saharan Atlas (Algeria). *In*: ROURE, F. (ed.) *Peri-Tethyan Platforms*. Editions Technip, Paris, 79–118.

WEIJERMARS, R. 1986. Flow behaviour and physical chemistry of bouncing putties and related polymers in view of tectonic laboratory applications. *Tectonophysics*, **124**, 325–358.

WILCOX, R.E., HARDING, T.E. & SEELY, D.R. 1973. Basic wrench tectonics. *American Association of Petroleum Geologists Bulletin*, **57**, 74–96.

WILKERSON, M.S., MARSHAK, S. & BOSWORTH, W. 1992. Computerized tomographic analysis of displacement trajectories and three-dimensional fold geometry above oblique thrust ramps. *Geology*, **20**, 439–442.

Preliminary microfocus X-ray computed tomography survey of echinoid fossil microstructure

S. R. STOCK[1] & A. VEIS[2]

[1] *Institute for Bioengineering and Nanoscience in Advanced Medicine, Northwestern University, Chicago, IL 60611–3008, USA (e-mail: s-stock@northwestern.edu)*

[2] *Department of Cell and Molecular Biology, Northwestern University Medical School, Chicago, IL 60611, USA*

Abstract: Microfocus X-ray computed tomography (μCT), a high resolution variant of medical CT, was used to non-invasively examine Jurassic echinoid fossils with spine, demipyramid and test plate fragments serving to assess the extent to which the microstructure remained unaffected by diagenesis. The sizes of the calcite-crystal stereom remains dictated the resolution that could be obtained. The smaller diameter spines were imaged with $9 \times 9 \times c.\ 25$ μm voxels; the larger demipyramids and test plate fragments were imaged with 13×13 μm and 17×17 μm voxels, respectively, in the plane of reconstruction and with a proportionally larger slice thickness. The stereom structure was not seen in the μCT slices of the fossils. This is not surprising because, for the ossicles studied, the stereom pore dimensions are, for the most part, smaller than the resolution of the μCT system. Tide marks and other low absorption features were found in the spines and appear to be related to diagenetic changes. What appear to be cleavage cracks were observed in the demipyramid and some of the pores for tube feet could be seen in the test fragment.

Understanding patterns of mineralized tissue formation and organization in modern echinoids benefits greatly from comparison with fossil predecessors. The survey of sea urchin fossils reported below is motivated by an ongoing study of mineral distribution in teeth of the short-spined sea urchin *Lytechinus variegatus*. Underlying the interest in sea urchin tooth mineral is the hypothesis that many of the tooth-associated proteins are highly conserved, in an evolutionary sense, between the echinoderms and mammals (Veis *et al.* 1986). While the available protein-derived data on sea urchin spicules support this view (Kitajima *et al.* 1996; Ameye *et al.* 1999, 2000), it is nonetheless remarkable given that the mineral associated with the proteins changes from high magnesium calcite in sea urchins ($Ca_{1-x}Mg_xCO_3$, with $c.\ 0.05 < x < 0.45$, depending on species and ossicle type, with no Mg ordering evident in X-ray diffraction patterns for any x) to calcium phosphate-based minerals in mammals. This paper focuses on fossil sea urchins and assesses the use of microfocus X-ray computed tomography (μCT) in palaeontology.

Fossils do not invariably preserve the microstructure of mineralized tissue. Several stages of diagenesis affect what remains and one view of the process (Towe & Hemleben 1976) is as follows. The first phase corresponds to magnesium loss to $x < 0.04$, with no significant change in calcite texture. Note that the incorporation of magnesium significantly strains the calcite lattice and if the organism employed magnesium-related microstrain to strengthen the mineral crystals (i.e. solution hardening as employed in engineering alloys), this information would be lost in the first, earliest stage of diagenesis. The magnesium can be removed from ossicles entirely or can congregate locally to form microparticles of dolomite (e.g. MacQueen & Ghent 1970; MacQueen *et al.* 1974; Dickson 2001). The second stage involves recrystallization of grains into a more tightly welded mosaic of equiaxed grains, but biogenic textures (outlines of plates and other ossicles) remain well-preserved. Crystallite size effects (i.e. diffraction peak broadening, see Cullity & Stock 2001) which persisted through the first stage of diagenesis, would be eliminated. In the third stage, the onset of a coarser mosaic of microspar calcite tends to obliterate any remaining structure.

The calcitic ossicles of the sea urchin may be divided into three groups: the plates of the test; the spines radiating outward from the test and

From: MEES, F., SWENNEN, R., VAN GEET, M. & JACOBS, P. (eds) 2003. *Applications of X-ray Computed Tomography in the Geosciences*. Geological Society, London, Special Publications, **215**, 225–235. 0305-8719/03/$15.

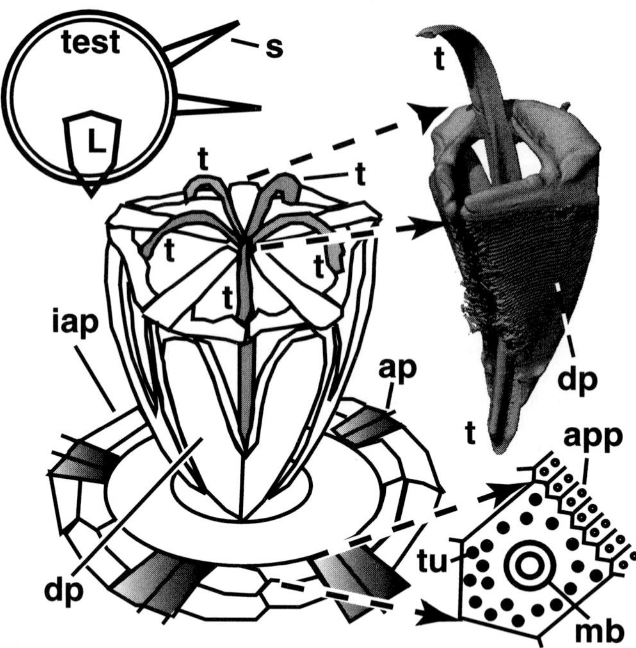

Fig. 1. Geometry of the Aristotle's lantern and portions of the test of a regular sea urchin. Upper left – side view of the sea urchin test showing spines, 's', and lantern, 'L'. Lower left – structure of the lantern, opening in the test through which the lantern protrudes during feeding and interambulacral, 'iap', and ambulacral, 'ap', plates surrounding the opening; due to their small size, the individual ambulacral plates are not shown but are indicated by shading; one of a pair of complementary demipyramids, 'dp', and all five teeth, 't', of the lantern are labelled. Upper right – three-dimensional rendering of a single pyramid of *Lytechinus variegatus* based on μCT data. Lower right – schematic of the fossilized test fragment from which the μCT data in Figures 6 and 7 were obtained; the interambulacral plate, the bordering ambulacral plates, the mamelon and boss, 'mb', the tubercules, 'tu', and ambulacral pore pairs, 'app', are shown in this view of the fragment from outside of the test.

the jaw structure, or Aristotle's lantern (Fig. 1). Except for the teeth, virtually all of the ossicle volume consists of more or less equal volume fractions of stroma (various soft tissues and fluids) and stereom (fine fenestrated or trabecular structure). The plates of the test form a spheroidal or ellipsoidal armored structure around the sea urchin's vital organs and the simple schematic of Figure 1 (upper left) shows the relationship between test spines, 's', and lantern, 'L', as seen from the side of the test. The pentaradiate lantern consists of five pyramids, to which the five teeth, 't', firmly attach and other ossicles that are not attached (Fig. 1, main drawing). Sets of interambulacral plates, 'iap', alternate with groups of ambulacral plates, 'ap', and surround the peristome and its opening through which the lantern protrudes and retracts during feeding. Because the interambulacral plates are relatively small on the scale of the main drawing, the shaded areas indicate where each set of these plates lie. The bulk of each pyramid consists of a pair of complementary demipyramids, 'dp', and the associated tooth (see Fig. 1 upper right corner, for a 3D rendering from μCT data). In the lower right of Figure 1 the mamelon and boss, 'mb', and tubercules, 'tu', are shown on the outer surface of an interambulacral plate. These are the mounting points for spines. An array of ambulacral plates and associated pore pairs, 'app', border the much larger interambulacral plate.

Sea urchin teeth are quite rare in the fossil record, but several studies (Märkel 1978; Smith 1982; Smith & Hollingworth 1990) using scanning electron microscopy (SEM) explicated the 3D spatial distribution of plates and prisms from polished sections, demonstrating that samples could be found where diagenesis had not progressed so far as to eradicate the microstructure. As demonstrated in a recent paper on teeth of the modern sea urchin *Lytechinus variegatus* (Stock *et al.* 2002), current X-ray methods (e.g. microfocus X-ray CT and synchrotron X-ray diffraction mapping) can add much to the understanding of the organisation of echinoid ossicles.

X-ray diffraction had been applied to echinoderm calcitic structures such as plates of the test and spicules (Towe 1967; Nissen 1969; Donnay & Pawson 1969; Berman *et al.* 1988, 1990, 1993), but previous X-ray studies do not appear to have observed microstructural variation nor to have studied recent or fossil teeth. Microfocus CT, a high resolution version of the medical CAT scanner (e.g. Stock 1999), appears to have been applied to echinoderms only by the authors of this paper and their co-workers (Stock *et al.* 2002).

The non-invasiveness of μCT and concomitant lack or minimization of sample preparation are important potential advantages. If fossilized ossciles are to be examined with microscopic techniques such as SEM, the filled pore space often requires chemical treatment to reveal the stereom (Smith 1980, 1990). Depending on what filled the stromal space, standard micropalaeontological disaggregation techniques (paraffin) can be sufficiently effective, but non-calcareous, strongly cemented infilling matrices often require treatments as aggressive as hydrofluoric acid. Furthermore, with μCT the ossicles of Aristotle's lantern can be studied *in situ*, that is, within the filled test without disturbing the ensemble.

Plates of the test, spines and even demipyramids (largest ossicle of the oral apparatus, i.e. the jaw) are much more common fossils than teeth. Surveying selected specimens of these types with μCT is, therefore, useful in establishing the likely outcome of such characterization on sea urchin teeth before extraordinary efforts are devoted to obtaining suitable samples of teeth. Accordingly, fossil echinoid test plates, spines and demipyramids were obtained on loan from the Field Museum of Natural History (FMNH), Chicago and examined with μCT, as described below.

Materials and methods

Spines, demipyramids and test fragments were examined as part of this preliminary μCT survey. Data are reported for three cidaroid spines (FMNH collection number PE53930, Jurassic), one demipyramid that is typical of the several other demipyramids examined (PE53943, Jurassic), and a large fragment of a test plate (PE53930, Jurassic). Species identities have not been established, but the test fragment appears similar to figures of *Miocidaris* found in the literature (Jackson 1912). A pyramid and a spine from an adult *Lytechinus variegatus* provided reference images.

Data for reconstructing slices of the sea urchin samples were collected at 50 kV using a commercial fan beam system (MicroCT-20, Scanco Medical, Bassersdorf, Switzerland). Integration times of 0.35 s, 500 projections per slice and 1024 samples per projection were used. Measurements on a millimetre-sized, inorganic calcite crystal produced values of the linear attenuation coefficient between 7.2 and 7.7 cm^{-1} and revealed the effective energy to be about 26 keV. A Scanco MicroCT-40 system was used to produce a reference image of a pyramid of the modern sea urchin *Lytechnius variegatus* (see Fig. 4c) for comparison with the fossil demipyramids.

The spines were positioned with their axes parallel to the rotation axis of the μCT system (i.e. the slices were normal to the spine axis) and reconstructed with 9 × 9 μm voxels (volume elements) in the plane of reconstruction and a *c.* 25 μm slice thickness. The larger size of the demipyramids and the test fragment dictated reconstruction with 13 × 13 μm and 17 × 17 μm voxels, respectively, in the reconstruction plane and the slice thicknesses were slightly larger than that of the spines. The major axes of the demipyramids were also positioned parallel to the instrument's rotation axis.

Results

Spines

Figure 2 shows μCT slices of two fossil cidaroid spines (a and b) and, for reference, an optical micrograph of a cross-section of a recent cidaroid spine and a μCT slice of the verticilate spine of the recent *Lytechinus variegatus* (c and d, respectively). In the μCT slices of all of the figures, the lighter the pixel (picture element) the higher the linear attenuation coefficient of the voxel. The images of the recent spines show what might be seen in the absence of stereom filling and diagenesis, as well as what the μCT apparatus was capable of resolving (Fig. 2c and d, respectively).

In Figure 2a, a tide mark, 'tm', of lower attenuation voxels parallels most of the outer contour of the spine. It may be that it follows, more or less, the cortex of the spine (see 'c' in Fig. 2c). Knob-like irregularities, termed granules in the palaeontological literature, ornament the outside of the spines (Fig. 2a); in Figure 2b the regularity of the granules is striking. The apparent size of the granules in a given slice has less to do with their actual size and more to do with the location of the slice. A collection of small, low attenuation features, 'v', in the centre of the spine in Figure 2a appear to correspond to portions of the medulla (see 'm' in Fig. 2c)

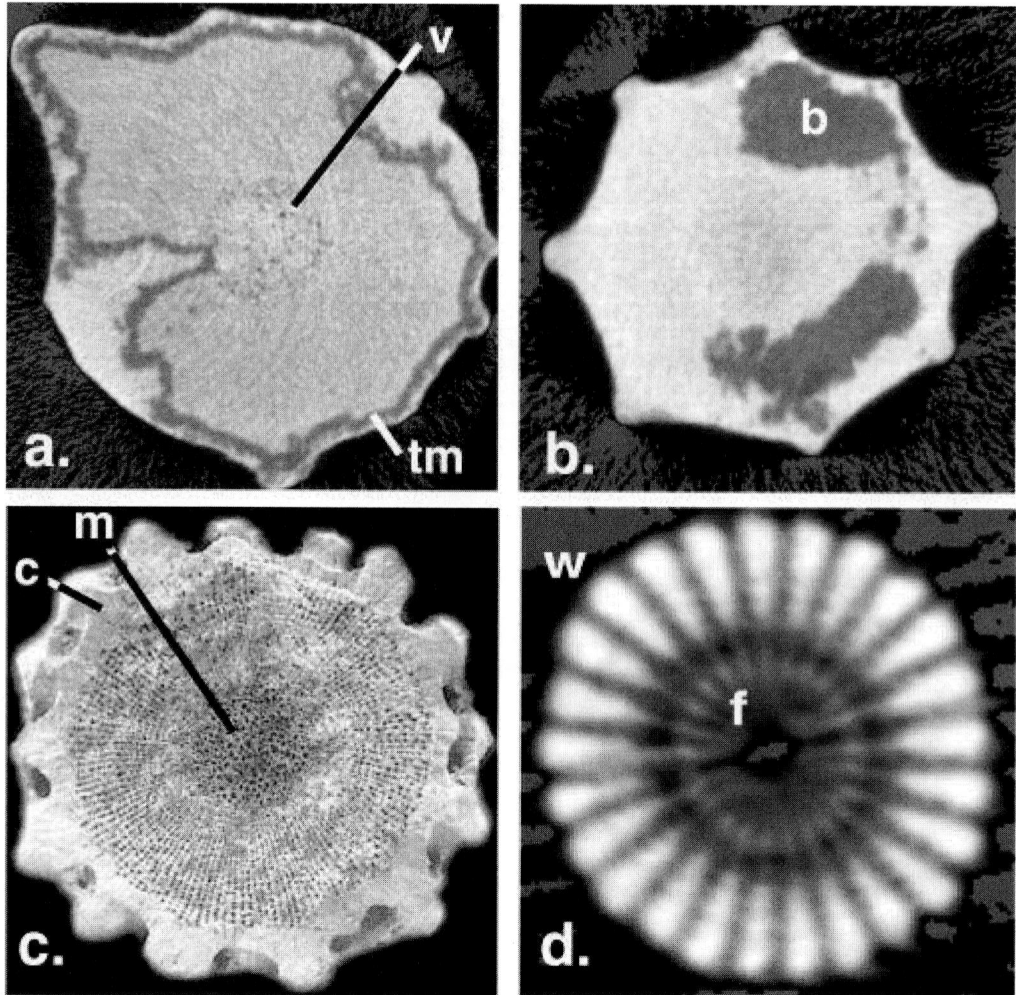

Fig. 2. Slices of two different fossil spines (**a** and **b**); the optical micrograph of a recent cidaroid spine, (**c**), and the slice of a recent *Lytechinus variegatus* spine, (**d**), are shown for reference. The maximum diameter across the spine is slightly more than 5.5 mm in (**a**), slightly more than 3.2 mm in (**b**) and 1.0 mm in (**d**); (**c**) was scaled to match (**a**). All labels are explained in the text. (**c**) is reproduced from the website *Echinoid Directory* (http://www.nhm.ac.uk/palaeontology/echinoids), with permission of A. B. Smith and the Natural History Museum, London.

that escaped filling subsequent to the demise of the sea urchin, or that were filled with low attenuation material similar to that in the tidemarks. Note that the fraction of the area of the spine spanned by void-like features, 'v', is about the same as that of 'm'. The spine in Figure 2b differs from that in Figure 2a: the low attenuation features, 'b', are blobs with no apparent connection with the geometry of the spine. Because the wedges, 'w', and fine stereom, 'f', are clear in Figure 2d, similar features should be resolved in reconstructed slices of fossilized spines.

Inspection of Figure 2a also reveals that the voxels within the tidemark appear slightly darker than those outside of it. Inside of the tidemark, the linear attenuation coefficients μ range between 5.7 and 6.7 cm^{-1}, excluding features 'v'. The tidemark itself consisted of voxels with values between 3.6 and 4.3 cm^{-1}. Outside of the tidemark, in what one assumes is the cortex of the spine, the spread in μ was 6.2 to 7.6 cm^{-1}.

Figure 3 shows three slices through a third fossil spine. The tidemarks seen in Figure 3 are similar to those in Figure 2a except that the

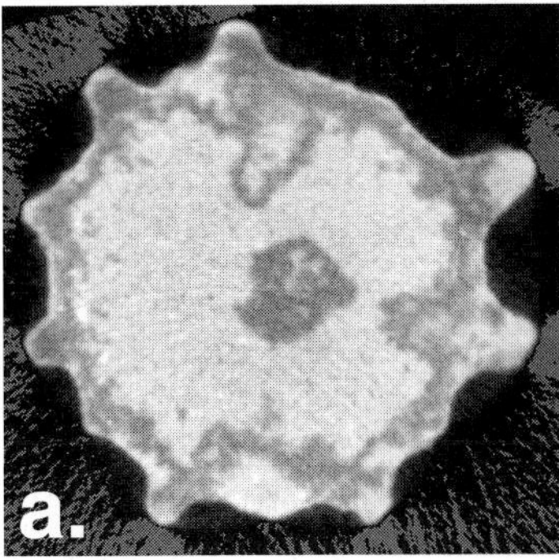

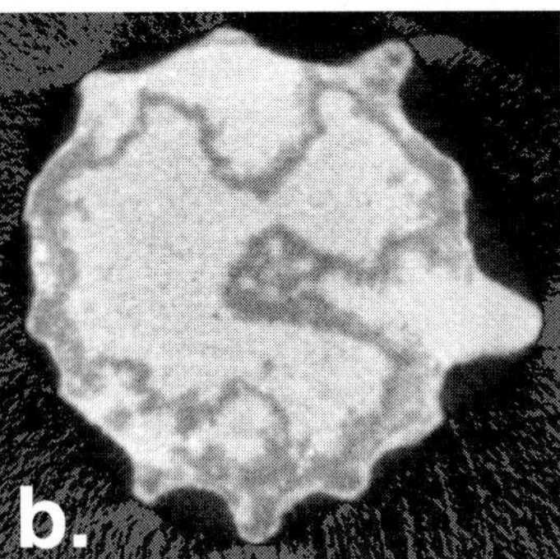

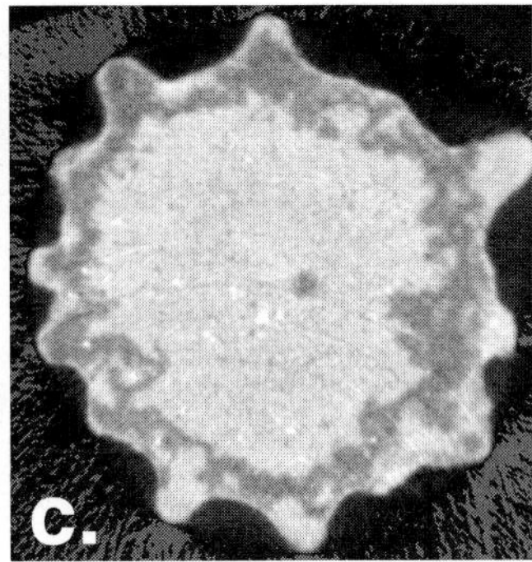

Fig. 3. Three slices through a cidaroid spine showing a complex 'tidemark' pattern of low attenuation voxels. The horizontal field of view in each slice is 3.68 mm. Just over 0.6 mm separates the slices shown in (**a**) and (**b**) and about 1.6 mm separates the slices in (**b**) and (**c**).

tidemark extends into the centre of the spine. There is considerable fine structure within the central low absorption zone. Small high attenuation particles (with μ approaching the limit of 8 cm^{-1} for the apparatus) appear in most slices of the fossil spines. There are more granules around the periphery of the spine than in the spines of Figure 2 a number similar to that of the recent cidaroid spine shown in Figure 2c. The ranges of linear attenuation coefficients in Figure 3 are similar to those in Figure 2a.

Demipyramids

Figures 4a and b show two slices through a fossil demipyramid; Figure 4c shows an enlargement of the area of Figure 4a indicated by '*' and Figure 4d shows a slice through a pyramid of the recent *Lytechinus variegatus* (the suture between demipyramids is labelled 's', the tooth, 't' and the intrapyramidal cavity, 'i'). In the tangential wall, 'tw', of the fossil, two sets of parallel, linear, low-absorption, very high-aspect ratio features (labelled '1' and '2' in Fig. 4c) appear in both Figure 4a and Figure 4b and in adjacent slices spanning several millimetres. On the plane of the slices, the angle between lines '1' and '2' is about 110°. Note that this is not the interplanar angle but rather the angle between the traces of the respective planes in the illustrated cross-sections. Along the radial wall, 'rw', of the demipyramid, the feature labelled 'z' (Fig. 4b) seems to be a remnant of glue (presumably used to mount the specimen for display). The area 'q' is seen in many slices and has the same appearance as a filled intrapyramidal cavity. If this is true, then the pyramid is not of a cidaroid sea urchin.

The linear attenuation coefficients vary from 5.2–5.8 cm^{-1} in the zone between 'q' and the external surface of the radial wall, 'rw', to 6.2–6.8 cm^{-1} at the end of the radial wall and to 5.9–6.7 cm^{-1} between the 'cracks' in the tangential wall, 'tw'. Outside of the ossicle the values are 0–0.4 cm^{-1}, within the boundary of 'q' 0.8–1.5 cm^{-1} and within the crack-like features as low as 3.4 cm^{-1}.

Figure 5a shows a 3D rendering of the portion of the fossil demipyramid containing the array of low absorption, high-aspect ratio features (Fig. 4). The labels '1' and '2' are placed on the plane of the rendering parallel to the slices and indicate the same orientation of low absorption linear features (shown lighter than the surrounding sectioned surfaces), as in Figure 4. The slightly irregular, blocky nature of the features appear to be an artefact of the variable contrast along their length and the requirement to select a single threshold level for production of the rendering. Selection of a lower threshold breaks the features into small, disconnected segments, which defeats the purpose of producing the rendering. The other visible side of the volume in the rendering demonstrates that the low absorption features are, in fact, planar objects intersecting the slices. It would certainly be more interesting to produce a rendering showing the planar features rather than the surrounding material, but attempts to do this without heavy-handed imaging tricks failed due to the nature of the contrast (described above).

The rendering in Figure 5b covers the volume around the feature labelled 'q' in Figure 4a. Note that it is at about one-half the scale of the rendering of Figure 5a. The front surface and viewing direction for the rendering of the volume intersecting 'q' is (approximately) the plane indicated by the 90° arrows in Figure 4a. The boundaries of 'q' run parallel to the outer surface of the demipyramid and through the entire

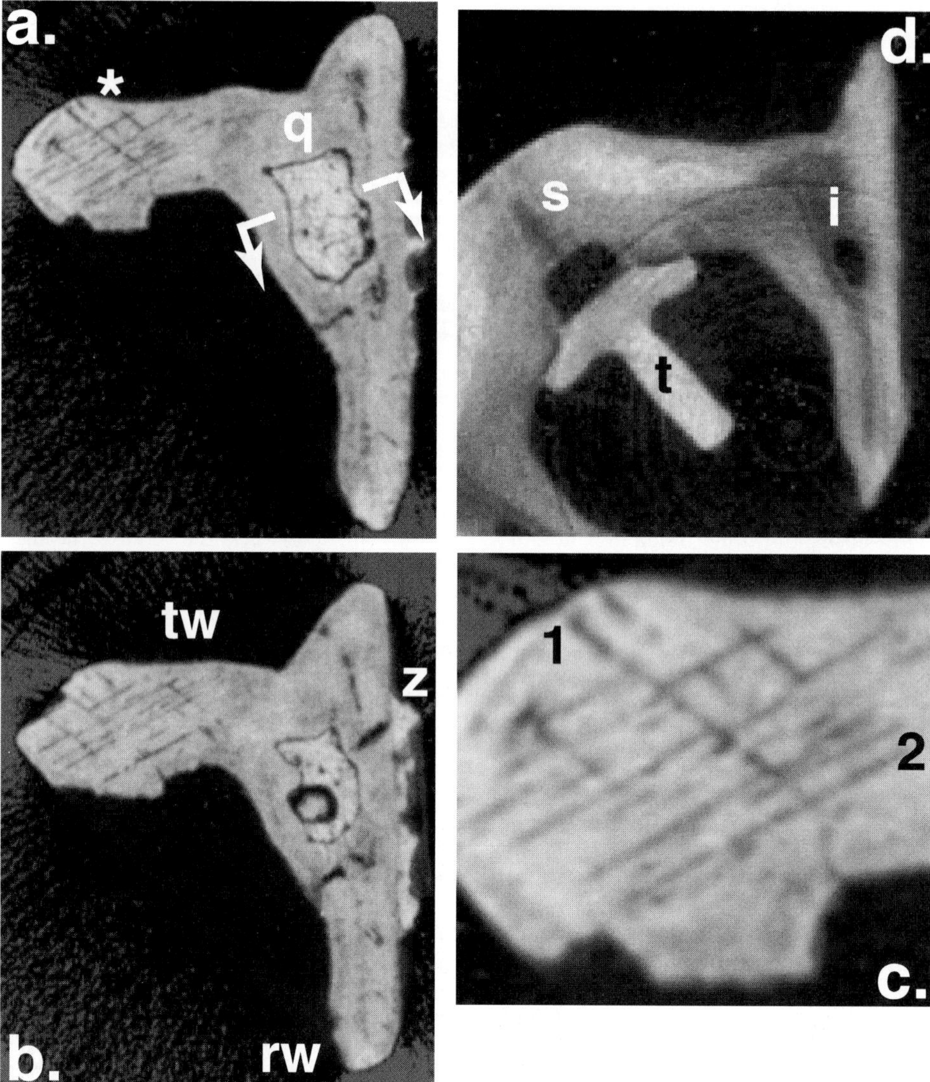

Fig. 4. Two slices through a fossil pyramid, separated by 0.95 mm (**a, b**), an enlargement of the area marked in image (**a**) by the asterisk (**c**), and a slice through an intact recent pyramid and enclosed tooth, (**d**). The horizontal field of view is 6.25 mm in (**a**) and (**b**) and 4.05 mm in (**d**). The arrows with a ninety degree bend mark the position of the vertical sectioning plane for the rendering of Figure 5b. Note that the slice shown in (**c**) was collected with a cone beam system (Scanco MicroCT-40) instead of the fan beam system used to collect the other data, and the values of the linear attenuation coefficients differ slightly from those obtained with the MicroCT-20 system.

volume imaged. Visual observation of the demipyramid reveals that 'q' is open in the volume above the area that is covered in these scans and extends to the surface where in life the epiphysis would presumably sit. It is also interesting that there is significant open volume in the bottom portion of 'q'. Physical examination of the sample reveals that this does not communicate with the exterior of the fossil.

Plates

Five slices through the plate fragment, spaced 100 μm apart, are shown in Figure 6. The outer

Fig. 5. Renderings of volume of a fossilized demipyramid containing cracks. (**a**) shows the volume near the asterisk in Figure 4a and (**b**) shows the volume indicated in Figure 4b, with the front surface of the rendering being a plane through the mostly filled cavity, 'q'. Note that the rendering in (**b**) is rotated 180° from the orientation of the demipyramid in Figure 4 i.e. so that the volume is viewed from the direction of the arrows in Figure 4b. The lighting in (**a**) and (**b**) was chosen to optimize crack visibility.

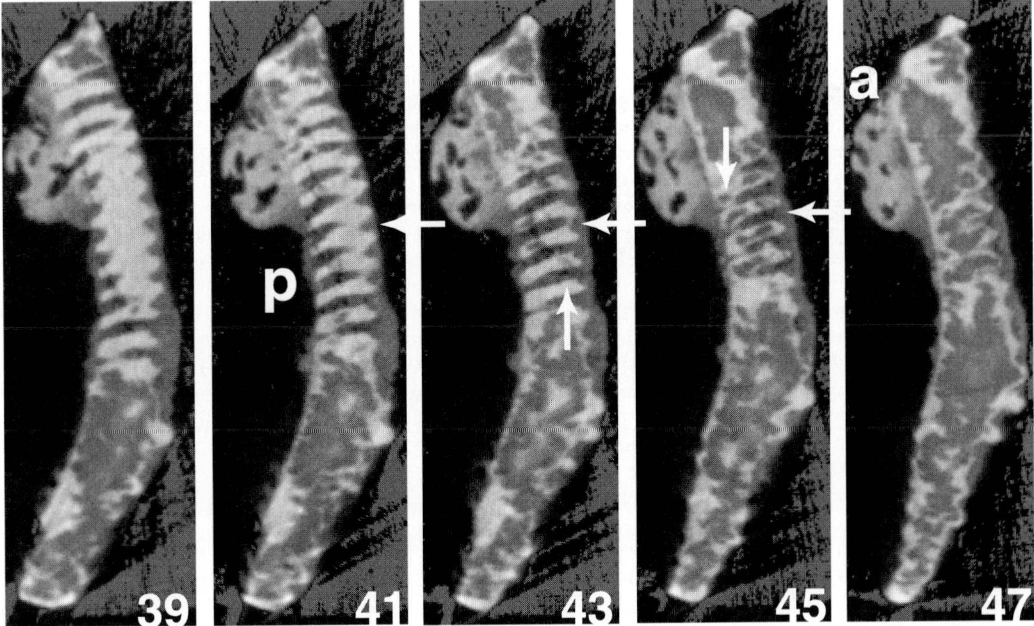

Fig. 6. Five slices through the plate fragment. The horizontal field of view in each image is 1.74 mm and the slices are 100 μm apart.

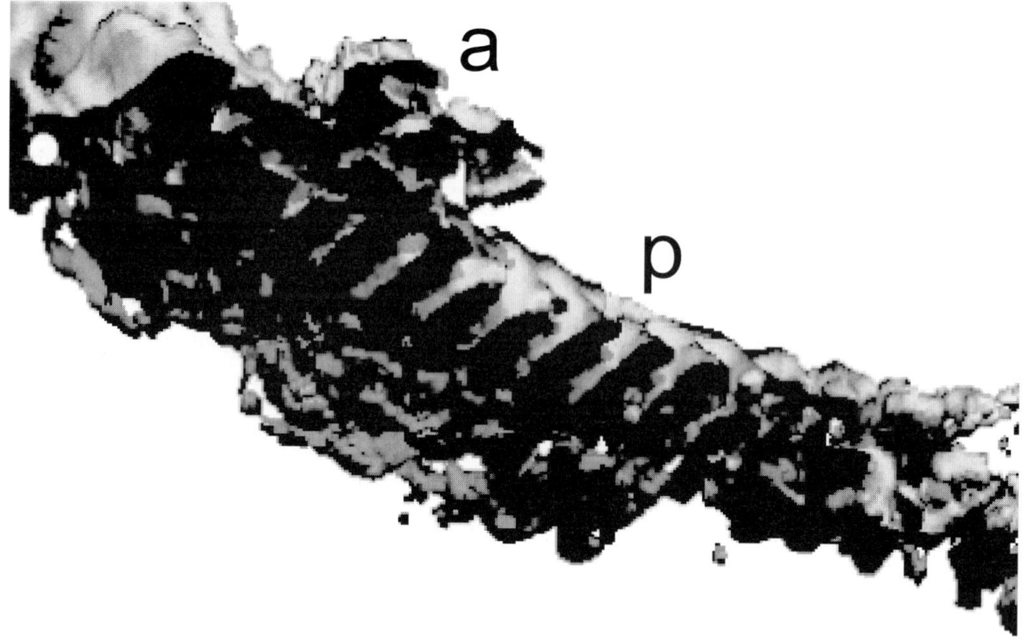

Fig. 7. Three-dimensional rendering of the volume of the plate shown in Figure 6.

surface of the test (i.e. the convex side of the fragment) is on the right-hand side of each image and the thickness of the plate, away from obvious fracture surfaces, varies between 1.3 and 1.5 mm. The feature 'a' on the interior of the plate (slice 47) may be a later accretion. It is interesting to note what appear to be voids in 'a'. At 'p' a series of perhaps a dozen pores in the high attenuation material of the fragment are intersected by the slices. Presumably the fragment contained portions of a number of ambulacral plates bordering a much larger section of an interambulacral plate and the slices happened to cut across the row of ambulacral plates and their pores (see the schematic in the lower right corner of Fig. 1). The interior of the pores appears to have been filled with lower attenuation material (three horizontal arrows, right side of slices 41–45), although a small number of voids may remain in the pores (e.g. upward vertical arrow in slice 43). One of the voids in 'a' extends into one of the pores in the test (slice 39). Within the area of the plate well away from the pores, contrast is variegated. Near the pores, the situation appears more complex, with uniform high attenuation in slice 39 and small areas of low attenuation in the pore walls in the middle three slices of Figure 6 (e.g. downward arrow in slice 45).

Within the low attenuation regions of Figure 6 the linear attenuation coefficients vary from 2.8–3.3 cm^{-1} (uppermost grey region in the fragment) to 3.3–3.7 cm^{-1} (the largest grey region, in the bottom third of the fragment) to 3.4–4.4 cm^{-1} (accretion 'a' just above the label 'p' in slice 41). The small, very low attenuation features scattered in the sample have attenuation coefficients in the range of 0.7–1.0 cm^{-1}, while the empty space outside of the sample has values of 0-0.4 cm^{-1}. High absorption voxel values covered a range of 6.0–6.4 cm^{-1} through 6.5–6.7 cm^{-1} with small areas having values of 7.4–7.7 cm^{-1}.

A 3D rendering of the slices in Figure 6 appears in Figure 7 with the low attenuation voxels stripped away. The volume is viewed from the left of and below the slices in Figure 6 and the white circle on the sectioning surface indicates the uppermost position. The rendering's bottom (solid black in the figure) is slice 35 and slices with higher numbers lie in back of this plane.

Discussion and conclusions

The fraction of the spine diameter that the tidemark (Fig. 2a, Fig. 3) has penetrated approximately equals the ratio of thickness of the dense cortex to the total diameter in a modern cidaroid spine (Fig. 2c). This may be nothing more than coincidence, but the correspondence is suggestive. The cortex of cidaroid spines is one of the few regions of polycrystalline calcite in regular echinoids (Märkel *et al.* 1971) and the volume fraction of calcite (mineral) in this portion of the spine is much higher than in the interior. This is consistent with the higher values of the absorption coefficient outside versus inside the tidemarks. It is unlikely, of course, that the original structure of the spine was preserved, but after decay of the stromal tissue and before any diagenetic removal of calcite, the cortex has a much lower potential than the spine interior for filling with other mineral species. Unless minerals containing elements with atomic numbers higher than that of Ca filled the stromal space, the original difference in attenuation coefficients would persist. The presence of the blobs (Fig. 2b), the presence of the tidemark (with significantly lower attenuation than the material inside of it) and the tidemarks' irregular geometry suggest that the processes during diagenesis were more complex than outlined here.

The relatively large range of linear attenuation coefficients in the three distinct regions within the slice shown in Figure 2a suggests that there is considerable structural heterogeneity (a mixture of relatively high and relatively low absorptivity features), at a scale somewhat smaller than could be resolved. The range of voxel values of $1.0\,cm^{-1}$ inside the tidemark and about $1.4\,cm^{-1}$ outside this band are substantially larger than that seen in the pure calcite crystal ($0.5\,cm^{-1}$) and this is consistent with the interpretation of unresolved structural heterogeneity.

The extreme values of μ found in Figures 2 and 3 and the range of values observed are consistent with the presence of calcite accompanied by a lower attenuation material. Combined with the presence of tidemarks, this suggests the presence of at least one additional diagenetic phase that has been incorporated into the original stereom structure. Whether or not it is present only in the original stromal space can of course not be answered by measurements with the present spatial resolution.

X-ray diffraction can guide identification of the phases present, if not their spatial distribution, and would be invaluable in interpreting the μCT results on the spine fossils of Figures 2 and 3. Synchrotron X-ray data of the same spines reveal large volumes of single crystal material (low Mg calcite), possibly some dolomite epitaxially-oriented with respect to the calcite, and polycrystalline SiO_2 (quartz) (S. R. Stock, A. P. Wilkinson, P. Lee, unpublished data). The data suggest that the tidemarks and other internal low absorption features in Figures 2 and 3 are the result of siliceous diagenesis.

No tidemarks or variegated contrast were observed in the demipyramids examined. This suggests that the demipyramid experienced a very different diagenetic environment than the spines.

It is difficult to imagine the array of low absorption features in Figures 4a and 4b as being anything other than cracks produced during diagenesis after deposition. The well-defined angle between the two sets of linear features is certainly what one expects of cleavage cracks in a single crystal. Given the 3D data set, this hypothesis can be tested directly using the relative positions of the 'cracks' on several slices to infer the orientations of the normals of planes '1' and '2' and the associated interplanar angle. Manipulations of this sort are covered in standard X-ray diffraction and crystallography texts (e.g. Cullity & Stock 2001) and will not be covered here. The interplanar angle was determined to be 105° and equals the angle between the expected calcite cleavage planes for inorganic calcite: {211} family of planes with indices given for the rhombohedral unit cell (Warren 1969). In terms of hexagonal indices, this family of planes is {10.4}. The reader should note that the Miller-Bravais indices for the cleavage planes are sometimes given in the hexagonal system but referred to the cleavage rhomb (Gaines *et al.* 1997, pp. 428–431) and this designation of the cleavage planes is {10.1}.

The interpretation of '1' and '2' as {10.4} cleavage cracks implies that the demipyramid at the time of death of the sea urchin, or after diagenesis, was single crystal material. This is in accord with earlier observations on fossil sea urchin ossicles: the stereom structure of high Mg calcite with single-crystal orientation was present prior to the Jurassic (Smith 1984) and the original single crystal stereom can be preserved free of diagenesis or can be augmented during diagenesis with calcite precipitated in crystallographic continuity with the original stereom (Smith 1980, 1990). The presence of cleavage cracks also implies that the organic constituents intrinsically included in the biologically-generated calcite (Berman *et al.* 1988, 1990, 1993) must have been eliminated prior to cleavage. Otherwise, such calcite fractures conchoidally (Merker 1916).

X-ray diffraction would be able confirm that the 'cracks' indeed follow {10.4} for the sample in Figure 4 but such data have not been obtained. However, the orientation of the calcite *c*-axes has been determined for ossicles of a number of recent sea urchins. The crystallographic orientation of test plates have received the most attention (Raup 1966), but Märkel (1979) reported *c*-axis orientations for pyramids, epiphyses and rotulae of two non-cidaroid and one cidaroid species. According to Märkel's diagrams, both types of pyramids have c-axes in or near the plane perpendicular to the main axis of the lantern and the c-axes are intermediate in orientation between parallel to the suture between demipyramids and the outside of the radial demipyramid wall. The angles between this approximate c-axis orientation and the experimentally-determined normals to the crack planes are consistent with the {00.1}, {10.4} interplanar angles for calcite. This identification of features '1' and '2' of Figure 4 as {10.4} cracks must remain tentative until/if diffraction data can be collected. The same could be done with optical microscopy, but this would entail sectioning the sample.

The test fragment (Figs 6 and 7) appears to have been greatly affected by diagenesis, but some structure has been preserved (pores for tube feet). The diameter of the pores appears to be about 100 μm consistent with data in the literature (Smith 1978). The variegated contrast is extremely complex, much more so than in the spines or pyramids, and interpretation is presently impossible.

It may well be that X-ray μCT with voxels a factor of two or three smaller than were possible for the present study would reveal considerably more detail. Newer μCT apparatus (e.g. Scanco MicroCT-40) allows a somewhat smaller voxel size and synchrotron μCT permits spatial resolutions below 2 μm on samples of about 2 mm maximum diameter. Examining larger samples requires cutting the sample (probably objectionable to the owners of the fossil), obtaining an X-ray detector with many more pixels (very expensive), stitching radiographs together to provide coverage of the entire sample at high resolution (severe registration problems exist), using region of interest reconstruction techniques (recording the views of the region of interest at the required resolution and using low resolution data over the entire sample cross-section to correct for the missing portions of the high resolution views) or increasing the field of view by about a factor of two by placing the sample off the centre of the mechanical rotation axis and rotating from 0° to 360° instead of from 0° to 180°. All of these require effort to implement and so it is not surprising that they have not yet found wide use.

The authors thank W. Taylor and S. Lidgard of the Field Museum, Chicago, for their help in locating suitable specimens and for helpful discussions. The authors are also grateful to A. Smith of the Natural History Museum, London, for lengthy electronic correspondence on echinoid fossils. Authors also thank S. Nagaraja of Georgia Institute of Technology for producing the 3D μCT rendering of the pyramid of the modern sea urchin *Lytechnius variegatus* (Fig. 1, upper right). The authors are also grateful to the reviewers for providing helpful suggestions for improving the original manuscript. This work was partially supported by US NIH , National Institute of Dental and Craniofacial Research grant DE-01374 (AV) and by NSF BES grant 9977551 (SRS).

References

AMEYE, L., HERMANN, R., KILLIAN, C., WILT, F. & DUBOIS, P. 1999. Ultrastructural localization of proteins involved in sea urchin biomineralization. *Journal of Histochemistry & Cytochemistry*, **47**, 1189–1200.

AMEYE, L., HERMANN, R. & DUBOIS, P. 2000. Ultrastructure of sea urchin calcified tissue after high pressure freezing and freezing substitution. *Journal of Structural Biology*, **131**, 116–125.

BERMAN, A., ADDADI, L. & WEINER, S. 1988. Interactions of sea-urchin skeleton macromolecules with growing calcite crystals – a study of intracrystalline proteins. *Nature*, **331**, 546–548.

BERMAN, A., ADDADI, L., KVICK, Å., LEISEROWITZ, L., NELSON, M. & WEINER, S. 1990. Intercalation of sea urchin proteins in calcite: a study of a crystalline composite material. *Science*, **250**, 664.

BERMAN, A., HANSON, J., LEISEROWITZ, L., KOETZLE, T.F., WEINER, S. & ADDADI, L. 1993. Biological control of crystal texture: a widespread strategy for adapting crystal properties to function. *Science*, **259**, 776.

CULLITY, B.D. & STOCK, S.R. 2001. *Elements of X-ray Diffraction. 3rd Edition.* Prentice-Hall, Upper Saddle River, New Jersey, USA.

DICKSON, J.A.D. 2001. Transformations of echinoid Mg calcite skeletons by heating. *Geochimica et Cosmochimica Acta*, **65**, 443–454.

DONNAY, G. & PAWSON, D.L. 1969. X-ray diffraction studies of echinoderm plates. *Science*, **166**, 1147–1150.

GAINES, R.V., SKINNER, H.C.W., FOORD, E.F., MASON, B. & ROSENZWEIG, A. 1997. *Dana's New Mineralogy. 8th Edition.* Wiley, New York.

JACKSON, R.T. 1912. Phylogeny of the Echini, with a revision of palaeozoic species. *Memoirs of the Boston Society of Natural History*, **7**, 1–491.

KITAJIMA, T., TOMITA, M., KILLIAN, C.E., AKASAKA, K. & WILT, F.H. 1996. Expression of spicule matrix protein gene SM30 in embryonic and adult mineralized tissues of sea urchin *Hemicentrotus pulcherrimus. Development Growth and Differentiation*, **38**, 687–695.

MACQUEEN, R.W. & GHENT, E.D. 1970. Electron microprobe study of magnesium distribution in some Mississippian echinoderm limestones from western Canada. *Canadian Journal of Earth Sciences*, **7**, 1308–1316.

MACQUEEN, R.W., GHENT, E.D. & DAVIES, G.R. 1974. Magnesium distribution in living and fossil specimens of the echinoid *Peronella lesueuri* Agassiz, Shark Bay, Western Australia. *Journal of Sedimentary Petrology*, **44**, 60–69.

MÄRKEL, K. 1978. On the teeth of the recent cassiduloid *Echinolampas depressa Gray,* and on some Liassic fossil teeth nearly identical in structure (Echinodermata, Echinoidea). *Zoomorphologie*, **89**, 125–144.

MÄRKEL, K. 1979. Structure and growth of the cidaroid socket-joint lantern of Aristotle compared to the hinge-joint lanterns of non-cidaroid regular echinoids. *Zoomorphologie*, **94**, 1–32.

MÄRKEL, K., KUBANEK, F. & WILLGALLIS, A. 1971. Polykristalliner Calcit bei Seeigeln (Echinodermata, Echinoidea). *Zeitschrift für Zellforschung*, **119**, 355–377.

MERKER, E. 1916. Studien am Skelet der Echinodermen. *Zoologische Jahrbücher – Abteilung für allgemeine Zoologie und Physiologie der Tiere*, **36**, 25–205.

NISSEN, H.U. 1969. Crystal orientation and plate structure in echinoid skeletal units. *Science*, **166**, 1150–1152.

RAUP, D.M. 1966. The Endoskeleton. *In*: BOOLOOTIÁN, R.A. (ed.) *Physiology of Echinodermata*. Wiley, New York, 379–395.

SMITH, A.B. 1978. A functional classification of the coronal pores of regular echinoids. *Palaeontology*, **21**, 759–789.

SMITH, A.B. 1980. *Stereom Microstructure of the Echinoid Test*. Special Papers in Palaeontology No. 25. The Palaeontological Association, London.

SMITH, A.B. 1982. Tooth structure of the pygasteroid sea urchin *Plesiechinus*. *Palaeontology*, **25**, 891–896.

SMITH, A.B. 1984. *Echinoid Palaeobiology*. Allen and Unwin, London.

SMITH, A.B. 1990. Biomineralization in Echinoderms. *In*: CARTER, J.G. (ed.) *Skeletal Biomineralization: Patterns, Processes and Evolutionary Trends*. Van Nostrand Reinhold, New York, Vol. 1, 413–223 and Vol. 2, 69–71.

SMITH, A.B. & HOLLINGWORTH, N.T.J. 1990. Tooth structure and phylogeny of the Upper Permian echinoid *Miocidaris keyserlingi*. *In*: *Proceedings of the Yorkshire Geological Society*, **48**, 47–60.

STOCK, S.R. 1999. Microtomography of materials. *International Materials Reviews*, **44**, 141–164.

STOCK, S.R., DAHL, T., BARSS, J., VEIS, A., FEZZAA, K. & LEE, W.K. 2002. Mineral phase microstructure in teeth of the short spined sea urchin (*Lytechinus variegatus*) studied with X-ray phase contrast imaging and with absorption microtomography. *Advances in X-ray Analysis*, **45**, 133–138.

TOWE, K. 1967. Echinoderm calcite: single crystal or polycrystalline aggregate. *Science*, **157**, 148–150.

TOWE, K.M. & HEMLEBEN, C. 1976. Diagenesis of magnesian calcite: evidence from miliolacean foraminifera. *Geology*, **4**, 337–339.

VEIS, D.J., ALBERGER, T.M., CLOHISY, J., RAHIMA, M., SABSAY, B. & VEIS, A. 1986. Matrix proteins of the teeth of the sea urchin *Lytechinus variegatus*. *Journal of Experimental Zoology*, **240**, 35–46.

WARREN, B.E. 1969. *X-ray Diffraction*. Dover, New York.

Index

Page numbers in italic, e.g. *214*, refer to figures. Page numbers in bold, e.g. **216**, signify entries in tables.